CREATING DIGITAL CONTENT

JOHN RICE

BRIAN McKERNAN

McGraw-Hill

New York • Chicago • San Francisco • Lisbon

London • Madrid • Mexico City • Milan • New Delhi

San Juan • Seoul • Singapore • Sydney • Toronto

McGraw-Hill

A Division of The **McGraw-Hill** *Companies*

1 2 3 4 5 6 7 8 9 0 DOC/DOC 0 9 8 7 6 5 4 3 2 1

ISBN 0-07-137744-1

The sponsoring editor for this book was Stephen S. Chapman, the editing supervisor was Daina Penikas, and the production supervisor was Pamela Pelton. It was set in Fairfield by Patricia Wallenburg.

Printed and bound by R. R. Donnelley & Sons Company.

This book is printed on recycled, acid-free paper containing a minimum of 50 percent recycled, de-inked fiber.

CONTENTS

INTRODUCTION

Humans have been making pictures for millennia. Roughly a century ago they learned how to make them move. As the decades progressed, movies, television, and—more recently—home video and the Internet have enhanced communication and defined human experience as never before. Color, sound, and other innovations continually improved the moving image, and today it is a global *lingua franca* that transcends international borders, cultures, and traditional languages by being visual.

As the twenty-first century begins, humankind's other major information technology—the computer—is revolutionizing the moving image. The technologies of computers and television have been on a collision course for some time, sharing a purpose that includes the precise capture, storage, manipulation, transmission, and presentation of information. Both even use cathode-ray tubes (CRTs) for display. Computers, however, process information digitally—as a series of 0s and 1s—an efficiency that provides many advantages when applied to moving-image technology. And as computer microprocessors and related components become ubiquitous in everything from cars to toasters to toys, they grow in power and decrease in price.

The digitization of moving-image technology is by no means complete; broadcast television's transition to digital is moving slowly, but it is already having a major impact. Video and film production are going digital, and in the process gaining improved tools for creative expression and affordability. New computer-based storage, playback, and display devices are expanding uses for moving-image content. And the Internet is providing an instant, global means of content distribution.

In such a digital environment, distinctions between specific forms of moving-image media begin to blur or disappear altogether. A stream of digital data comprising moving images and their accompanying audio can be scaled for multiple uses, depending on how much picture and sound information is necessary. Movies, TV shows, videos, and even simple computer presentations become, in truth, a quantity of data existing as binary digits, or *bits*. The moving images that have traditionally been defined according to the technologies used to create and display them are now liberated for multipurposing; they have become, in reality, "digital content." This new world that moving-image creatives find themselves in at the dawn of the twenty-first century is what this collection, *Creating Digital Content*, is all about. The impact of the digital content creation age is only just starting to be felt. This book is intended to introduce the reader to a broad range of existing areas that are being affected, and, in so doing, enhance understanding of this fundamental industrial transition, business opportunity, and communications revolution.

Although we do not anticipate that the many areas encompassed by digital content creation will see overnight change (e.g., feature filmmakers won't suddenly become indistinguishable from corporate video professionals), we do feel it's imperative that those in the video, motion-picture, broadcast, news, entertainment, Internet, and education sectors understand the nature of the changes wrought by digital content creation revolution. First and foremost is that program production is being democratized, with increasingly more affordable—and yet powerful—technologies. And whether for entertainment, education, business, or other purposes, digital content can be consumed via a growing array of devices. This situation opens up potentially rich new possibilities for content creators, but it also increases competitive pressures. Ultimately, it will be creative talent and innovative thinking that will determine success. Digital is a great equalizer that reduces the cost of entry, but we should remember that Shakespeare did quite well without a laptop.

Digital content creation technology has many intriguing possibilities. Assuming there is sufficient data, content can be repurposed for multiple displays: theatrical movie screens, consumer televisions, small computer windows, or even hand-held devices and cell phones. As data, moving and still images are assets that can be managed—stored, cataloged, indexed, and repurposed with minimal or no loss in quality. Archived news footage can be retrieved instantly for broadcast. Images or portions of images can be digitally "cut and pasted" for economical repurposing in alternate versions or to customize advertising for different demographics. A high-resolution three-dimensional computer graphic can be used to mold a solid object; a two-dimensional image from a digital movie or TV show can be output in printed form as a billboard, a book, or even a t-shirt.

The uses applied to digital content creation tools will be determined by the creative talents using them and by current and future market forces. As time goes on, HDTV, interactive cable, and greater Internet bandwidth will present further challenges and opportunities to moving-image professionals. We have assembled a collection of essays by a broad array of uniquely qualified experts who have graciously provided the background, tutorials, and analysis necessary for understanding many of the changes brought about by the digital content creation age. We are grateful to these contributors, all of whom tackled topics worthy of separate books in their own right.

It is our hope that this edition will enable moving-image professionals to adapt and prosper in these changing times, and to participate in the evolution of what is, in the end, storytelling, one of humanity's oldest and most important forms of communication, welcome to the Age of Digital Content Creation.

John Rice, Brian McKernan
August 2001

FOREWORD

THROW ANOTHER ANALOG ON THE DIGITAL FIRE: CONFESSIONS OF A DIGITAL CONTENT CREATOR

PETER BERGMAN

For 25 years I had no trouble describing what I did or who I was. I was a comedian, an artist turning out high-end comedy on records and stage with my band of brothers, The Firesign Theatre.

But those halcyon days are over thanks to the advent of the digital revolution. Ah, the digital revolution, operating at 800 million clock cycles a second, processing 11 trillion machine decisions every 20 minutes, and producing four ex-billionaires every six months. I should have seen it all coming with the arrival of Bill Gates, the Kubla Khan of the Nerds. He does possess the perfect digital name. Think about it: Microchips are nothing but millions and millions of little gates opening and closing, and he's billing us for it.

In this strange new cyber landscape, peopled with the Princes of Packet-share, the Esquires of E-Commerce, the Browser Barons, and the Dukes of URL, I could no longer compete as a simple artist; I've had to reinvent myself as a "digital content creator."

Me digital? How could I, the last baby born before the boom—how could I, the youngest member of the generation whose oldest member is Bob Dole, ever hope to be digital?

I certainly didn't grow up digital. The only computer I encountered as a schoolboy was a picture of the Univac on the front page of *Life* magazine. It looked like a giant black toaster that took up three rooms. In college I was never in the same room with a computer, which explains why I failed to grasp Shakespeare's underlying inherent interactive message when I studied him in freshman English:

2B~<2b

The first great Boolean inequality in modern literature, and I missed it!

I had a chance to go digital back in the late '70s when a buddy of mine built an 8080 Imsai computer out of a kit. That was no help. I've never been able to build anything out of a kit. I tried twice. The first, a Heathkit voltmeter, which never registered a single volt. The second, a pair of Heathkit walkie-talkies that neither walked nor talked.

Like it or not, I had to get on board the digital train before it left me at the station with the rest of the Luddites, blind men all, tapping their way into the future on their manual typewriters. Fortunately, I woke up just in time and joined the swelling ranks of the early adapters. Now, I buy the newest and fastest computer the minute it comes on the market. I run home immediately and set it up; and then to nip that nagging feeling of buyer's remorse right in the bud, I turn on my old computer and pour what's left of my Big Gulp into the mother board of that slow, obsolete, useless Pentium-assed dinosaur that I paid beaucoup bucks for six months ago.

So now I'm a bona fide digital content provider and although this brand-new job description is great for picking up Web Mistresses in Silicon Valley, it does have its inconvenient side when it comes to actually delivering comedy in the digital environment. A caveat here: If you, gentle reader, are tasked with designing nothing more entertaining than the home page for a plastics extrusion company in Indianapolis, then you may skip the upcoming paragraph on the perils of fashioning comedy bits out of bytes. In fact, you can blow off this entire chapter and move on to the nether realms of this worthy tome, where weighty thoughts and globs of impenetrable computer code await you.

Now that those humorless drones have fled, we can return to the challenge of delivering entertainment in the digital mode. Five years ago I created *PYST* a best-selling CD-ROM comedy game, based on a parody of the much better selling adventure game, *MYST*. I learned straightaway that the sense of timing I had developed performing before a live audience was of no use to me in the realm of digital interactivity. On stage, I could develop a joke, nurse the audience, hold them until I was ready, and then wham!—hit 'em with the punch line. Not so, when the audience is on the other end of a CD-ROM or out there somewhere floating on the Internet. There's no real time and space connection between me, the provider of digital ha-has, and the bozo or bozoette stroking the mouse. I lay out the joke, and they can wait for bloody ever until they decide to click for the punchline.

To entertain and amuse via the digital domain is no longer an issue of split-second timing, but of providing your interactive audience with humorous situations and icons that intrigue and inspire them to click on and experience the joke. In *PYST*, for example, there's a screen that depicts the lawn in front of the main temple. No longer the idyllic greensward portrayed in *MYST*, I had transformed it into a decidedly low-end trailer park. Click on the dog and he relieves himself on a marble plinth. Click on the dilapidated Airstream and hear the TV inside the mobile home blare out a promo for Ken Burns' new 12-part documentary on "Hopscotch" or an infomercial for the latest in careers, reading books backwards for blind Satanists.

Five years ago when I crafted *PYST*, I also launched RFO.NET, a Web site that offered a rich menu of streaming and downloadable comedy routines, which included the talents of John Goodman, The Firesign Theatre, and a bevy of Los Angeles humorists. Back then, the equation was fairly simple: "me" plus "media" equals "multimedia," but in the ensuing half-decade the variables have increased enormously. To bring

RFO.NET up to speed, I have had to offer streaming video files, chat-room access, Shockwave animation, and an ever-increasing assortment of bells and whistles that I have to ring and blow just to avoid being wiped out as I surf the bitstream in the digital pipe.

And that's just the Web. All you emerging digital content providers have to wrestle simultaneously with the ever-expanding palette of digital content distribution channels—live, tape, broadcast, narrowcast, satellite, and the burgeoning intranets, which are hungry for formats that will keep the worker bees stuck to their computer screens while the boss reels out the company line.

Then there's the matter of the competition. Any ordinary citizen in this post-modem world can put his or her hand on his or her PC and declare, "I am a digital content provider!" Example, my neighbor, Dot Broadband. Once a typical Stepford Wife, she has up-stepped herself into the digital domain, Web casting "Dot's Home Page Cooking" from her kitchen and making real bread along the way.

That being said, there's a ton of good work waiting out there for you real professionals; and nowhere else has the rapid onrush of digital content creation posed so many challenges and opportunities than in Hollywood, where the Hot Shots and the big-time Suits gamble umpteen millions of dollars on the latest blockbuster. That's where the most glamorous cyber jobs reside; so I think it only meet that I let you in on the latest digital dish from Tinsel Town.

In Hollywood, the stars are lining up on both sides of the digital divide. Case in point, my lunch last week with Braveheart and Ms. Icepick at Baldy's Le Dome, the glitter eatery on the Sunset Strip.

Braveheart was real excited with his latest cutting-edge encounter with digital filmmaking.

"Ever since I laid in that all-cyber Irish army in *Braveheart,*" he said, "I've learned there aren't any actors, just gigabit parts. Now, I'm in the middle of lensing the sequel and thanks to the digital pipe they laid in my back-yard, I don't have to go over to the set any more to be me. I just show up in my Home Entertainment center in this blue jumpsuit, and the studio remotely renders my costume, hair-do, and skin color. Then, I motion-capture the action and I'm wrapped for the day. It's so liberating. This fall, I'm making three movies simultaneously—a spy-snuffer with Sean Connery, a Tibetan Western with Claire Danes, and an animated, space algae musical for Pixar."

"Yeah," I said. "Where there's Jobs, there's work."

Ms. Icepick sat like an alabaster goddess waiting for Braveheart to finish his digital rhapsody. Then, she spoke, "Once they make an algorithm out of my mouth, you can kiss these lips good-bye." She sank back, deep in thought, sipping her double sissy espresso, leaving us for a moment. When she came back, she stood up and said, "Take a look at these legs."

We did.

"The only way you can catch these pins is with silver. You know, real motion picture film. These gams only steam when you pull them out of the soup. Digital don't do it."

"Well," says Braveheart, "Claire Danes and I are going to ride digital yaks in our Western. They say it's the best ride in town."

Ms. Icepick looked like she was about to dispute the claim when The Terminator ambled over. He and I go way back to his gym rat days in Venice, when he didn't know whether he was Conan or goin,' so we speak straight to each other.

I say, "How much of you that we see on the screen these days is new media, and how much is real wetware?"

"What it's all about," says The Terminator, "is that I don't have the juice any more to pump myself up ten suit sizes when it's superhero time. I let the megachips do the dirty work. They beef me up with bits, so I don't have to break any barbells getting in shape. You know what SGI means to me? Save Gym Injuries."

"Right on," says Braveheart, pulling out a couple of chairs for The Terminator to straddle. "Last week I was checking out *Lethal Weapon Four* and, right away, I notice that my eyes are a full turquoise glint factor down from *Lethal Weapon Three*. I call the IT honcho at the studio and tell him straight, 'You get your wunderkinds to code me up the aquamarine peepers I had in *Mad Max*.' Check out *Lethal Weapon Five*; old Blue Eyes is back."

This was too much for Ms. Icepick. She left her seat like wind filling a new sail and was out the door and into her stretch-Humvee before the paparazzi could find an f-stop.

B-Heart and The T-Man started to arm wrestle for a thousand bucks a pop. The biceps and the stakes were too big for me, so I did a slow fade over to my virtual office at the end of the bar, where I nursed an enhanced Bloody Mary and pondered on how to carry on this chapter.

Then it came to me. Enough already with the practical, let's get philosophical. Here goes. Most of us reading this tract were born in the age of the analog and are fated to live out our lives in the age of the digital. We all grew up with photographs as the standard representations of reality. *Snap! Pop!* There it is, a three-inch-square, full-color moment of our past, frozen forever.

Photographs are analogs of our very selves, lifelike simulacrums of the history we have accumulated. Paste them in a scrapbook and return to them from time to time for a shot of happy memories. We wouldn't dream of capriciously tearing up those precious snapshots, unless, of course, they depicted an ex-boyfriend or ex-wife whose image we wanted to irrevocably expunge from our lives. These images are whole, integral, fixed, analog.

The digital representations of our lives are another thing entirely; they are ultimately deconstructable. *Click!* There's your picture on the computer screen. Hit a key and it dissolves into a billion bits. You want it back the way it was? No problem. Hit a key and there it is, as good as ever—maybe better. Don't like your hair color? Add some highlights with the swish of a mouse, and while you're at it, tighten up your tummy and beef up the beefcake on your significant other.

Of course, the downside is that we are forced to constantly compare ourselves and compete with the images of the touched-up people and touched-up places that bombard us from every angle in our digitally done-over, multiscreen society. "Ah," one wonders wistfully, "If only I had the picture-perfect complexion of that digitally altered model and lived in a place as peaceful and scrupulously manicured as that digitally altered neighborhood, how happy I would be."

In the good old days, as a simple analog vaudevillian, I was limited in the havoc I could wreak; but now as a digital content creator I have at my ready disposal an enormous bag of sorcerer's tricks and a vast audience of citizens ready to be totally disconnected from reality and community in return for being totally entertained from cradle to grave. Life in the digital world is becoming one great special effect.

Sounds scary? It is; but, no surprise. Western civilization has been running ahead of its capacity to socially cope with the technology it produces ever since the Spinning Jenny wrenched little Jenny from the English countryside and sent her spinning into the gin-soaked streets of London. We're simply experiencing the dislocation from the familiar and the alienation from the real at the speed of light, aided by a gaggle of Spin-Meisters and Little Piggy Marketers, who in a world teetering on the edge of ecological chaos, admonish us to close our eyes and "Just Do It."

So, fellow digital content providers, it has been given to us to wield the sword that cuts and pastes reality. Let us use it wisely, for the whole world—geezers, wheezers, hackers, and slackers—is waiting and watching.

FOREWORD

WHAT HAPPENED TO CONVERGENCE?

NOBUYUKI IDEI
CHAIRMAN AND CHIEF EXECUTIVE OFFICER,
SONY CORPORATION

For years, people have been predicting the convergence of media. It has yet to happen. Several years ago, Bill Gates said that PCs and TVs would converge. That did not happen. WebTV, which allows you to send e-mail using a TV, did not become a hit, although TV sales are as strong as ever, especially large-screen models.

Bill Gates also predicted that we would play games on our PCs. That did not happen on a large scale either. In fact, as of the beginning of 2001, Sony's PlayStation and PlayStation2 have sold more than 90 million units worldwide. And Microsoft is developing its own game machine, the X-Box.

When I tried to analyze why the PC and TV have not converged, I realized that the two have totally different characteristics. I call this the "30 Degree Principle."

When you watch TV, you usually lean back about 30 degrees and relax. TV is a passive medium and very easy to use. TV programming appeals to our emotional senses.

This commentary is excerpted from a speech given to the Pacific Telecommunications Council in Honolulu in January 2001. Although Mr. Idei's were made to a gathering of manufactures and business leaders, we felt there is relevance in these ideas for anyone in the digital content arena.

When you are using a PC, you usually lean forward about 30 degrees and focus. PCs are an active medium, somewhat complex to use. PCs involve logical thinking.

Although most of us would be happy when our Internet search produces a result within 10 seconds, no one would tolerate such slowness when TV channel surfing. TVs and PCs are so fundamentally different, the two shall never converge...at least not in the current environment.

INTERNET'S IMPACT ON CONVERGENCE

Development of Internet development has shown three distinct phases. The first period of the Internet was characterized by the emergence of cyber companies. Some of these companies created totally new businesses, such as America Online (AOL) and Yahoo!, whereas others were cyber forms of old businesses, such as Amazon.com or E-Trade.

The second period of the Internet was characterized by the comeback of real companies. Established companies, such as Merrill Lynch and Barnes & Noble, realizing the impact of the Internet on their business, began staging a comeback by establishing their own cyber presence. The Empire Strikes Back, so to speak.

The third period of the Internet has been characterized by the convergence of cyber and real companies. This period was set off by the announcement of AOL and Time Warner's merger in January of 2000.

PARADIGM SHIFT

The Internet has caused a fundamental paradigm shift from the Industrial Age to the Information Age.

The United States, which was the first to embrace the Information Age, has reaped the greatest benefits. According to *Digital Economy 2000*, the U.S. economy expanded for 10 years in a row and is only recently beginning to slow. World Competitive Scoreboard ranks the United States number 1, whereas Japan is number 17.

FROM NARROWBAND TO BROADBAND

Now we are entering into the fourth stage of the Internet era, the transition from narrowband to broadband Internet services.

Broadband encompasses multiple distribution channels (telephony, cable, satellite, wireless), multiple access avenues (PCs, PDAs, TVs, refrigerators, and other non-PC devices), and multiple content choices (text, images, music, video, one-way and interactive, personal, and mass media), delivered at 30 to 100 Mbps. A CD takes 23 hours to download on current narrowband networks, but using 30 Mbps, it would take just 3 minutes. (1–5 Mbps is not broadband, only high-band.)

The characteristics of broadband communications are that it is a *pervasive network* (always on, everywhere, like oxygen), and that it can provide *customized information*. Information comes to you, not the other way around.

BROADBAND AND CONVERGENCE

Broadband will trigger the first true beginning of convergence, but unlike past predictions of convergence, I do not think it will be PC-centric.

We see the early signs of convergence with the Web-capable i-mode cellular telephones in Japan. In less than two years, there are now more than 17 million subscribers.

The strong popularity of Sony's Video Audio Integrated Operation (VAIO) computers is also a hint of the type of convergence to come. When we introduced VAIO in 1997, we tried to create a totally new concept in personal computing. VAIO is an entertainment rather than productivity tool. Since introducing the VAIO 505, an ultralight notebook PC, we have expanded the subnotebook market almost 15 times, and Sony has consistently commanded more than half of the market share in this category.

STRATEGIES IN THE BROADBAND ERA

Using broadband technology, TV, games machines, PCs, and other information devices will be connected to the network.

Because of this trend, Sony's strategy has changed. Sony has for many years been an audio visual (AV) champion in the manufacturing era, but we know the era of the stand-alone AV product is over. Now, Sony is pursuing a comprehensive strategy to become a Personal Broadband Network Solutions Company. We are doing so by making easy-to-use network gateways. (Currently, our four main gateways are digital TVs, PCs, the PlayStation, and mobile products.)

By using our assets in movies, music, and games, we are also developing unique network content for these network gateway products. And to differentiate ourselves from our competitors, we are creating new network services and applications.

We call this integrated business model strategy the Sony Dream World. Our goal is to provide new forms of entertainment lifestyles for the broadband age.

IMPACT OF BROADBAND ERA

I recently read a little book called, *Who Moved My Cheese?*, by Dr. Spencer Johnson. It talked about the difficulties of adapting to change. For the telecommunications, broadcasting, media, and computer industries, now is a time of great change.

I would like to ask you, "Where is your new cheese in the broadband era?"

Today we have cable, satellite and terrestrial broadcast, public telephone networks, mobile networks, and home networks. Up until now, each was in a distinctly different industry. In the coming broadband era, I think these networks and industries will no longer remain separate. For example, there was once a clear distinction between common carriers and IP carriers. One handled voice, the other handled data. In as soon as five years, the two will converge.

In addition, companies like Sony, Cisco, Hewlett-Packard, and others involved in the development of network-related hardware may become a new breed of communications company.

This is both a threat and an opportunity. Rather than fight for a smaller piece of the same old cheese you have been eating for years, the challenge is to take this opportunity and start thinking about new business models and new alliances with other companies, industries, and nations.

CHAPTER ONE

A DIGITAL PRIMER, SCHUBIN-STYLE

MARK SCHUBIN

Editors' note: Mark Schubin is a engineer, writer, historian—and for anyone who has seen him in the aisles of trade shows or giving a technical presentation, one of the most memorable people you will ever meet.

It is a simple matter of fact that when you have a question at any technical level, you call Mark (the editors of this book included). So, it seemed to be a simple choice to ask Mark to start off this book with a Digital Primer. Easier said than done.

Mark has been writing about the technology of the television industry, in which he works, for over 25 years. In fact, his work has appeared in every issue of *Videography* magazine since it premiered in 1976. Perhaps, Mark suggested, the best primer could be found in his volumes of writings.

So, this is Mark's primer, derived from four of his past *Videography* columns (with the date of their original appearance indicated).

NUMBERS, PLEASE (JUNE 1988)

If you're confused about the brave new world of digital video technology, you're not alone. Fortunately, the digital domain isn't as mysterious as it may seem.

At the 1973 convention of the National Association of Broadcasters (NAB), the introduction of what appeared to be the first digital video product, a timebase corrector (TBC), occurred. Fifteen years later, not only was every TBC on display a digital one, but there were also digital graphics stations; digital audio and video recorders using tape, discs, and solid-state memories; digital video and audio effects devices; digital test equipment; picture monitors with digital inputs; digital audio and video transmission systems; and even headphones and cameras with the word *digital* on them.

Has television gone digital? If so, is that a good thing? Read on. You should be able to follow the processes described here—digitization is not really that complicated. Here's what digital means to the world of television equipment:

There are five obvious aspects to the technology of television. There are, of course, the mechanisms necessary to capture or create images and sounds (*acquisition*); there are systems to store them (*storage*); there are ways of changing them (*processing*); there are means for getting them to viewers (*distribution*); and, finally, there are devices that change video and audio signals back into images and sounds (*presentation*). There's also an unobvious aspect to the technology of television, one into which such devices as sync generators, VU meters, and oscilloscopes fall, which will be called *operation* here.

Digitization has not affected all of these aspects equally, and that's as it should be. Digital simply means numerical. Anything that can be expressed in numbers can be expressed as digits. Your birth date, your telephone number, your driver's license number, the number of books in your library, and the number of hours of television you watch each year can all be expressed digitally.

On the other hand, certain things do not seem to lend themselves to digital expression. Can your home, your car, your personality, or your face be reduced to numbers? Can images and sounds? The answer to that last question appears to be a qualified "Yes." The reduction of images and sounds to numbers (digitization) may have been going on longer than you think.

There are three steps to the process of digitizing something. First, it must be *sampled*. Second, it must be *quantized*. Third,

JURASSIC VIDEO

Need a date for the origin of digital? How about 395,000,000 years ago. That's about 205 million years earlier than the beginning of the Jurassic period. *Crossopterygii*, lobe-finned fish, predate not only mathematicians but also dinosaurs.

Sometime in the Devonian period, it seems, a fish was born with bony fins that eventually evolved into paws, feet, and hands. The fin bones of *Crossopterygii* included those that would eventually form what the ancient Romans would call *digiti*, fingers and toes. There were certainly many genera that evolved between the Devonian *Crossopterygii* and the Roman *Homo*, but while experiments have proven that modern animals have a sense of number, there's no evidence that they can count. Humans can, often assisting themselves by touching parts of their bodies.

There have been cultures that have counted knees, elbows, and other readily identifiable body parts. Today, we usually count only on fingers, but the French *quatre-vingts* and the English *score* indicate that it wasn't all that long ago that we stopped using our toes, too.

Ten fingers led to the base-ten number system, but, other than its human-hands orientation, there's not much to recommend it. Near Bombay, the base five Maharashtra number system can be counted on one hand. The ancient Sumerians used a base-60 (sexagesimal) system, vestiges of which remain in the way we tell time and use a compass. Sixty has the advantages of being evenly divisible by all six of the first positive integers.

In the Eighteenth century, the French mathematician Joseph Lagrange proposed a number system based on primes, indivisible by anything but themselves and one (but he was also responsible for the metric system's decimal base). His contemporary, naturalist George Buffon, argued

*This was a sidebar to the column "The Great Finger Twirl," which appeared in *Videography*, September 1993.

in favor of shifting to base-12 (duodecimal) to retain at least divisibility by all positive integers through four; the Duodecimal Society of America keeps the latter idea alive with its numerals from zero through nine plus X and E (the last two pronounced "dek" and "el").

Buffon and Lagrange both had the advantage of printed numerals, said to be Arabic, though, in fact, the symbols we use for four and five were developed in Europe, our zero came from India, and, except for seven and eight, the other symbols can be traced to other non-Arabic cultures. About the only thing Arabic numerals have in common with each other is that they're not finger-based, like Roman numerals (*I* is the shape of a finger, *V* is a hand with the thumb spread, and *X* is two hands wrist-to-wrist.

The term *Arabic numerals* is simply one of many we use to honor abu-Ja'far Mohammed ibn-Musa al-Khowarizmi, an early-Ninth-Century mathematician in the court of Mamun in Baghdad. Latin translations of his works brought not only Arabic numerals but also much of medieval Europe's mathematical knowledge. The first word of his book *Al-jabr w'al-muqabalah* became what we now call "algebra"; the last part of his name is familiar to those working in digital video as "algorithm."

Even today, however, machines cannot always read handwritten Arabic numbers accurately. A modern European might write a one in a manner looking remarkably like an American's seven. Accordingly, Europeans place a dash in their sevens, but Americans don't. Computers are easily confused by those variations and others. Is a dash very close to a seven part of the numeral, or is it a minus sign? Is a strange-looking cross a crudely drawn four or should it be a plus sign? The only number system clearly distinguishable to a machine is one consisting only of two states: on or off, yes or no, present or absent, 1 or 0.

No one can say who first considered a base-two (binary) number system. The concept appeared in al-Khowarizmi's work, but it had certainly also been known to earlier thinkers. Long before the first electronic computer graphics or music devices, machines, operating on binary mathematical principles, created both graphics and music. One such music machine, an ancient form of an organ, was excavated at a Roman archeological site just north of Budapest.

It wasn't until the Twentieth century, however, that binary mathematics was applied to the recording, transmission, or manipulation of sounds or pictures. The problem was largely technological. First, there weren't even electronic sounds (let alone pictures) to record, transmit, or manipulate until the end of the Nineteenth century. Second, the circuitry required to digitize even a simple telephone call didn't exist until shortly before that feat was achieved, in 1939 (it was the analog-to-digital converters that slowed things down; digitally generated speech and music predated the digitization of sound, just as computer graphics predated the digital video timebase corrector and international standards converter).

Scientists working on digital signals weren't even able to use today's common term for the little pieces of information they dealt with until 1948. In July of that year, in the *Bell System Technical Journal*, Claude Shannon, considered by many the creator of information theory, credited J.W. Tukey with suggesting a contraction of the words *binary digit* into *bit*. But the word *digital* definitely comes from the Latin *digitus*, which means fingers and toes, and they all come from a fish that lived hundreds of millions of years ago.

it must be *coded*. Sampling is the process of examining a signal at distinct moments in time, quantization is the process of assigning a numerical value to each sample, and coding is the process of changing that numerical value from everyday numbers to computer numbers.

The coding process is really almost trivial. Suppose some digitization process seems to require the ability to assign 100 different numbers to samples. In digitization, the quantity of numbers affects the quality of signal. Those numbers can be expressed in base-two (binary) arithmetic by the digits 0000000 through 1100011. As many as 128 numbers can be coded into seven *bits* (a contraction of the word *binary digits*), 256 into eight, 65,536 into sixteen, and so on. The advantage of such coding is that computer circuits can deal more easily with the ons and offs of base-two arithmetic than with the ten conditions required for base-ten arithmetic. The disadvantage is that 1100011 takes up seven digits, 99 only two. The first process, sampling, may not seem as simple as coding, but it can be proven mathematically to be 100-percent reversible (which means that a signal can be sampled and then reconstructed with no degradation). Real-world processes (the kind you experience every day) are continuous; numbers are not. Therefore, to digitize a real-world process, such as vision, it must be broken up into moments of time before if can be assigned numbers.

The idea of breaking vision up into time intervals dates back at least to 1821, but the ancient Roman philosopher Lucretius may have set forth the principles even earlier, in his master work, *On the Nature of Things*. By 1829, Joseph Plateau had already published a paper on the "fusion frequency," the basis for the frame rates of motion pictures and television. Surprise! Television frames are sampled in time; television has always been a sampled process.

In fact, television images are sampled not only in time but also vertically, by the scanning lines. This idea was the basis of a patent awarded for a facsimile transmission system in 1843 and had been known earlier. Therefore, the idea of sampling television images in their one remaining dimension, the horizontal, is not a radical one.

There's a simple law that allows the sampling process to be perfectly reversible. The signals must be sampled at more than twice their highest frequency. Although a proof of that law would involve more complex mathematics, you might get a feel for it by considering yourself an alien astronaut sent to the Earth to determine the length of an earth day. Your spacecraft has no windows, and you have only a limited amount of alien atmosphere in your space suit. If you ventured out every hour, you could determine after a couple of days that an earth day lasts 24 hours, but you'd also run out of alien atmosphere. If you ventured out every 24 hours, you'll never learn the length of an earth day.

If you ventured out every 23 hours, you could incorrectly determine that an Earth day lasts 24 23-hour periods, and, if you happened to glance at a clock, you'd think it ran counter-clockwise. That's what's referred to in digital video as an *alias*, something that is not what it appears to be. Clocks don't run counter-clockwise. Earth days don't last 552 hours.

If you ventured out every *12* hours, and you happened to start at midnight, you'd quickly determine that an Earth day lasted 24 hours. Venturing out every 12 hours is sampling the earth's rotation at exactly twice its frequency.

If you ventured out every 12 hours but you started at six a.m. in the Spring or Fall instead of midnight (and you were facing north or south) you'd see twilight at dawn followed by twilight at dusk followed by twilight at dawn, and so on. You could incorrectly determine that an Earth day lasts forever and never gets too bright or dark.

That's why the law of sampling requires that the sampling rate be *faster* than exactly twice the highest frequency, or, to put it another way, that the highest frequency to be sampled be less than half the sampling rate. If you venture out faster than every 12 hours, you'll eventually determine the correct length of an Earth day. Back in the world of electronics, since there are usually spurious signals above the highest desired frequency, a filter—called an *antialiasing filter*—is usually imposed prior to sampling, to eliminate undesired elements.

Unlike coding, sampling, therefore, places a restriction on the quality of the digitization process. It can never handle any frequencies greater than one-half the sampling frequency. The quantization process introduces restrictions, too, and is the only one of the processes that is not perfectly reversible.

Consider your digital characteristics again. Your telephone number is a number, plain and simple. But the number of hours you watch television in a year needs to be quantized. First, a number of quanta need to be selected. There are 8,760 hours in a year. Let's say those hours are the quanta, the number of numbers we will deal with in this process.

Now, suppose you watch 1,460 hours, 36 minutes of television a year. There's a 1,460-hour quantum and a 1,461-hour quantum, but no 1,460-hour-36-minute-quantum. If you're placed in the 1,461-hour quantum, there's an error of 24 minutes; if you're in the 1,460-hour quantum, there's an error of 36 minutes. These errors are correctly called *quantization error*. They are what prevent quantization from being perfectly reversible (in other words, free from degradation). Some people refer to these errors as *quantization noise*. In fact, they're usually much more a distortion than a noise, a problem that can be eliminated, strangely, by intentionally introducing a small amount of real noise (sometimes called *dither*) into the digitization process. Either way, the number of quanta determine the maximum signal-to-noise ratio of a digital process.

Given all of these potential problems, why would someone want to digitize video or audio signals? Well, in the aspects of storage and distribution, it seems a wise move. Transmitting or recording a television signal is much like the traditional children's game of telephone. By the time the information gets where it's going, it has become garbled. Numbers, however, can be continuously regenerated with no change. A copy of an analog tape is said to be *down one generation*; it has picked up noise and distortion. A copy of a digital tape is a clone; correctly copied, it is indistinguishable from the original.

It may seem surprising then that it wasn't until 1986 that the first digital videotape recorders went on sale, and that their sales, for years, were all but insignificant compared with sales of analog videotape recorders. Part of the problem is the number of bits demanded by the sampling, quantizing, and coding process, and part of the problem is the restrictions caused by quantization and sampling.

Consider all of the various analog videotape format improvements. Low band, two-inch quadruplex tape, the first format to be introduced (in 1956), begat high-band and then super high-band. Betamax begat SuperBeta and Super Highband Beta; VHS begat VHS HQ; U-matic begat U-matic SP; Betacam begat Betacam SP; and Hi-Fi audio revolutionized Betamax and VHS.

Analog formats can be improved; digital formats cannot. The maximum playback frequency of a digital format is determined by its sampling rate, and its maximum *signal-to-noise* ratio is determined by the number of quanta (in effect, determined by the number of bits per sample). Had a digital videotape recorder been possible in the early days of quadruplex recording, it almost certainly would not be used in today's high-quality analog world.

That there was no digital videotape recorder in those days is attributable to the demands of sampling and coding. To record the highest legal broadcast video frequency of 4.2 MHz (4.2 million cycles per second) demands a sampling frequency of more than 8.4 MHz. Multiply that by eight bits per sample and you get 67.2 Mbps (million bits per second), more than six times the highest frequency being recorded on analog tape (modern standard-definition equipment samples at the even higher rates of 13.5, 14.3, or 17.7 MHz). The same problem kept digital video out of its other obvious aspect, distribution. A television channel is 6 MHz wide. How could no less than 67.2 Mbps be squeezed in there?

It may not be surprising, then, that what appeared to be the first digital video product showed up in the operations aspect of television, rather than in any of the more obvious aspects. Timebase correctors need to correct timing errors. Digital samples are easy to clock at a steady rate. Therefore, why not a digital timebase corrector?

The TBC led to the frame synchronizer, and it, storing an entire picture digitally, led to the first digital video effects—picture freeze, and quarter-size compression—moving digital video from the operational aspect into the processing aspect. Quarter-size compression led to variable compression, push-on and push-off, mirror images, and image multiplication. Next came rotation, then perspective, and finally curved effects, such as page turns and sphere wraps—first introduced in Mirage, what Quantel called "the ultimate illusion."

Chyron, meanwhile, called its Scribe "the ultimate character generator." Character generators predate digital TBCs, which is why the TBC was not really the first digital video product and why acquisition was actually the first television aspect to use digital technology. The history of character generators and electronic graphics systems is lengthy. Suffice it to say that the use of digital video in this aspect has grown as fast as has its use in effects.

In fact, a curious situation has occurred. The lines between the concept of creating images and the concept of processing them have blurred tremendously. One of the most common terms on the show floor at the 1988 NAB convention was "workstation," an area where someone can create or acquire images, manipulate them, animate them, record them, edit them, add text (if desired), repeat any of those processes, and finally squirt them out into a videotape recorder, a videodisc system, a film recorder, a hard-copy printer, or even a live production.

The recording aspect of the workstations (a natural for digital video, given the multiple generations that may be involved) brings even the storage aspect into play—everything except distribution and presentation.

Even the latest videotape formats that might be used to feed or be fed from workstations seem instantly accommodated. No sooner did JVC announce the first professional S-VHS recorders than Pinnacle showed up, right at JVC's press conference, with an operating Super-V1000 workstation, ready for the new format.

But television consists of both video and audio. Has audio been left behind in the application of digital technology? Hardly. Although the three steps of digitization (sampling, quantization, and coding—collectively known as *pulse code modulation*, or PCM) were actually invented for audio transmission in 1939, the use of digits in recording and distribution has been slow to be accepted in audio, too.

If the data rate of digital video prevented its use in recording, however, digital audio was able to make use of *analog* video. A typical digital audio rate for stereo (48 kHz sampling × 16 bits × two channels), 1.536 Mbps, while extreme for an audio recorder, was a cinch for a video recorder, allowing digital audio to be coded as a video signal and recorded on even analog consumer formats. Thus, digital audio was able to make inroads into the storage aspect in the 1970s; all CDs were mastered using digital audio recorded on 3/4-inch U-matic videotape.

Still, digital audio for video followed much the same path as digital video. *Time code*, a digital signal, and one that might be considered part of the operational aspect, made audio sweetening a reasonable process by freeing VTRs from audio duty. Next, seemingly, came the first digital audio processing equipment—*delay lines*. These were used for much the same reason that digital circuitry was used in TBCs, to clock and delay signals. From the delay lines came other processing equipment: reverberation systems, pitch changers, and the like.

Meanwhile, just as character generators and graphics systems evolved separately from processing equipment in video, so did audio synthesizers evolve separately from audio processing equipment, eventually adding "sampling" capabilities similar to the image capture capabilities of videographics systems. And, just as the lines blurred between acquisition, storage, and processing in video workstations, they've done the same in audio workstations.

But is there digital outside of workstations? Sure. In the storage aspect there are now digital videotape recording formats, in addition to magnetic and optical disk systems and even no-moving-parts recorders. Digital audio is similar.

Outside of workstations, digital video and audio processing equipment continues to flourish, and the distribution aspect, recently used to reach audiences directly (via DTV, digital cable, digital satellite, streaming media, and even DVD), also flourishes in inter- and intrafacility transmission. There are digital audio satellite links, and digital video coaxial cable and fiber optic links, to mention just a few products serving this aspect.

In the presentation aspect, Texas Instrument's Digital Light Processing (DLP) video projectors are truly digital displays. Only the non–image-creating acquisition aspect (cameras and telecines) remains analog. Charge-coupled device (CCD) cameras aren't truly digital devices because they have neither quantization nor coding. All they do is add horizontal sampling to the vertical and time sampling that television already has. So-called "digital" cameras may use digital processing *after* their acquisition chips, but the chips themselves are analog.

So, is this headlong charge into digital a good thing? Clearly, some digital equipment costs a lot and some costs a little. Some do things that couldn't be done before; some do things that could be done before, but better. And, since all timebase correctors are digital, virtually everything anyone has watched on television has, for decades, been digitized at some point.

For a given highest frequency and a given signal-to-noise ratio, there's nothing inherently wrong with digital recording or transmission. Analog-to-digital converters, digital-to-analog converters, and digital processing, however, can really mess up signals. It is now well known that all CD players do not sound alike and that at least some CDs sound worse than their phonograph record counterparts. Many of the causes have been identified and corrected, usually by adding the correct amount and type of digital noise to eliminate distortion. Similar steps have been taken in digital video.

The future of television production and postproduction will undoubtedly be still more digital. It's virtually impossible today to imagine professional videotape recording without either digital timebase correction or digital storage, television titling without character generators, postproduction without digital video effects, or consumer Hi-Fi without CDs.

If the process of digitization introduces flaws it's not unique, and there's every reason to believe someday those flaws will be eliminated. The process of NTSC color was introduced in 1953. It has flaws: the rainbow hues of cross color that appear in pinstripe suits; the dots of cross luminance that crawl around colorful graphics; the smear of reduced color detail; and so on.

Through diligent effort, television engineers came up with NTSC encoding systems that reduce or eliminate those flaws, 35 years after the introduction of NTSC color. Perhaps digital's flaws will be eliminated a "bit" faster than that.

ARE BITS BITS? (FEBRUARY 1996)

Chicken parts can be stored in a wide range of containers. So can digital videography parts. Or can they? There's no question that videography is getting ever more digital. Prior to the advent of the character generator, essentially nothing associated with television or video was digital except the channel numbers of broadcast stations. Then came digital timebase correctors, frame synchronizers, digital effects systems, broadcast electronic "paint" systems, digital videotape and videodisc recorders, digital nonlinear editing systems, digital processing circuitry in cameras, and more.

The CBS technologist Frank Davidoff once predicted, back when few people foresaw an inexorable change in the industry from analog to digital, that the then few digital islands in an analog sea would eventually change to a few analog islands in a digital sea. Today, when it's possible to create, modify, distribute, and store programs entirely in the digital domain, that digital sea, like the salty one in the movie *Waterworld*, seems ready to rise to cover most of the analog islands left. Only cameras, microphones, displays, and soundmakers seem destined to retain some analog circuitry (but see the sidebar, "Digital Is Analog").

The flood tide of digital seems an obvious development when one considers the technology's two massive benefits, both derived from the "atoms" of digital chemistry, binary digits or *bits*. By definition, bits can have only one of two states: 0 or 1, on or off, positive or negative, high or low, broad or narrow, etc.

That two-state nature leads to the two digital benefits. First, a digital signal must seemingly be incredibly robust. Whereas an analog video signal might have to represent a subtle color shade and could easily cause the hue, color saturation, or brightness of that shade to change because of poor recording or transmission or noise or interference, a digital video signal would represent that same shade as some combination of bits, each of which could be only on or off. It would take a tremendous amount of noise or interference to change an on to an off or vice versa.

Thus, digital signals are said to be perfectly recordable and transmissible. Copies of analog recordings are somewhat degraded from the originals; copies of digital recordings can be perfect clones, indistinguishable from the originals.

Second, all bits must seemingly be identical, since all can only be on or off. Any circuit that can process one form of bits must be able to process *all* forms of bits. Thus, general-purpose computers have been turned into video paint systems, digital

DIGITAL IS ANALOG

Perhaps the single most bizarre aspect of the digital revolution is that it is being made possible by analog circuitry. Manufacturers will claim that a particular type of digital recorder works better than another. If both use only 1s and 0s, how could there be any difference? Transmission system proponents advocate one form of digital signal distribution over another, claiming superior performance. Again, if bits are just bits, how could one be better?

Unfortunately, although the world of videography is becoming ever more digital, the *real* world, above the level of quantum mechanics of subatomic particles, is essentially analog. It's easy to describe a bit as being either on or off, but how can such conditions be transmitted or recorded?

Consider a simple example. A person acting as a digital transmitter reads a list of 1s and 0s and "transmits" them to a person acting as a digital receiver in another room. The transmitter flips a light switch on or off at the beginning of every minute, depending on the bit to be transmitted. The receiver glances at a clock and marks down a 1 or a 0, depending on whether the light is on or off at the beginning of the minute.

If the two clocks are perfectly synchronized, and the transmitter flips the switch at the beginning of each minute, and the receiver notes the condition 30 seconds later, all is well. Suppose, however, that the receiver checks the condition at the beginning of each minute, too. Sometimes the transmitter will have flipped the switch by then, and sometimes not. Errors will be introduced.

Suppose the clocks aren't synchronized. As one clock speeds up or slows down, errors will be introduced.

Suppose bright sunlight makes it difficult to see whether the lamp is on or not. Suppose a power brownout dims the light. Suppose the whole process gets sped up. Suppose it's now a bit per second instead of one per minute. Suppose it's faster still. It will take time for the filament in the light bulb to cool down and extinguish or to heat up and illuminate. In fact, the power fed to light bulbs in the United States normally is switched on and off 120 times a second, but, at that rate, the filament never cools down, and the illumination is constant.

Digital video signals tend to require exceptionally speedy switch flips. The video signal of a U.S. analog television broadcast never alternates between positive and negative any faster than 4.2 million times a second. The common digital version of the same signal has bits that occur roughly 114.5 million times a second. To transmit the digital signal in the same space (bandwidth) as the analog would require 28 bits to be carried in each cycle (positive/negative alternation).

One bit has only two possible conditions, on or off. Two bits, however, have four possible conditions, on, mostly on, mostly off, or off. Three have eight. Twenty-eight bits have 268,435,456 possible conditions. Imagine the human digital receiver having to write down not merely whether a light is on or off but which of 268,435,456 levels of brightness it has, a situation that completely eliminates any advantage that the robustness of digital signals is supposed to provide.

Therefore, digital video signals are processed to reduce the number of bits they contain ("compression"), require larger-capacity transmission and storage channels than their analog counterparts, and/or use more than two states per bit in those channels.

Even with all of that, digital storage and/or transmission remains an analog process involving circuitry that takes time to change states and causes variations during the changes. How much magnetic flux constitutes an on state; how little an off? How high a voltage is a 1; how little a 0? How long is a clock period? And what happens when lightning strikes near a receiver, a tape coating is uneven, or an optical disk gets dirty?

These anomalous events can confuse digital receivers. Suppose our human digital receiver hears a sneeze from the north at the same time that the lamp is off. The system could get overloaded. Instead of a small entry of off, the receiver could have *bit off northerly kerchoo*.

audio workstations, nonlinear editors, color correctors, effects processors, noise reducers, and more.

Similarly, the bits of videography can be stored in any container that can hold any other form of bits: silicon-based random-access memory (RAM) chips or cards, magnetic bubble memories, ancient bead-and-wire core memory, magnetic tape, optical disks, magnetic disks, optical tape—the possibilities seem endless. But are they really?

As one saying has it, to err is human, but to really mess up requires a computer. A copy of a digital recording *can* be a perfect clone, but, if there are problems during the copying, it can also end up in worse shape than an analog recording. It takes a lot to change an on to an off, but it doesn't take quite as much to change a low high into a high low or a broad narrow into a narrow broad.

Poor reproduction could change an analog pink to rose or red or, perhaps, orange. An error in the least-significant bit representing the pink in a digital recording would probably be unnoticeable; an error in the most-significant bit could turn white, black. An error in the bit representing the start of a frame could render the whole frame unwatchable.

These problems, however, are probably the least to be worried about relative to digital storage options. A horrible error in a video recording could theoretically render a picture unwatchable, but an error in an electronic funds-transfer operation could cost billions of dollars. If digital storage devices are secure enough for the latter, they must surely be good enough for the former.

Raw video or audio data is well protected by layers of error correction coding, and digital storage systems have sometimes proved themselves in the financial industry (e.g., Ampex's DST) before being used in the videography industry (e.g., Ampex's DCT). Ordinary bit errors are no more significant in videography than in any other field—in fact, they're usually much *less* significant. The same is true of data encryption to prevent stored digital information from falling into the wrong hands.

There are, however, a number of areas where videography is different in its requirements for video storage from such other computer-using industries as finance and education. One is in the vast quantities of information that moving images can require. A sum in the trillions of dollars, even including error protection, can be dealt with in less than 200 bits; a typical two-hour movie, at full resolution, could represent some 31,104,000,000,000 (see the sidebar, "Grouchibytes").

Another area of video storage difference is in data-transfer rate, the speed at which bits can be recorded or played from a storage device. As noted previously, ordinary composite (uncompressed) digital television typically has a data rate exceeding 100 million bits per second; digital high-definition television recorders operate today at rates exceeding a billion bits per second. Non–moving-image data travels very successfully at much slower data rates. Even the somewhat pictorial World Wide Web is typically accessed at data rates of only 28,800 bits per second or less.

The massive storage and high transfer-rate requirements referred to apply to all or most forms of moving-image digital storage (bit-rate reduction, commonly called *compression*, can reduce the requirements perhaps a hundredfold, but that's still a lot of bits). The postproduction process imposes still more requirements.

Tape, for example, even in its current, surface-recording form, is a tremendously efficient digital recording medium. Well over a trillion bits can be recorded on some cassettes. Unfortunately, getting from a bit near the front of the tape to a bit near the end of the tape requires spooling the tape from one end to the other, a process that doesn't mesh well with the requirements of nonlinear editing. It's possible to use multiple tape decks to improve access time, but it still won't be near-instantaneous.

Then there are videotape's so-called "stunt" modes. All current videotape recorders use *helical scanning*: The tape is wrapped around a spinning head drum in a path comprising a portion of a helix. When the tape is stopped, the drum can continue to spin and provide a still image. At non–play speeds, helical scanning can offer, in both forward and reverse, slow-motion, fast motion, and high-speed searching with visible images.

There is an even more efficient way to record data on a tape by recording *into* the tape instead of on its surface. This more-efficient technique (in terms of information-packing density) is

GROUCHIBYTES

As the detail of electronic imagery and the need to manipulate it in real-time increase, so do demands on storage capacity. In Kodak's Cineon electronic film system, a single 35 mm film frame is given approximately 3,000 × 4,000 picture elements of resolution, or 12 million points. Each of those points is assigned values of red, green, and blue components, and those values might be chosen from a range of perhaps 4,000 possibilities, or 12 bits worth. Thus, a single 35 mm film frame could generate 3,000 × 4,000 × 3 × 12, or 432,000,000 bits, not counting any error correction or other coding.

At this point, absolute numbers become unwieldy, so it becomes appropriate to use mathematical prefixes. The same number of bits may be expressed as 432 megabits (Mb) or 54 megabytes (MB). (It is worth noting that, while standard mathematical prefixes increase by factors of 1,000, in the computer industry prefixes sometimes increase by factors of 1,024. For the special case of the first such factor, the distinction is normally noted by case as follows: 1 kB is 1,000 bytes; 1 KB is 1,024 bytes. In this section of the book, for convenience, factors of 1,000 will be used).

That 54 MB, however, represents only a single frame. There are calls for moving imagery to have no less than 72 frames per second. That's 3,888 MB per second, allowing the next prefix to be invoked: 3.888 gigabytes per second or roughly 4 GB/s.

There are 3,600 seconds in an hour, which means 14,400 GB of storage for this hypothetical uncompressed high-resolution imagery. The next prefix makes that 14.4 terabytes (TB). A typical, uncompressed movie could easily fit on three 10 TB disk drives (whenever such drives are introduced).

A week has 168 hours. That would require 2,419.2 TB of storage. Moving to the next prefix changes that to a "mere" 2.4 *petabytes* (PB). There are 52 weeks in a year, and programming services like DirecTV may offer some 200 channels. That would be 24,960 PB of high-resolution programming for DirecTV in one year. Moving to the next prefix makes that a "manageable" 24.96 *exabytes* (thus the name of that storage equipment manufacturer) or roughly 25 EB.

DirecTV, however, is only one programmer. Suppose a thousand programmers decide to offer the same amount of programming. That would be 25,000 EB or just 25 *zettabytes* (ZB). And, if surround video someday requires a thousand times more information, there still won't be any problem. One more official metric system (SI) prefix allows those 25,000 ZB to be expressed as just 25 *yottabytes* (YB).

It's easy to be lulled into a sense of simplicity by the use of these prefixes, but 25 YB is, in reality, 200,000,000,000,000,000,000,000,000 bits. And if those programmers decided to store ten years of material instead of just one?

The official prefixes run out after 24 zeroes (*septillions* in American terminology, *quadrillions* in British), though they also cover quantities smaller than one. A *millihelen*, for example, would be a thousandth of a Helen, or the amount of beauty, according to Greek mythology, necessary to launch a single ship.

Micro- is the prefix for millionth, *nano-* for billionth, *pico-* for trillionth, *femto-* for quadrillionth, *atto-* for quintillionth, *zepto-* for sextillionth, and *yacto-* for septillionth. The last few, no doubt, inspired one person to suggest that the next prefixes be *harpo-* (for octillionth) and *groucho-* (for nonillionth). In the other direction, *harpi-* would connote octillions and *grouchi-* nonillions.

Many in the digital storage industry have already adopted these new prefixes, although metric officials have not yet leapt at the chance. Thus, there is absolutely no need to worry about storage capacity. A single one grouchibyte memory will easily handle all of the world's videographic needs until long after we're all dead.

called *perpendicular recording*, but it has a flaw relative to the storage of video information: It normally requires recording heads on both sides of a tape, something not possible when tape is wrapped around a spinning drum. Perpendicular recording works fine for recording financial data, but it was not designed with stunt modes in mind.

Disk drives are well suited to the random-access requirements of nonlinear editing, although their capacities are more limited. A television commercial or even a scene of a movie might not require an exceptional amount of disk capacity, but a documentary, with hundreds of hours of material to choose from, could.

Videography, however, also introduces other issues relative to disk drives. Disks spin, and always in the same direction. Video postproduction sometimes requires imagery to move backwards; financial data never need be read in reverse. Editing, depending on system design, can also result in fragments of data being recorded all over a disk, reducing already limited access speed and transfer rate.

No-moving-parts memories can be used in conjunction with disks to help alleviate some of these problems. By themselves they offer certain advantages—usually instant access, high transfer rate, no concerns about direction—but poor capacities, although they are sometimes used (notably by EVS) as slow-motion recorders and even for broadcast spot playback.

The concept of mixing digital storage technologies extends beyond RAM and disk. Robotic tape libraries are being used to provide what one manufacturer, Storage Tek, referred to as *near-line storage*. Because the tape drives feed only disk drives, they need neither video transfer rates nor stunt modes. And, because the robots can select from hundreds of cartridges or cassettes in large libraries, users have access to vast amounts of information—capacities measured in terabytes (trillions of eight-bit bytes). The sorts of tape cartridges developed by Digital Equipment Corp. and IBM for non–moving-image data are now finding themselves in the heart of a number of video facilities.

Conversely, the helical-scanning technologies developed for videotape recording are proving popular in non–moving-image applications. Not only are D-1, D-2, D-3/D-5, DCT/DST, and

Digital Betacam cassettes and transports being used, but even Sony's postage-stamp-sized NT ("Scoopman") digital audio cassette formed the basis of Datasonix's Pereos data recorder.

Completing the circle, some moving-image applications use non–moving-image digital tape recorders based on moving-image versions. Cineon workstations have used DST cassettes; competitor Quantel's Domino has used D-1. The Fox Movietone newsreel library was transferred to a unique electronic format but was recorded on a data-recording version of D-1 cassettes.

Thus, while there are considerations that make recording digital video not necessarily as simple as recording other forms of data, there are systems, subsystems, and techniques that seem to eliminate or reduce concern about those considerations. John Watkinson, who literally wrote the book about digital video (*The Art of Digital Video*, Focal Press, 1994), once expressed incredulity during a keynote speech to a winter conference of the Society of Motion Picture and Television Engineers that anyone still made special-purpose recorders for specific forms of video. In Watkinson's view, all recording should be done in non-specific "bit buckets," with a computer figuring out what got recorded where and when.

Buckets are also used to contain chicken parts at fast-food takeout restaurants. A commercial for a fast-food chain some years ago showed a wary customer asking a counter clerk what kind of parts comprised an order of "Chicken Parts." The clerk's response became a line used widely: "Parts is parts."

Despite video's peculiar requirements, it seems that bits *are* bits—at least when they're not bridle parts in horses' mouths, drills, tobacco holders, tong jaws, key fronts, cow markings, watermark makers, eighths of dollars, pieces of material things, or moments of time. And one who works with bits, whether on a computer or in digital videography is, of course, bitter for it.

THE THIN NEW LINE (FEBRUARY 1997)

Q: What single invention may be more responsible than any other for the existence of videography?

A: The telegraph.

Q: What invention may be driving most of today's developments in videography?

A: The telegraph.

Q: And what invention could change the world of videography in the future?

A: The telegraph.

These answers, if true, certainly seem odd. Despite AT&T's name (American Telephone and Telegraph) telegraphy is relegated to museum status today. Even historians of the subject may be more interested these days in the last telegram sent, rather than the first.

Telegraphy, however, has particular relevance to videography—a relevance that, remarkably, seems to be increasing rather than decreasing. After all, a definition of the common electrical telegraph may be "something that transmits information digitally, using bit-rate-reduction (compression) techniques." Digital? Compression? Perhaps it would be best to start at the beginning.

Videographers of the 1990s were by no means the first people to communicate digitally. Smoke signals and drum beats may be considered early forms of digital communication. The smoke was either visible (on) or not (off).

There's no question that binary digits (bits) were organized into groups (bytes) for information transmission no later than 150 B.C. That's when the Greek historian Polybius (born in Megalopolis, by the way) wrote about a system of signaling involving ten torches per letter of the alphabet. By 1605, using common binary arithmetic techniques, Francis Bacon reduced that to five bits per letter, the minimum required—even today—for uncompressed digital transmission of the 26 letters of the European alphabet.

How far could things be transmitted? Well, about as far as you could see five torches and unambiguously know their positions. That wasn't good enough for Robert Hooke, who wanted to send messages over greater distances. He proposed, in 1684, a series of towers on hills. People in one tower would look at the next through a telescope to see what it was signaling. They would then send the same signal to the next tower. The process could be repeated indefinitely.

There's no indication that Hooke's system was ever built in his lifetime, but Claude Chappe came up with a version in 1790 that eventually covered all of France and spread from there across the globe. Today, we refer to that form of signaling, with arms swinging into various positions, as *semaphore*. Because messages could travel hundreds of miles in mere minutes, however, Chappe originally planned to call his system the *tachygraphe* (speedy writing); Miot de Melito suggested *telegraphe* (distant writing) instead, which Chappe adopted by 1792. A traveler today encounters many sites called *Telegraph Hill*; these are all reminders of systems of visual telegraphy.

Even before Chappe, people tried applying electricity to telegraphy, but the only versions of that form of energy then available were either lightning (powerful, but not very reliable) or static. Nevertheless, it's possible that electric signaling was demonstrated before 1736, and messages may have been sent electrically from Madrid to Aranjuez (about 26 miles, or the distance of a marathon) by the Spanish royal family beginning in 1798.

Two years later, Alessandro Volta brought the world into the "Volt Age" with the battery, and reliable sources of continuous electrical current became available. Almost immediately, people began trying to apply electrical current to the problems of telegraphy. Don Francisco Salva, who had built the Madrid system, suggested in 1804 using electrolysis of water into hydrogen and oxygen bubbles as a means of telegraph message reception.

Indeed, just a few years later, Samuel von Soemmerring built a multi-wire telegraph using bubbles in a water tank as the indicator. Each letter and number had a wire, and bubbles emanating from that wire indicated transmission. In a way, it could be considered an early TV set: People stared intently at a glass faceplate to watch information transmitted electrically.

That's not how the telegraph created videography, however. Baron Pavel Lvovich Schilling, a diplomat in the Russian embassy in Munich, happened to see von Soemmerring's demonstration and was fascinated by the idea of electronic telecommunications, if not by the implementation (which, in contemporary drawings, looks like a component of NEC's recent Fish Club, a thin bubble tank intended to be placed in front of an HDTV monitor to make it look like an aquarium).

Hans Christian Oersted, a Danish physicist, discovered in 1819 that an electrical current passing through a wire could deflect a compass needle. That was all Schilling needed. In 1825, he created and demonstrated a single-wire telegraph system, utilizing a binary system of compass-needle deflections as the receiver's indicator.

He went a step beyond Bacon's five-bit alphabetic code, however. The letter *E* is the most common in English. It is quite wasteful to transmit an *E* with the same number of bits as a *Q* or *X*. Schilling, therefore, used (and may well have invented) a new form of coding, one in which the most common letters used the fewest bits and the least common the most (think of the letter values on Scrabble tiles). Today, we refer to that form of data-rate compression as *variable-word-length coding*; then it was just Schilling's telegraph code (the later Morse Code is a minor variation).

Schilling died in 1837, but not before influencing Professors Carl Friederich Gauss and Wilhelm Weber, as well as William Cooke (who became a partner of Charles Wheatstone in a successful telegraph business) and Gustav Fechner. Cooke and Wheatstone received a patent for their telegraph system even before Morse's 1837 "successful experiment" (see the sidebar, "What Hath *Who* Wrought?"), and one of their telegraphs was instrumental in capturing a criminal before Morse's famous "first" transmission of 1844 (news of a pickpocket escaping by train was wired to the next station, where he was arrested).

In fact, telegraphy had spread worldwide long before 1844. By 1839, a telegraph cable was successfully laid by the East India Company underwater across the Hooghly River in Calcutta. In 1850, a telegraph cable crossed the English Channel. The first transatlantic cable was completed in 1858 but failed roughly two weeks after it went into operation.

WHAT HATH *WHO* WROUGHT?

Ask someone who invented the computer, and you'll probably get a shrug. Most people can't name an inventor of television, either. No one seems to have a clue as to who invented the magnetic disk drive. Almost anyone, however, can easily tell you who invented the telegraph: Morse.

The same people may have learned in elementary school that Eli Whitney invented the cotton gin, but few remember what a cotton gin is, and there's nothing to associate Whitney's name to it. For Morse and the telegraph, there's both Morse Code and the awesome sentence he came up with as the first telegram. When President Kennedy initiated telephone service via geosynchronous satellite in 1963, he said, "What hath God wrought?"—the exact words transmitted by Morse in 1844. After all, satellites were just the latest way to turn the planet into a global village, a process that began with Morse.

There are just a few problems with this glorious history: Morse didn't come up with that sentence, it wasn't the first telegraph message, he didn't come up with Morse Code, and he didn't invent the telegraph. As for his feelings about the community of humanity, he ran for public office (and lost) as a candidate of the Native American party, an anti-immigrant, anti-Catholic organization.

Morse *was* a good artist, regarded highly enough to make a living and to be appointed a professor at the University of the City of New York (now NYU). He also founded the National Academy of Design and served as its first president for 16 years.

His credentials were strong enough that he was able to successfully petition Congress for funding to develop telegraphy, an idea he'd heard about. When Annie Ellsworth, daughter of the U.S. Commissioner of Patents, brought him the news that President Tyler signed a bill granting Morse $30,000 (a precursor of more recent government grants for the development of advanced television), he offered her the opportunity to come up with the sentence he would transmit, and she provided "What hath God wrought?"

It was by no means the first telegraph message Morse transmitted. *The Journal of Commerce* carried a story some seven years earlier: "Successful experiment with telegraph September 4, 1837." In the code Morse devised, that sentence was sent as 215 36 2 58 112 04 01837. Morse's own code was numerical and required a decoding book, "the numbered dictionary," to use. What we call Morse Code today was actually developed by Morse's "assistant," Alfred Lewis Vail (and was preceded by such earlier variable-word-length codes as Schilling's).

Vail (Morse's financial backer, researcher, and mechanic) may also have been responsible for many other innovations in telegraphy that Morse "invented" (electrical telegraph transmissions were actually fairly common long before Morse's "first" message). In a sworn 1849 deposition for a case before the Supreme Court, Joseph Henry, the American scientist once considered a friend of Morse, had to testify that "I am not aware that Mr. Morse ever made a single original discovery, in electricity, magnetism, or electromagnetism, applicable to the invention of the telegraph."

Revisionist history? Perhaps "wrought" irony.

During the laying of the second transatlantic cable, therefore, a great deal of care was taken. In 1866, an electrician, Joseph May, carefully measured the selenium resistors being used, but something strange happened: When the sun shone, May got one reading; when a cloud passed overhead, he got a different one.

May reported the findings to his supervisor, Willoughby Smith. Smith performed a series of experiments and confirmed that selenium was photoconductive, a fact he published in 1873, getting a lot of people to start thinking about television.

What was required for television? A light-sensitive electrically varying material (May's and Smith's selenium), a means of image scanning (published by Wheatstone in 1827), a way to get the signals from the camera to the receiver (telegraph wires), and a way to make them visible when they got there (various telegraph indicators were used or suggested). Alexander Bain, by the way, created the first image transmission system (a fax, not a TV set) based on telegraphic work he'd been doing with Wheatstone, and received a patent for it in 1843, almost a quarter-century before the discovery of photoconductivity.

What about sound? Whether you choose Alexander Graham Bell or Elisha Gray as the inventor of the telephone, both were actually trying to come up with ways to cram more messages down telegraph lines (compression?). Edison's phonograph was designed for the same purpose (record telegraph messages, speed them up, and replay them through the line, then reverse the process at the receiving end); he had previously gotten rich from an invention that allowed multiple messages to travel simultaneously on the same telegraph wire.

The phonograph, in turn, became the first video signal recorder in John Logie Baird's Phonovision. Magnetic recording? It was developed by a Danish telephone engineer, Valdemar Poulsen. Baird even dabbled in magnetic disk recording and non–real-time applications. Broadcasting? Needless to say, Guglielmo Marconi was working on a method of wireless telegraphy.

The international standards organization (the ITU) that created the famous Recommendation 601 used in component digital video equipment and facilities worldwide is the exact same organization founded in 1865 as the International Telegraph Union. Today, the *T* stands for telecommunications.

As for all of those names listed earlier, Oersted's is now used as the unit for measuring coercivity on videotapes, and Gauss's measures their retentivity. The Weber, a unit of magnetic flux, surfaces when video engineers argue over the audio levels on reference tapes. Weber and Fechner worked on the basic laws governing human vision (leading to such concepts as a video camera's *gamma*), and Fechner performed the first known research into aspect-ratio preferences.

Thus, it's pretty clear that telegraphy had a major impact on the development of videography. What about question two, however? Is telegraphy affecting today's developments in videography?

THE FIRST TELEGRAPH CABLE

Long before Volta's battery, Chappe's semaphores made the word *telegraph* synonymous with speed. Thus, *The Daily Telegraph* was so named to indicate the speed with which it could deliver the latest news, and a fast train was sometimes called *The Telegraph*. One stagecoach driver, John Cable, was so quick that his conveyance bore the *Telegraph* name, and he was so closely associated with his job that people referred to him as Telegraph Cable. This was reported in a learned Nineteenth-century volume as the first "Telegraph Cable" by Willoughby Smith, the same person whose report of the photoconductivity of selenium started the video revolution.

This proves that not only digital bit-rate reduction has venerable roots; so do unsuccessful scientific attempts at humor.

An affirmative answer to that one may be a bit more of a stretch, but, again, telegraphy is a form of digital transmission. Bell's and Gray's plans for harmonic telegraphs are little different in theory or form from the wavelength-division-multiplex schemes used in the latest video fiberoptic systems.

Schilling's variable-word-length coding was the ancestor of all current data-compression schemes, without which there would be no Betacam SX, DCT, Digital Betacam, Digital S, DV, DVCAM, DVCPRO, or HD D-5. There would be no JPEG, MPEG, or M-JPEG. There would be no DirecTV or DISH and no DTV. Nonlinear editing might have been a curiosity sold only in 1971 and not since. Today's giant-capacity 23-GB hard drives would hold only about 11 minutes of component digital video. Direct satellite broadcasts would offer only about 24 channels per satellite, and antennas would probably be considerably larger than 18 inches in diameter.

The cover story of the December 1996 issue of *Wired* magazine was devoted to the latest descendants of the early underwater telegraph cables and how they help make the Internet work. Today's cable-laying ships use differential global positioning satellite (GPS) systems to identify positioning to within one centimeter and bow and stern thrusters to try to maintain it; the first transatlantic cable was laid from a wooden ship, the *Agamemnon*, that, in the words of Arthur C. Clarke, the man who first proposed the geostationary orbit for communications satellites, "would not have looked out of place at Trafalgar" (*How the World Was One*, Bantam Books, 1992).

Still, the techniques of submarine cable laying haven't changed very much in the last 150 years, and, despite roughly 35 years of geosynchronous communications satellites, the cables continue to be laid. Why?

The answer to that question might offer a clue to why the telegraph could be the answer to the third question that began this column: What invention could change the world of videography in the future?

From a sending and receiving standpoint, it's hard to imagine a technology simpler than the telegraph. The sender is, in essence, a switch, and the receiver might do nothing more than click. They could even be the same device, connected to the same wire. Even in the nineteenth century those simple devices, however, could utilize a global telecommunications network involving news, finance, commerce, and more. One could even send a message in English and have it delivered in Chinese characters.

Clearly, to make this happen, the telegraph network had to have a lot more than simple switches, clickers, and wires. It did. In its first year of operation, Western Union expanded from less than 600 miles of telegraph wires to 100,000 miles. It had thousands of offices; employees serving as translators, engineers, researchers, technicians, and operators; and enough money to stimulate innovation even outside the company.

In contrast, a typical video postproduction facility consists of a lot more than just a switch or clicker. It has videotape recorders, digital disk recorders, image manipulators, sound processors, image and sound generators, and more, not counting controls and monitors. It is becoming ever more computer oriented, and, with computers obeying Moore's Law of increasing in power and decreasing in price roughly fourfold every three years, they are able to do more and more. In 1971, a postproduction facility's computer could control three video storage devices and, perhaps, a switcher; 25 years later, the computer could *be* the video storage devices and manipulate the images, too.

Moore's Law notwithstanding, video postproduction facilities often complain—justifiably—about the high cost of hardware and software. Yes, a simple personal computer can do some form of video and audio postproduction, but the latest and greatest effects, color correction, and such, always seem to require more expensive systems.

Could it ever be possible that a video postproduction facility would become telegraphlike? It would keep the "clicker" (picture and sound monitors and other indicators) and the "switch" (control panels and keyboards), but that would be it. The rest of the operation would be handled somewhere else—somewhere connected by a thin new line to the postproduction facility.

Hitherto, this idea has not worked in part because the connection costs would have exceeded all of the existing hardware and software costs. A full-bandwidth (4:2:2) Recommendation 601 component video signal, for example, requires 270 million bits per second (Mbps) for transmission. Commonly available digital transmission lines offer less than 45 Mbps, necessitating seven such lines per 4:2:2 data stream. A typical three-device dissolve would require some 19 such lines, each having connection costs of thousands of dollars per month, not counting whatever the central processing facility might charge. And then there's the problem of getting material from clients and getting it back to them.

The client issue is potentially quite serious in still more ways. In the 1990s, the facilities of the Detroit chapter of what was then the International Teleproduction Society (ITS) arranged for Ameritech (their phone company) to provide 270 Mbps interconnections for costs approximating those of 45 Mbps. The secret was suggesting the use of so-called "dark fiber" (fiber optic strands not connected to anything else), thus saving Ameritech the cost of video compression and decompression equipment at each end.

Quizzed about the same possibility in 1997, ITS members in New York did not think it was necessarily a good idea. They didn't want clients being wooed by other facilities. One member thought a connection to a West Coast facility would be fine, but not one to a West Side facility a few blocks away ("My clients aren't going to travel 3,000 miles to get a better deal; crosstown is a different story").

As long as high-bit-rate (wide bandwidth) connections are relatively expensive, they will remain merely an adjunct to existing facilities. Those prices, however, may be coming down. Writing in *Forbes ASAP* of December 5, 1994, George Gilder (author of the books *Microcosm* and *Telecosm*) said "bandwidth will expand from five to 100 times as fast as the rise of microprocessor speeds." Microsoft's Bill Gates was quoted in *PC Magazine* of October 11, 1994 as saying, "We'll have infinite bandwidth in a decade's time."

Perhaps the bandwidth won't really be infinite, and perhaps it will take longer than a decade, but, if and when high-bit-rate connections ever become comparable to telephone calls in costs and locations served, video postproduction will almost inevitably change. Clients wouldn't necessarily ever have to deliver or pick up their material; they could either allow a distant post facility to access their server directly (everyone would, of course, have servers by this time), just as the billing computer calls satellite pay-TV subscribers today, or "call the material in" about as simply as a reporter calls in a news story.

Editors (the humans) would have their monitoring and control panels at their South Pacific island hideaways. Their control computers would establish the necessary virtual paths: video and audio connections from the client's server to the massive ITS processor in Virginia, monitoring and control connections to the island, and, of course, hundreds of billing connections.

The same sort of virtual paths are used today to, say, order flowers. A toll-free call connects you to a telemarketing center. One call from there contacts an actual florist to get the flowers delivered (or perhaps just a distributor who makes still more calls). Other calls debit your credit card and credit the florist and the telemarketer. There might even be calls for debits and credits for a delivery service.

If and when 270-Mbps circuits become as common and as inexpensive as ordinary telephone lines, all the rules will change. Given that the cost of a local phone connection has not dropped in recent years, however, it might take more than technological advances to bring about the change.

The future is unpredictable. Banks scoffed at telegraph lines because they had messengers (using the lines for security alarms was a clever marketing ploy to win them over). Investors couldn't understand the value of a photocopier given the existence of carbon paper. And 1997 was the birth year of the HAL 9000 computer that was a major character in *2001: A Space Odyssey*.

Back in the late 1960s, *2001* anticipated a computer that could not only understand ordinary speech but even read lips, something unlikely to be achieved by 2001. On the other hand, that computer was a mainframe, something unlikely to be found in any spacecraft, now or in the future.

Did the concept of laptops, notebooks, and personal digital assistants elude *2001*'s author, the same Arthur C. Clarke who is considered the godfather of the Age of Satellite Communications? Or did he leap beyond mere small computers to the Age of Infinite Bandwidth, an age when a network of mainframes could rule the universe?

Hmmmm. Maybe that's what HAL was trying to do.

CHAPTER TWO

ANY CONTENT, ANYWHERE, ANYTIME

CRAIG BIRKMAIER

Flashback: The scene is a living room in a Long Island bedroom community in 1958 (*family rooms* won't be invented until the Seventies; come to think of it, neither will the term *bedroom community*).

Dad: "Who's on *Ed Sullivan* tonight?"

Son: "Just a few circus acts and some old singer."

Mom: "Sure wish we had one of those new color TVs so we could see *Disney* and *Bonanza* in living color."

Sound offstage: telephone rings.

Son: "It's Larry. His family's invited us to come over and watch their new color TV."

Mom: "Is there room for all of us in that little den?"

A few hours later....

Dad: "I think we'll wait on buying a color TV. The color isn't very realistic and those sets are expensive. Besides, there are only a few hours of color programs a week."

The Fifties—what a time to grow up. Television pictures coming through the air were still a novelty back then, still considered magical. Good thing, because my parents didn't like to go to the movies. We were lucky; I lived in a suburb of New York, which was the broadcast center of the universe. I once sat in the "peanut gallery," the audience section for a live telecast of *Howdy Doody*. To complete this Norman Rockwell portrait, one of the Mouseketeers, Kenny, lived up the street.

Not only did I have three networks to watch while growing up, there were three independent stations as well, including WOR-TV, Channel 9. Every night they broadcast the *Million Dollar Movie*, which was the same film every night for a week. I sat through *King Kong* five times.

Back then, television could be characterized as "What You See Is All You Get." We didn't have many choices, but nobody cared. Having four or five choices was better than none! Watching TV was a *synchronous* event bringing families and nations together—a collective experience.

Back to the future: a Nineties family room in Anytown USA.

Daughter: (bordering on a scream) "He changed the channel! I was watching a cartoon!"

Son: "The cartoon was over. Besides she's had the TV for an hour."

Mom: "Where's the *TV Guide*?"

Dad: "Don't bother, the Braves are playing the Dodgers, and I haven't had the TV all week."

Mom: "Well, that's 'cause you're never here!"

Dad: "Hey, I have to make a living. The kids can watch the TV in the playroom."

Daughter: "No fair. There's nothing to lay on in there."

Dad: "Use your imagination. If you're nice I'll let you watch *The Little Mermaid*—again—after the game."

Thirty-four years—and at least that many channels later, television has evolved. Now television is a case of "What You See Is What *They* Want." Cable TV has expanded our choices; broadcasting has been replaced by multicasting. The VCR permits time-shifted viewing—provided one can figure out how to program the thing. And there's the corner video store; just a short drive to find out that they don't have the movie you want to see. In reality, television has not changed all that much.

There's more choice, but it's still difficult to access the content you want, when you want it. You turn to the Weather Channel to get the local forecast or see what conditions will be like in the city you're traveling to, then you wait until *they* get around to telecasting the information you need. Same thing with the news channels. There are music video channels for every taste; too bad *they* never play the songs you want. Shopping networks are interesting, but what are the odds that *they* will offer the product you're looking for when you tune in? And sports channels always have the game you're looking for...right?

Here we are in the twenty-first century—500 channels, but nothing to watch. You could surf the Web, but let's face it, compared to TV, the Web-surfing experience is overrated; like doing the waves in Florida compared to the real thing at Waimea Bay, on the north shore of Oahu.

And then there's the question of a "surfboard." You could get one of those complicated computers, or one of those set-top systems like WebTV. You know the idea: Take your basic TV—a simple appliance that even a three-year-old can learn to operate in about two minutes—and give it a wireless pointing device and a keyboard. What a concept! Now it can display fuzzy text that the three-year-old and a significant percentage of high school graduates—raised on a steady diet of TV—can't even read.

But wait there's more: How about those pixelated 256-color images that don't even move most of the time, or when they do, you start looking for the crank on the side of the box so that you can turn it faster. It's hardly "surfin'!" You want big-screen pictures with enough resolution to see the details, a human voice, great sound instead of text, and high-fidelity color that will really make you want to buy stuff. You want to use the Internet as a back-channel to complement the big *broadcast pipes* so that you can search for and subscribe to content from *anyone* on *any* network. You want access to all kinds of content: traditional linear entertainment programs; interactive video including entertainment, news, sports, documentaries, and how-to programs; electronic magazines with rich-digital media content; and interactive product experiences—rich-media content about products you are actually interested in buying.

You want to use The Net for transactions, to buy stuff, so that the people who create this ocean of content can be compensated directly. But most important, you want a "surfboard" that saves all of the stuff you subscribe to, so that you can consume it asynchronously, when *you* want.

We survived Y2K. Now, if we can survive the last-gasp efforts of the well-entrenched gatekeepers who control the mass media, the balance of power may finally shift from the privileged few to *you*.

WHAT YOU WANT IS WHAT YOU SEE

We are a product of our environment. As a product of the first television generation, and as a parent of the next, it is impossible to ignore the impact that television has had in shaping our lives. We carry much of this baggage with us when we make the decision to become content creators, as I did in 1963, at the highly impressionable age of 15.

You cannot be a digital content creator without first understanding your role as a content consumer. And, hopefully, as a content creator you're teaching kids (yours or someone else's) to be media-literate; it gives a whole new meaning to the concept of responsibility. Having lived and worked through four decades of revolutionary change in the process of creating television and, now, digital media content, it is humbling to recognize that the most dramatic changes in the way we create and consume content still lay ahead. And it is daunting to realize that the process of learning our craft can never stop. The digitization of media has enabled the opportunities to create and proliferate content to grow exponentially. Although the principles of good storytelling have not changed in thousands of years, the tools for expressing those stories are evolving at an ever-accelerating pace. (Incidentally, practically all content involves some kind of storytelling, from the shortest clip to the longest movie.) Thanks to digitization, itself driven by an exponential increase in digital processing power, these tools are becoming evermore affordable and accessible to the next generation of content creators.

The evolution of digital media is far from complete, but the directions in which it is taking us are already apparent. The days of limited choice—What You See Is All You Get—are ancient history. Media today is characterized by abundant choice, but control still rests in the hand of the privileged few. Driven by the technologies of the digital revolution, we are rapidly converging on a new media paradigm: What You Want Is What You See (WYWIWYS). A key element of this new paradigm is the shift from *synchronous* (real-time) consumption of media content to *asynchronous* consumption. To fully appreciate how liberating this can be, consider the role that e-mail has played in enhancing our ability to communicate with one another on a global basis. By forwarding and storing messages, it is not necessary for two people to link up in real-time. And unlike the conversations that finally take place after a game of phone tag, the content of the message is preserved. It can even be shared by forwarding it to others, and large groups can converse by broadcasting messages and responses to everyone on an e-mail list.

The same advantage can apply to digital media content. When media is forwarded, or broadcast, then stored locally, it can be consumed at any time. You can program your local data-caching device—for example, a personal video recorder (PVR) or a PC—to *subscribe* to content that is of interest, and to create filters to look for specific kinds of content when it is broadcast. You then have a broad selection of content that you want to watch, available on demand, in addition to live synchronous broadcasts and the searchable content of the Internet. Again, it's a case of WYWIWYS. This fundamental shift is shaking the very foundations of the mass-media empires that have been assembled over the past century. Much of their power flowed from the need for synchronous consumption of media and tight control over the channels of distribution.

In the days when there were only four or five channels, before the home VCR existed, content was fleeting; watching TV was strictly a synchronous experience. And because a handful of programs had to appeal to the masses, the content was typically targeted at the lowest common denominator. These limits on the channels of distribution imposed limits on competition; the mass media evolved into an oligopoly with tight control over distribution and content.

The growth of cable increased programming choices and opened up new opportunities for content creators. But cable franchises were structured as a monopoly, and the industry quickly learned the value of controlling access to these systems. They created new networks to fill up the bandwidth, and restricted access to independent content producers. The gatekeeper mentality was not only preserved, it was strengthened.

The concepts behind the Internet created a very different environment for the evolution of digital media. The value of the network is directly proportional to the number of *connections*, rather than the number of *viewers*. The Internet supports multiple topologies: many-to-many, many-to-few, and few-to-many. Anyone can become a content creator and distributor on the Internet. If you want to compete with the mass media, however, the trick is in attracting the many. If you only want to reach a highly targeted audience, however, it is relatively easy to succeed.

One of the key concepts behind the success of Internet applications such as the World Wide Web is that servers are always on: You can access information on demand. Unfortunately, this is also the biggest problem with the Web. Each thread of information flowing across the Internet backbone is one-to-one; if many people want the same information at the same time, bandwidth is consumed when duplicating the threads. Broadcast techniques such as IP Multicast have been developed for the Internet, but many of the existing routers do not support this protocol. And most consumers do not have the high-speed connections to the Internet that support high-quality video streaming and other forms of rich-media content—yet. Thus, there are significant advantages in using digital broadcast networks (DTV, cable, and DBS) to push popular content to the masses, where it can be stored locally for consumption. In essence this mirrors popular content at the point of consumption—at the edges of the network—freeing up the Internet backbone for other important tasks like transactions, searching, and accessing information that is not for mass consumption. By filtering the content you want from an ocean of bits, WYWIWYS becomes a reality.

Asynchronous consumption of content also acts as a bandwidth multiplier. In the world of synchronous media, certain time periods are far more valuable than others. For TV it's "prime time," for radio it's "drive time." In the world of asynchronous media consumption, however, *anytime* is "prime time;" your content can be delivered at 3:17 A.M. Sunday morning. We can begin to look at the value of a network on a 24/7 basis versus the 3/7 basis of "prime time." All that remains is to get the gatekeepers to make some of this bandwidth available to independent content producers. That is inevitable; it's just a matter of time.

THE TOOLS REVOLUTION

Although the digitization of media and video-compression techniques are beginning to bypass those gatekeepers who have limited access to the channels of distribution, far more profound changes are taking place in the tools used to create digital media content. To the casual observer, it may appear that the most significant changes are related to the cost of these tools. In the short period of two decades, the cost of the hardware and software required to create high-quality, professional content has plummeted. A broadcast-quality camera, field recorder, studio deck, computerized editing system, video switcher, DVE, character generator, audio mixer, and supporting gear could easily cost $500,000 to $1,000,000 in the early Eighties. Today, with a three-chip DV camcorder, a notebook computer, and some software, it is possible to produce high-quality content with effects that were impossible to conceive with older, more traditional tools. An aspiring content producer can get started for an investment of about $10,000. A professional "boutique" project studio business can be fully equipped for about $100,000.

To focus on the cost of the tools, however, is to miss the larger implications of what has happened to the content-creation business. Tools have always been a very small part of the cost to create high-quality entertainment content for the masses. This market depends on star power to attract large audiences, and large audiences produce significant revenues, the lion's share of which flows to the stars and the producers of the content. It's not uncommon for the cost of a top-rated episodic television show to exceed $1,000,000; a blockbuster movie can cost more than $100,000,000 to produce. Removing cost barriers to content creation has had a major impact on the creation of new markets for content creators, especially outside of the traditional entertainment and broadcast markets. Corporate, institutional, and educational applications for traditional video and new forms of interactive digital media content continue to grow; event videography (of weddings, bar mitzvahs, etc.) has become a major business; and now, the ability to create digital media content is available to the masses. With a camcorder and a PC anyone can use video to tell his or her story.

Fortunately, for professionals trying to make a living in the content-creation business, there is little reason to fear the cannibalization of their business from below. Success in this business does not flow from the tools or even the ability to master their use; success flows from the ability to tell stories and to express content in appropriate ways across multiple distribution media.

The real revolution in the tools of content creation has strong parallels to that which has already been described for content consumption. The digitization of media assets has placed tremendous control in the hands of content creators. In the world of analog video, the ability to manipulate images was at best fleeting and imprecise; an often frustrating real-time synchronous process. In the new world of digital content creation, the ability to manipulate every element of a digital media composition is precise. Media assets are digital files that can be replicated perfectly and moved across networks, within a facility, or around the world. Real-time processing is not a requirement for most applications; the asynchronous nature of today's content-creation techniques provides levels of control that were impossible using the traditional tools of video production.

Another aspect of the digitization of media assets to have emerged, and one that is having a far more profound effect on the content-creation business, is that digital content-creation tools have become the gateway to a wide range of opportunities to enhance the storytelling process. Digital media enables consumers to interact with stories and information, and digital media can be distributed to a wide range of devices via multiple networks, while maintaining consistent quality and branding.

When the "desktop video" revolution began, many believed that this was nothing more than the digitization of analog video storytelling. What really happened, however, was the transformation from manipulating physical media (film, tape, etc.) to the manipulation of digital media—*bits*. This transformation has eliminated many of the barriers that once existed between various content-creation markets. A consistent set of tools can be used to manipulate assets, which can be expressed in multiple physical media, like print, outdoor and point-of-purchase display, product packaging, video, audio, and direct consumption via a digital media appliance such as a PC or PDA. The manipulation and re-expression of common digital assets supports the delivery of a consistent message and branding to anyone, anywhere, anytime. The key elements of the convergence of media and content-creation tools from multiple media include:

- The use of media objects stored as digital media files
- The use of software-based tools that employ consistent user interface design
- The ability to cross-reference (hyperlink) content delivered across multiple media; this in turn supports the ability to access *depth* information and/or similar content
- The ability to localize and customize content through the intelligent assembly of media objects into digital media compositions

Many of these elements are coming together via the World Wide Web. Hypertext Markup Language (HTML) is a relatively simple way to create object-based compositions; although it is based on the assumption that text is the most basic object type in the composition, the principles are equally applicable to advanced forms of digital media content that rely on audio and video to carry the story. HTML adds tags that define relationships among various objects in the Web-page composition, as well as links to other pages (or points within a page). Objects can include text, graphics, still images, audio, video, and animations.

These objects are delivered to a browser, which composes them based on the HTML tags. As the Web has evolved, more advanced techniques have been developed to control the composition and to customize the presentation based on user preferences. These preferences may be specified in browser settings, or they may be based on past interactions with the server—in essence the server recognizes the user and customizes the content based on a profile it has created from past behaviors.

Web browsers also rely on *local cache* to minimize the need to request frequently used objects and large objects (large files) that require a significant amount of time to download. If an object is not cached locally, the browser requests it from the server(s) that are contributing objects to the composition. It is important to note that objects may come from multiple servers. For example, it is becoming commonplace to mirror large video files in servers all around the world—servers that are close to the edge of the network. When you request one of these objects—for example, a streaming video file—the request is routed to the closest mirror server. In most cases this bypasses network congestion and distributes the load across multiple servers. It is only a small leap to the mirroring of content locally. For example, if you subscribe to a Web-based service like *The Wall Street Journal Online Edition*, the content could be broadcast when it's updated and mirrored in the cache of hundreds of thousands of subscribers. When a component of a page is updated, only the changes would need to be broadcast to all of the subscribers. The broadcast medium could be the Internet, DBS, cable, or terrestrial data broadcasts.

Going a step further, it is quite feasible to localize and personalize all kinds of content in using local composition techniques and objects played from cache. For example, the name and address of a local merchant in your neighborhood could be overlayed on a television commercial. Or a TV commercial targeted at your specific demographics (or recent actions) could be inserted into a program you are watching.

The ability to interact with, and personalize, content is likely to have a profound impact on the future of digital media content creation, not to mention electronic commerce. The ability to manipulate physical media using a computer is just the tip of the iceberg.

A NETWORK OF NETWORKS

Another popular misconception is that the Internet is an isolated infrastructure, like cable TV or the phone system. Some believe that today's communications infrastructures will retain their unique identities, that consumers will continue to expect them to be different and disconnected. Others, including myself, believe that all communications networks are converging and that formerly disconnected technologies will interoperate with one another.

As indicated earlier, the real value of networks is realized from their interconnections. The term *Internet* is a contraction of inter-networking. The value of the Internet and other communications infrastructures such as cable or broadcast TV increases when they are interconnected. Just as today's Web servers pull content from multiple interconnected servers scattered around the country or world, it is quite feasible to pull content from multiple interconnected infrastructures. For example, cable or broadcast TV can deliver high-quality digital media experiences to a local cache. When this content is viewed, the appliance may use an Internet back-channel to update the information in the cache, to connect to links embedded in the content, or to facilitate transactions, like buying the product that is being advertised.

Because of its "open architecture" and the culture in which it has evolved, the Internet has become a crude prototype for the digital media infrastructures of the future. Internet standards are beginning to influence other digital media infrastructures, just as those media have influenced the content delivered via the Internet. It comes as no surprise that many of the entrenched media interests have been trying to resist the Internet tide by keeping various aspects of their infrastructure isolated, and by developing proprietary standards that keep their customers locked up inside a walled garden.

The impact of closed or proprietary systems on content creators is enormous. Multiple versions of the same content may be required, not to mention the need for proprietary tools to create each version. This approach inhibits the pace of evolution; the need for interoperable solutions is more obvious every day, and slowly the barriers that have been erected to delay convergence are falling.

INTELLIGENCE, INTERACTIVITY, INTEROPERABILITY

There is a sense of inevitability about the future that may be impossible for the vested media interests to forestall. Control is shifting to the consumer, and once enough consumers realize the power that they can wield, continued resistance will be futile. History teaches us that blocking and delaying tactics are usually followed by sweeping changes. Content creators would be well advised to anticipate these changes so that they can benefit from them, rather than being swept away by them.

When digital video is delivered through wide-bandwidth switched networks, the number of choices is likely to increase exponentially; "channel surfing" will not be an effective method for exploring this digital world! This is where the "I" words become important for the content creators and armchair quarterbacks of the future:

- *Intelligence.* That old remote control is useless in the face of potentially hundreds or thousands of viewing possibilities. An intelligent intermediary will be required to help you sort through these possibilities. Once that intelligence is available, you can teach it to be your personal intermediary—to look for the kinds of programming and information that you are interested in.

- *Interactivity.* Teaching the system to look for information is, of course, a form of interactivity—but it need not stop at finding the programs or information you want. By comparing the vast amount of information that will pass through the digital network with your user profile, the receiver could accumulate information that you are interested in, essentially synthesizing programs for you. The ability to access information on demand may be the most compelling use of interactivity. Conversely, the ability to eliminate the unwanted will be the most important form of interaction to future advertisers; the most important interactions may be those that bring people who want to buy things together with the people who want to sell those things. Feedback from people who have already bought those things may strongly influence those purchase decisions.
- *Interoperability.* Integration with the local network within your home and the telecommunications network will open up new possibilities for your family-room display. Want to check on the baby, see who's at the front door, or make sure Johnny is doing his homework? The local network will allow you to monitor these activities, perhaps opening a window on the family-room display. You'll also be able to handle telecommunications—both audio and video—from that easy chair. And it may be just as easy to do these things from the office or halfway around the world. Interoperability will be one of the most difficult capabilities to purchase, because entrenched interests will resist and thwart it. Herein lies one of the most important responsibilities of the content creator.
- *Invest wisely!* Invest in yourself, by embracing change and the skills that will enable you to survive the digital transition, and hopefully thrive as a result. Invest in the tools that will enable this future. Investing in that which is comfortable—from vendors who are resisting change—is risky. The result may be that you have the wrong tools and the wrong skills, and no future. Invest in your content-creation education. Look at the content that has been successful in the past, and ask yourself why that is so.

These are indeed interesting times, but for the digital content creator the opportunities have never been greater. When any content can be consumed anywhere, anytime, anything is possible.

CHAPTER THREE

CONVERGENCE PROGRAMMING IN SPORTS

MICHAEL SHAPIRO

The road map to the future of the emerging media landscape is unfolding in both convergent and divergent directions. At present, we are experiencing growing acceptance and understanding of the differing functionality between traditional broadcast media and the new media-interactive applications available through the Internet. Generally, we utilize the television for video presentation and the PC for multimedia presentation. Television currently offers limited interactivity in that it can be turned on and off, channels can be changed, and, through digital cable and satellite delivery systems, limited two-way communication is available. The Internet offers more pervasive data manipulation, complete two-way communication, personalization, community functions, and commercial applications.

Sports programming has been at the vanguard of experiments with synchronized content display (the so-called "simulcast") whereby televised events on the field of play are supplemented with relevant materials such as statistics,

This chapter was excerpted from Mr. Shapiro's remarks at the United Nations World Television Forum 2000.

commentary, and analysis on the Internet. Typically, the sports broadcaster displays a graphic mention that directs the audience to log onto a Web site for more information. This methodology is in its infancy in terms of providing a valuable service to the audience and a viable traffic generator for the Web. It is functionally awkward and readily superseded by the next stage of emerging media applications: the interface of Internet protocols with television and the increased ability to provide full-motion video on the Internet. These applications offer short-term solutions to the growing consumer demand for greater choice, greater interactivity, and greater personalization in their entertainment appliances. The television is generally a poor device to access text because it is best viewed from a distance. The computer is a great device for text but less than ideal for viewing video. Bandwidth access into the home and consumer-friendly pricing for services and hardware also certainly influence the market penetration for these applications.

As such, it is assumed that over the next five years the marketplace will be primed for the opportunity to acquire appliances that integrate Internet protocols with high-definition digital video and audio. These so-called "convergence" appliances will require big "pipes" into the home over cable, or through satellite and telephony, as well as the set-top box products that will integrate Internet signals with broadcast signals. While convergence moves forward toward consumer availability, however, divergent appliances will begin to permeate the marketplace as well. These appliances will offer multiple entertainment environments (wall-sized entertainment systems, for example) and intelligent interface devices—such as those used in the "smart refrigerator" that notifies your grocery store when you're low on milk or eggs, or the "smart oven" that scans your cupboards for the ingredients for a preprogrammed recipe.

In the realm of sports programming this convergence/divergence dichotomy is an important distinction because programming rights, production devices, and the appliances receiving the program distribution will affect content strategies. Convergence production will greatly differ among appliances, bandwidth availability, and access platforms. Productions must be designed for multiple platforms of distribution to wireless devices, convergence and divergence appliances, broadcast television, print, radio, and many other access points.

THE RIGHTS ENVIRONMENT

Today's sports programmer must understand these distribution options in order to navigate successfully through the complicated rights environment created by the growing interconnection of interactive and traditional media. As the Darwinian realities of the Internet economy become more pressing, the pressure toward proof of sustainability will require more independent Internet content companies to prove their financial viability. The survivors will need to be well funded, with a clear path to profitability and a strong strategic alliance among broadcasters or delivery providers. As consumer demand for interactive television grows, broadcasters will need to integrate facility with interactive media as a production requirement to their programming portfolio. As such, they will not only create production models that integrate interactivity but they will need to negotiate for rights to do so. In the not-too-distant past there was a proliferation of Internet companies actively bidding for interactive programming rights, resulting in ever increasing rights fees. Because many of them have been forced under by changing economic conditions in the industry, only a select few players are left in the field to compete for rights that have become increasingly more difficult to "monetize" (i.e., develop sufficient revenues to offset production and rights fee costs). The remaining players are those most closely linked to the broadcasters either as wholly owned subsidiaries or through equity purchase.

This broadcaster–Internet connection is integral to understanding the future direction of sports programming in a convergence world. Today, the rights to premier sporting events command enormous sums for the ability to televise, promote, and sell on-air advertising in connection with the event. But because interactive coverage on PCs attracts different audiences (at times) and competes for an audience at other times, broadcasters can't support these costs of rights and production without complete exclusivity. Thus, the integration of media is necessary because of the desire to package advertising sales and to retain the audience within a closed network of coordinated storytelling, sponsor benefit, and cross promotion.

Separation of rights is inefficient and ineffective. Although many rights holders may seek to retain their own interactive rights so that they may determine their ultimate value as a revenue source, the trend is to couple the interactive rights with the event broadcaster. In the recent example of the United States' rights to the Olympics, NBC paid the International Olympic Committee (IOC) $4 billion for the combined TV and U.S.-based interactive rights. NASCAR, on the other hand, divided its TV rights among several broadcasters, one of whom, AOL/Time Warner (Turner), acquired the interactive rights pursuant to a separate arrangement for the astounding sum of $100 million over six years. Many rights holders look at this deal with envy, but the truth of the deal is that, in fact, it's actually a promotion and advertising deal with an assumed value equation. The day of the huge interactive rights fee without a broadcast partner is over.

THE WEB EXPERIENCE TODAY

The current Web environment for sports is divided between delivering at-home programming over 28k and 56k modems (broadband penetration is growing but is by no means pervasive) or, in the case of work environments, higher speed T1 line. This dichotomy makes production options difficult, because the richer multimedia experience is largely unavailable to the home computer. As interactive media producers seek to expand their connection with their broadcast partners through the use of full-motion video and simulcast features, their efforts are limited to workplace applications. Broadband penetration and the growing trend toward programming for wireless devices will open the scope of opportunity for new production options but the core entertainment value will continue to be driven by the trends toward converged experiences.

THE CONVERGENCE EXPERIENCE

Many broadcasters and their interactive media partners have begun to develop convergence models which, in most cases, focus on creating greater viewing choice (access to multiple camera feeds, for example), immediate access to statistics and other text-based information, and greater reliance on data visualization. Sky Sports in the United Kingdom has conducted trials with football matches. In the United States, NBC has produced convergence trials with NBA games and college football through their WEB TV and Wink systems. Generally, these trials have utilized the inherent vertical blanking systems to combine the broadcast signal with a text delivery frame that can be reconfigured for interactive applications. Through this device the user is able to access on-demand programming, commercial buying options, and two-way communications through e-mails, chats, and other community services.

The forthcoming convergence of broadcast and interactive media will create exciting opportunities for consumers as well as for broadcasters. As bandwidth increases into the home, and the Internet protocol-driven set-top box becomes integrated into the television set, the availability of convergence programming for sports events will be an outgrowth of today's "simulcast" production experiences. Consumers will increase their demands to choose and control their entertainment experience. The consumer will be able to access compelling and engaging news, information, and entertainment experiences through one appliance or through multiple appliances that can be accessed no matter where they are or what they're doing. This new value equation will not only give the consumer more choice, but will also allow programmers to increase their revenue streams, promotional opportunities, and breadth of distribution.

CHAPTER FOUR

THE NEW DIGITAL CONTENT CONSUMER

The Accenture Media & Entertainment Industry Group

A landmark study by Accenture* reveals that consumers want total control over content—anyplace, anytime, across any channel, and in any context—and they can hardly wait for the day when technology makes them "large and in charge."

E-books may rock, but will the average consumer read them? Napsterites claim content is free, but music moguls contend that copyright is king. Hollywood holds onto its DVDs and hopes that MP4 won't make movies the latest contraband casualty. With each passing day, more toys pop out of the technological toy chest, delighting digerati and challenging established industry players.

*(Accenture, formerly known as Anderson Consulting, is a $10 billion global management and technology consulting leader.) Excerpted from *The New Digital Content Consumer: Large and In Charge—Strategic Implications of Consumer Preferences in an Era of Digital Choice*, by the Accenture Media & Entertainment Industry Group. *Study authors:* Ken Mifflin, Managing Partner–North America, Stategic Services, New York; Andres Sadler, Partner, New York; Rick Joyce, Partner, New York; David Brodwin, Partner, San Francisco; Roberta Glaser, Experienced Manager, New York. *Key contributors:* Carolyn Hudson, Kathryn Pierson, Rebecca Sobo, and Kimberly Yates. *For more information:* Susanna J. Deegan (susanna.j.deegan@accenture.com) 917-452-2919; Perzon Mody (perzon.mody@accenture.com) 44-207-844-3088.

A new wave of cultural exchange is breaking all around us, spurred by digital advances. In the evolving digital landscape, content can go anywhere, thanks to wireless technology. Content can get personal with the integration of consumer information, and even interpersonal with the help of networked devices. Content from different sources can become interactive via broadband channels. The cumulative impact of all this is a radical transformation in the consumer's experience of content itself.

But are we ready to play? And how much will consumers pay? Will technology drive the market, or will the digital consumer command and control it? And what precisely does the consumer want? It was to answer such questions that Accenture commissioned this study.

As publishers, music labels, and TV and cable channels collectively work toward their digital density, Accenture looked to the consumer, because that's who will ultimately decide winners and losers. This study interviewed more than 600 people, asking them to project their buying preferences onto the digital entertainment devices and digital content expected to be commonplace by 2005. Accenture showed them eBooks and explained how they worked; asked the consumers to imagine a new kind of digital audio device, capable of accessing almost any music, anytime, anywhere; and then asked them to envision a new type of interactive television service. Then we asked them what they thought.

THE CONSUMER CALLS THE SHOTS: OVERALL FINDINGS

Here are some fundamental results of the study:

- Regardless of the medium, consumers are saying, "I want what I want when, where, and how I want it." Not interested in tricky technical features or hard-to-understand functions, consumers want easy access to the content they know and love. Controlling the entertainment, information, or educational experience is the top issue for the digital consumer.

- With over half of consumers interested in using digital content, large, valuable mass markets can be achieved quickly, provided key requirements are met. Content companies can best develop the market by meeting consumer needs for simplicity, convenience, comprehensive choice, and experiential control. This can be a big wave, not a slow swell, if industry leaders create conditions that optimize the dissemination and adoption of digital content.
- Content companies can succeed by delivering on three imperatives:
  - *Consumerize*. Think about what the consumer wants, rather than what you have.
  - *Collaborate*. Work with others in your industry to create the right climate for digital adoption.
  - *Configure*. Rethink and reshape your content for the digital frontier.
- As the digital consumer assumes control over the entertainment experience, there will be a new focus on intent-based content consumption ("Tonight I'm going to catch the game we missed last Sunday."), as opposed to passive exposure ("I wonder what's on tonight?"). Choices centered on consumer preference will displace choices driven by media availability and formats.

ARE CONTENT COMPANIES READY TO LISTEN?

In this new environment, the *a priori* existence of content companies or media formats does not guarantee their continuation. Established companies will have to move fast to defend themselves against new digital challengers. As firms work through a range of unresolved issues, from interoperability across multiple platforms and devices to digital rights management, they will need to understand the true scope of the consumers' newfound power.

THE THREE C'S OF DIGITAL ENTERTAINMENT: CHOICE, CONVENIENCE, CONTROL

One of the most surprising findings from the study was the large number of people who consider themselves likely to buy digital devices and content. Everyone is interested in digital media and entertainment, even though the first generation of these devices lacks many of the features most desired by consumers. What is more, the number of likely purchasers does not vary by age group.

When consumers were shown a futuristic eBook reading device, or considered a television entertainment service that let them control the viewing experience, or envisioned a digital "audio device" that gave them the ability to listen to any music of their choosing, anywhere and anytime, the response was largely enthusiastic (see Figure 4-1). Based on consumer responses and all other available data on adoption trends, Accenture estimates that, with the right conditions prevailing over the next five years:

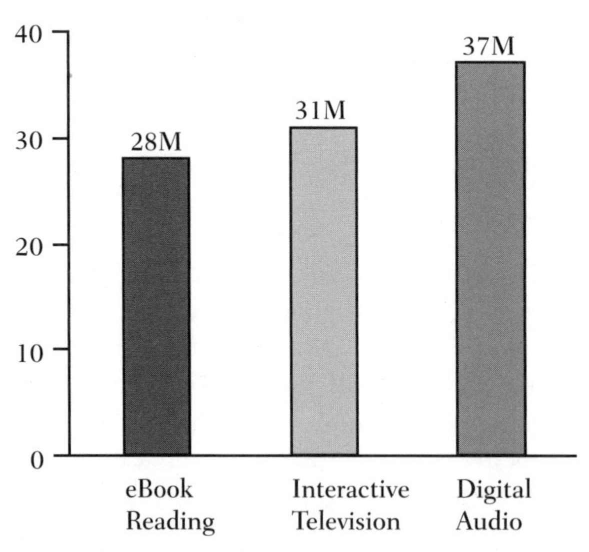

FIGURE 4-1 Number of U.S. consumers likely to adopt device by 2005 (in millions).

- 28 million consumers will purchase an eBook reading device
- 37 million consumers will use interactive digital audio
- 31 million consumers will subscribe to a multifunctional interactive television service

Based on their stated preferences, consumers want eBooks that are highly reliable. They want access to an extensive library of works on a lightweight device that has a back-lit screen displaying a clear, highly readable image. Color and connectivity are cool; features like phones, pagers, organizers, and video are far less attractive (see Figure 4-2).

In a digital audio device, consumers want high sound quality, access to virtually any music, a device that is virtually indestructible, and adequate storage capacity.

In interactive television, consumers are expecting video-on-demand, the flexibility to watch programs any time they want, the ability to watch past programs at will, and almost total control over live television, from pause to instant replay.

Key to the mass-market potential of these entertainment devices is their ability to excite consumers beyond the young "early adopter" trendsetters (see Figure 4-3). The appeal of these devices already touches the average consumer. This finding is quite surprising; conventional wisdom has it that younger consumers are more open to new technologies, and thus would be the early adopters of most new digital devices. Our study indicates that consumer interest in future digital entertainment devices is actually quite high across the age groups we surveyed, provided content is readily available and these devices have the functionality consumers want.

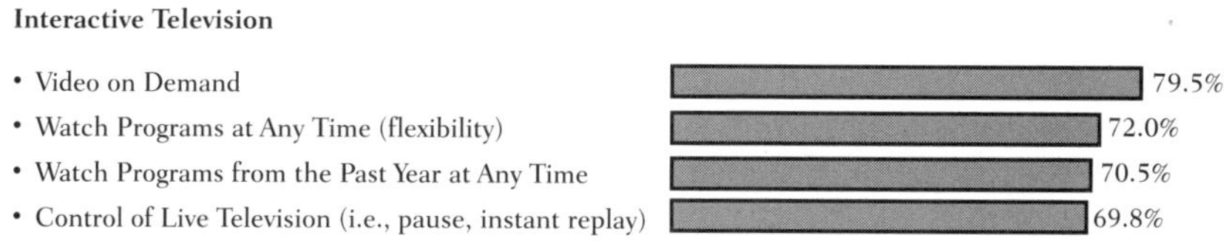

FIGURE 4-2 Critical device and content features.

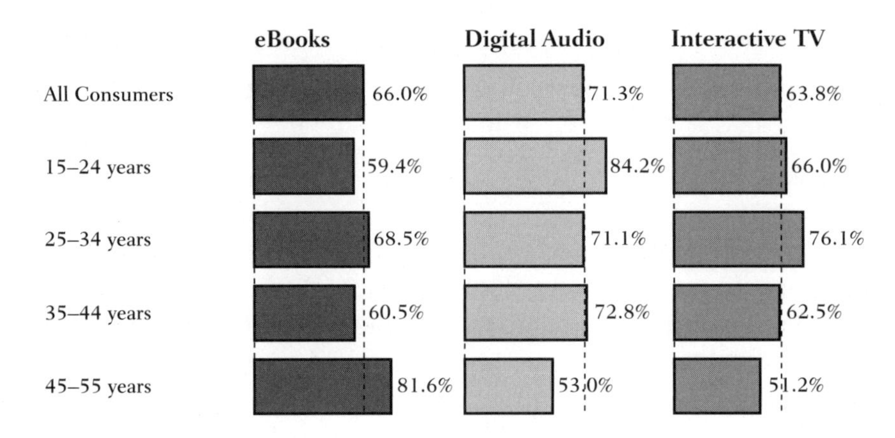

*Likelihood of adoption defined as expressed interest of 70% or higher; assumes device price is "fair and reasonable," and content is priced 20% less than today for audio and eBooks.

FIGURE 4-3 Likelihood of new device adoption by age group.

PURGATORY OR NIRVANA? THE DIGITAL DECISION

How content companies approach digital content will dictate their ultimate destination: Digital Purgatory or Digital Nirvana. Content companies can either work creatively and collaboratively to shape and optimize the digital content experience for consumers, or suffer as consumers experience the pain and frustration of conflicting standards and proprietary solutions. The choice and the future of digital devices and content lie largely in the hands of the people leading the content industries today.

In Digital Purgatory, proprietary solutions block interoperability and content is scarce. Complexity and restrictions constrain the marketplace, and the entire industry never escapes from its traditional media mindset. Vendors overlook consumers in the rush to secure advantageous positions in the new digital environment. Instead of working to develop integrated solutions that simply and conveniently provide consumers with broad access to secure content, some companies promote proprietary solutions with limited access to selected content—an approach that Accenture's research suggests will actually inhibit the growth of these new digital content markets.

Ultimately, and regardless of what digital providers may want, a de facto standard usually emerges, leading consumers back on the path toward Digital Nirvana. But until this happens, billions of dollars in potential revenues are needlessly squandered. We're suggesting that by addressing consumers' digital content needs, content companies can avoid the detour to purgatory altogether.

AN ALTOGETHER SUNNIER PICTURE

Visionary leaders can head straight for Digital Nirvana, in which open standards abound, content is abundant, convenience and control are paramount, and value-added products attract consumers.

By learning lessons from past mistakes—VHS vs. Betamax, stereo vs. quad, and DVD vs. Divx—content competitors can recognize success traits. New markets require cooperation. Win–win business models are superior to win–lose approaches, and satisfying consumer preferences is critical to long-term success.

When industry analysts look at the potential growth of the digital market over the next several years, their forecasts typically number in the millions of dollars. Based on Accenture's research, Accenture concludes that under the right conditions, market size could significantly exceed these expectations and may easily reach into the billions of dollars (see Figure 4-4).

For example, Accenture predicts that an "early market" comprised of Internet-savvy music fans will embrace legal digital music, leading to a market that could total $3.2 billion by the year 2005. The explosive popularity of Napster among early adopters proves that consumers across a broad range will play digital music files.

For digital TV, Accenture predicts a total industry revenue from DTV of $30.1 billion in the year 2005. This forecast includes revenues from basic service, premium channels, video-on-demand, other interactive services, related telecommunications services, and advertising.

Four key factors drive the rate of adoption of digital media:

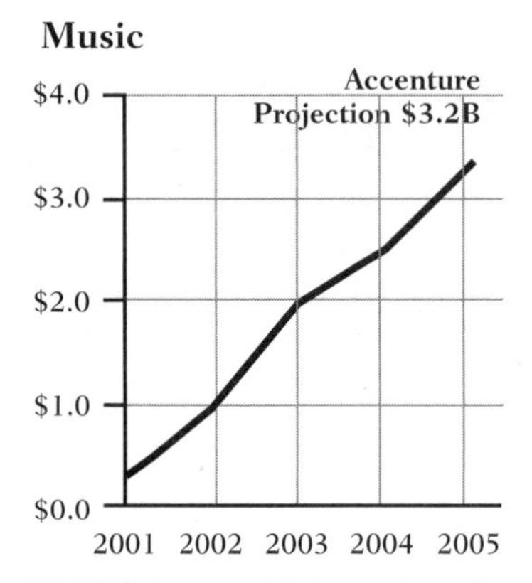

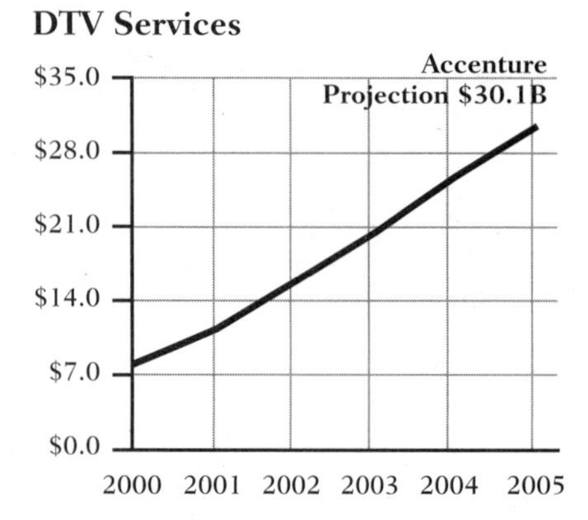

FIGURE 4-4 U.S. retail market forecasts for digital content (in billions of dollars).

- Widespread availability of content
- Quality and clarity of text, sound, images
- Simplicity and ease of use
- Durability and performance of devices

The stronger an industry performs against these four adoption criteria, the faster consumers will adopt digital devices and the digital content they convey. A mass market will, however, only develop rapidly if the environment created by industry leaders reassures consumers that digital content is widespread and that digital devices are risk-free. Content companies can create their futures most effectively by basing their market approach on a deep understanding of consumer wants and fears. The last thing in the world a consumer wants to do is purchase a device that is rendered obsolete by a new standard, or to own a device for which content is scarce.

What can content companies do in the face of these findings? Three core actions stand out: consumerize, collaborate, and configure.

CONSUMERIZE: START WITH THE CONSUMER AND WORK BACKWARDS

In coming to grips with the digital future, content companies naturally are tempted to start with what they know and with how their industry works today. But rather than taking existing industry formats and practices as givens, and then trying to adapt these to the new digital dynamics, content companies must adopt a more radical approach. Ignore the impediments imposed by established formats or existing content channels. Establish what the content consumer wants, and then rethink content creation, production, and distribution processes accordingly.

"Consumerize" implies thinking broadly about a legitimate transaction environment. For example, it may be necessary to rethink copyright-protection issues and begin to look beyond the purely technical solutions to a combination of approaches that collectively limit piracy, including convenient consumer processes, secure rights management technology, positive socioocultural influences, and effective enforcement.

COLLABORATE: DEVELOP WIN–WIN INDUSTRY BUSINESS MODELS

Consumers want integrated technology that is easy to use. They expect content and devices to be compatible, whatever the origin. They want a reasonable relationship between cost and value. They want content to have integrity. And they are looking for a low risk of obsolescence.

In addressing these expectations, the content companies of the digital economy must work together collaboratively to develop appropriate responses. Working together, industry leaders can:

- Collectively develop secure but accessible standards
- Cooperate to achieve interoperability, instead of competing to develop proprietary technology platforms
- Work to find win–win models
- Provide an upgrade path to avoid obsolescence

Can competitors cooperate? It is already happening in the publishing industry, where book publishers are collaborating to build the eBook market. The Association of American Publishers is helping to facilitate the creation of an open eBook market by recommending standards for digital rights management, numbering, and metadata. As a result, consumers will be able to read digital content with any reader software on any reading device. This level of industry cooperation is also likely to promote the availability of digital titles, as well as the development of new reading devices with attractive features.

CONFIGURE: REDEFINE OFFERINGS TO SUPPORT DYNAMIC PULL-DRIVEN SOLUTIONS

To realize the full value of all content, content companies should ask: How can we add value to the consumer experience? What are the components of the consumer's experience of my content? What other content do we have access to through our traditional business? How can it be mixed and matched to add value?

Think about giving consumers more control. How can consumers personalize their experience of content? Content companies should think about the consumer experience in terms of content modules and should build their ability to configure value-added products. A large, valuable market for digital content is available in the near term, if the industry takes an open, consumer-centric, value-added approach.

Digitization transforms media and entertainment more than almost any other industry—from creation through manufacture and retail into actual consumption. It drastically lowers scale barriers; enables radical new product functionality; alters pricing models; and eliminates historic constraints of time, speed, and distance. To avoid being commoditized or disintermediated, content companies must take advantage of new and larger markets, tapping into new media forms and broader distribution outlets that enable impulse purchasing anywhere. They can also benefit from the higher margins flowing from reduced or eliminated physical supply chain costs. To do so, they must seize the initiative and recognize that radical change is coming, whether it is comfortable or not. To wait passively for digitization to happen, as opposed to aggressively shaping it, is to consign oneself (and one's shareholders) to digital purgatory.

ENTERTAINMENT-ON-DEMAND FOR CONSUMERS-IN-COMMAND

The message on interactive television that emerges from Accenture's consumer research is clear: "I want what I want, when, where, and how I want it."

A three-way strategic debate is underway in the entertainment industry, a conversation involving consumers, content providers (entertainment companies), and the conduits (telecoms, cable, satellites). All share a strong belief that something big is about to occur in interactive television, yet no one is sure exactly how, or when, or even what will happen first. Our consumer research provides an important clue: Whichever combination of device, service, and content gives the consumer control over the entertainment experience will win. The research indicates that the consumer's needs determine adoption rates. So whether a company is a content provider or a delivery channel, business strategies should be driven by a deep understanding of these needs.

CONSUMERS WANT CONTROL

In Accenture's research, consumer respondents were given a description of a future interactive television service with several innovative features: Internet access, the ability to view and record television, on-demand capability, interactive programs, interactive commercials, promotions and on-line buying, video games, and a robust program guide.

Asked to rate the relative importance of these features, they responded in a surprising way (see Figure 4-5).

Our data reveals that consumers value "on demand" and "viewing/recording/television" features as essential or very important. All features associated with control of the viewing experience rated highly: video on demand; watching programs at any time, including being able to watch a program from the beginning while the end of the program is still being recorded; watching past television programs at any time; and control over live television, including the ability to pause and replay. There

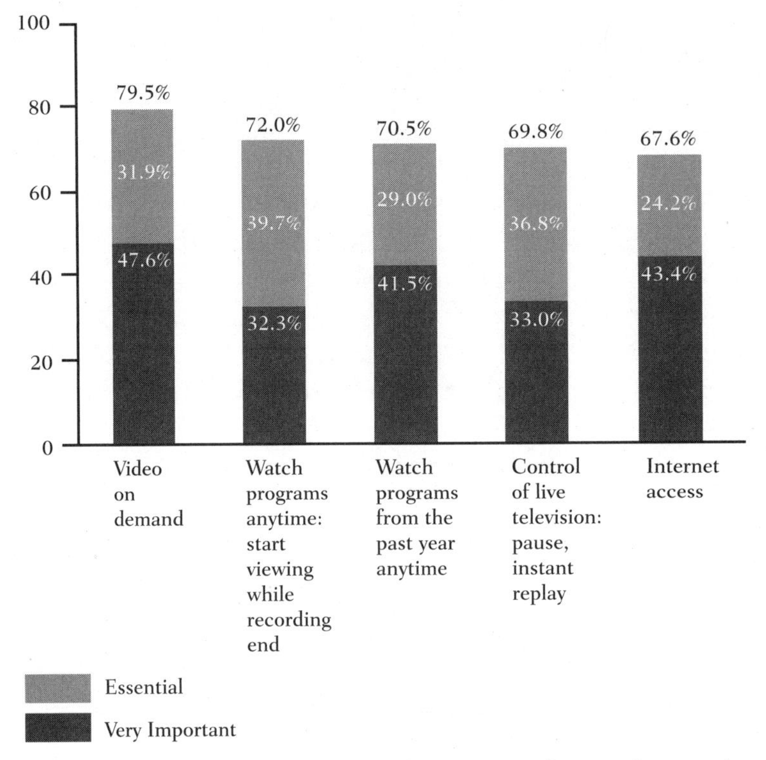

FIGURE 4-5 Relative importance of interactive television features (percentage).

was strong consensus around these features; two out of three consumers rate all of these features as "essential" or "very important."

It is an interesting fact that while 68 percent of consumers rated Internet access as essential or very important, there was a discernible drop in the "essential" rating (only 24 percent) for interactive television appliances, as compared with other highly sought features. This suggests that a significant portion of the consumer audience for interactive television already enjoys Internet access through another connection. This is confirmed by other readings that indicate that 87 percent of those who already access the Internet at home would be inclined to purchase an interactive television service. One conclusion to be drawn from this data is that Internet access may be a "nice-to-have" rather than a "need-to-have" feature in the earliest models of an interactive television service.

The good news from this research is that if you focus on the basics, there will be market uptake. The bad news is that exotic features do not generate much interest, making it more difficult to see a workable business model that leverages media convergence.

Accenture forecasts that the DTV marketplace will generate \$30.1 billion in revenue by the year 2005, with the largest sales flowing from premium channels (\$14.7 billion), and a sizable portion (\$3.2 billion) coming from video-on-demand (see Figure 4-6).

Content companies have an opportunity to assume a powerful position in the emerging broadband environment. Content could drive adoption of any given competing broadband technology over alternatives, because differentiated content has the potential to play a critical role in each step in the consumer's decision-making process. Consumer choices will include basic, premium channels and video-on-demand, games, enhanced television, commerce, and advertising. This content will be offered over DSL, cable, satellite, and terrestrial wireless by Internet Service Providers (ISPs), telcos, or third parties. There will be room in the marketplace for hybrids such as smart Tivo-like boxes, which cache material for playback without going all the way to full video-on-demand. Technological innovations such as MP4 will continue to change the marketplace.

In this environment, if you are a content company, the last thing in the world you want is to be beholden to a single channel.

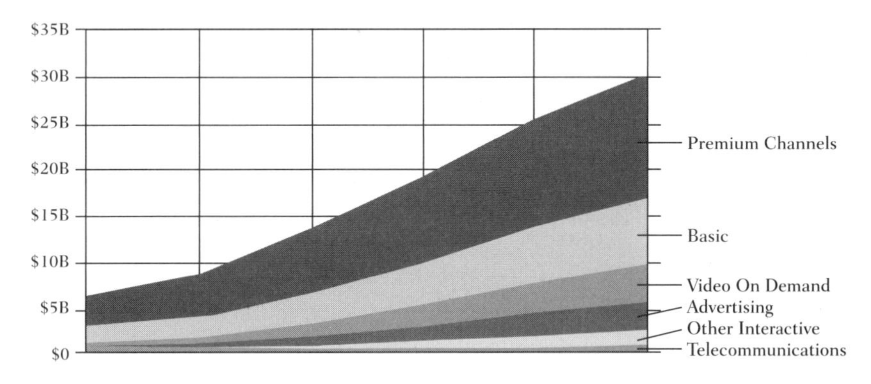

FIGURE 4-6 Accenture projections of DTV services revenues (in billions of dollars)

As a content provider, you cannot be put in a position where you are subject to distribution bottlenecks and pressure points. Ideally, you would like to play multiple channels off one another. So helping multiple channels survive is good business, and content companies should be prepared to make creative deals to reward innovators in the broadband world. For example, they can offer content at discounted prices to those who are first to market, until critical mass is achieved.

In the same environment, if you are a communications company providing the conduit, you typically face a multi-billion-dollar investment in infrastructure. You do not want to make this decision on the assumption that "if we build it, people will come." So the real issue becomes how do you deploy the capability you are building? When it's time to make tough decisions, you need a clear view of what the consumer will buy.

CHAPTER FIVE

TOOLS AND THE TRADE

DAVID LEATHERS

Creating digital content is a sequence of procedures. The number and complexity of those procedures is almost infinitely scalable. But whether you are making a simple home video or a Hollywood feature you will be using some form of technology, with—hopefully—some method to your madness.

The dimensions and requirements of productions and the tools available vary widely. There are tightly scripted projects with precise shot lists and there are guerilla documentaries. There are computer-generated images, film- and video-originated images, and ways of mixing any combination thereof. There are random image synthesis tools and precision camera-control rigs. The range is wide. The choices are many. There is a cornucopia of possibilities in the universe of content-creation tools.

The downside is that your desire to be creative can easily become overwhelmed by the technical details. It is part of the creative battle to avoid technological pitfalls. A careful end-to-end plan for the technological processes at the beginning is the only way to kept the technological tail from wagging the creative dog. Although there are many creative people, whether or not they are successful at executing creative digital media content depends on the goals, skills, attitudes, visions, and resources of those using the tools. Film, video, audio, visual effects, and interactive multimedia are all fundamentally technologies. They are also creative art forms and businesses. The fact is that the pure creative aspects of production do not take place in a vacuum. Technological and business realities are ever-present in the creative environment.

That being said, it is important to your project and your sanity to be realistic about both what you need to accomplish your goals, as well as the limits and possibilities of the technologies you select.

UNDERSTANDING ANALOG AND DIGITAL

Since the subject here is digital content, let's start with the basic concept that is behind digital media. Eyes and ears are analog by nature. We perceive the world through these analog sensors. Speakers, projectors, and monitors (video screens) are essentially analog devices. What we deliver, regardless of the technology used to create it, is going to be consumed in the analog universe in which we live. Before the 1970s, most production was done with analog tools. Since then, however, more and more digital technology has entered the picture, so to speak.

The concept of all-digital media technology is basically the same. An analog image or sound is "sampled" digitally. A simple example is the case of a black-and-white picture on a video screen. A digital sample of that image assumes that the image consists of a matrix of pixels. Sampling is the process of "encoding" the value of each pixel in the matrix as either black or white to a data file. That file is like a digital snapshot of a single field or frame of video, depending on the format. A new sample is encoded for every subsequent field or frame of video. As the video image changes over time, the corresponding encoded samples reflect the changing pixel colors on the screen. To play back the image, it must be "decoded." In other words the pixel information in the encoded file must be "read" and used to reconstruct the original analog image. The same kind of process takes place for audio. Audio is, however, sampled at much higher sampling rates, or frequencies. The digital audio standard sampling rate is 44,100 times per second for typical audio CDs (a digital technology). Remember, this is an oversimplified example. There are many variations in standards. Frame and field schemes vary with different video and audio formats.

So, why do we need to go to all this trouble? The reason is that once the media has been turned into digital data, the extreme power of computer and microprocessor technology can be applied to manipulate it. This results in both greatly reduced cost and greatly increased flexibility and power in how you can control and manipulate the images.

PROCESSES

The processes of creating most digital content are divided into logical steps: development, pre-production, production, post-production, and distribution. Digital technology is usually there at each step along the way.

Development

Definitions may vary somewhat, but the development process is the point at which the project is defined in terms of purpose, talent, budget, and script. This is the relatively low-tech part of the overall process. Not all projects involve scripts, but most projects involve some kind of written treatments and/or documentation before anything else gets done. Specialized word-processing programs are designed to facilitate writing scripts in a variety of formats, if needed.

Preproduction

Once the basic concept of the production is created using word processing software, spreadsheets, and other generic or specialized writing or business tools, the preproduction process begins. The script or concept may be "broken down" shot-by-shot to determine all the requirements to complete it. Sets, actors, gear, all have to be factored in. Shooting schedules and budgets are worked out, frequently employing specialized software packages to expedite the process and conform it to required standards.

Similarly, production design, visual effects, sets, and locations all are designed and tested using all kinds of specialized and general-purpose computer software; 2D and 3D graphics, compositing, image editing, and art and design software can all be used to advantage.

The degree of computer imagery and visual effects in a production varies from zero to 100 percent in today's world. Most productions, however, still use cameras and live talent for the majority of what they do. For the purpose of this chapter, we will concentrate on productions that are primarily based on camera-generated images

PRODUCTION

Production is the shooting phase. Until digital technology became pervasive in the media world, most high-end production was done with film cameras. Most video production was done with vacuum tube–based cameras. Now there is a much wider range of choices. In today's world, it is possible to convert film images to high-resolution digital video, and vice-versa, in a number of different ways, and the processes are getting better all the time. Feature films can be shot on a variety of digital tape formats and converted to film for distribution. Television projects are regularly shot on film and converted to digital video for postproduction and distribution. Parts of visual effects for feature film projects are shot on film, converted and combined with computer-generated effects and/or animation in the digital domain, and the results are recorded back to film. The processes are changing all the time, with more digital production, effects, postproduction, and distribution in the future.

This being said, this greater technological flexibility means that tools can be chosen for creative reasons. If you like the way something looks, and it's right for what you are doing, there is likely to be a technological way to get it into the mix. All you have to do is look at some the music videos being produced to see that images from video cameras, Super 8 film cameras, 16 mm, and 35 mm film cameras, as well as various computer-generated images can be combined with tremendous flexibility. There are many more subtle examples in feature films. The more content creators know about the characteristics of various tools, the more they will be able to make good creative choices.

FILM CAMERAS. The majority of high-end productions are still shot on film today, although the stage is set for a dramatic industry move to digital high definition video over the next few years. Film cameras have always been by far the most advanced form of image capture, and still are in many professional opinions. Decades ago, 35 mm film achieved the highest image resolutions attainable at that time; it's only now that 35 mm is finally being challenged by digital high definition video. Film-camera technology also provides some important advantages. The available lenses and lens accessories are extensive, having been developed over many years, and some represent the very highest quality technologies available. Also, the ability to vary the frame rate of film cameras makes the creation of dramatic slow- and fast-motion effects very practical. Nevertheless, the availability of top-quality lens systems and variable-frame-rate technology for digital high definition video is quickly closing the gap in capabilities, and the differences between film and digital is becoming increasingly less important.

ANALOG VIDEO CAMERAS. High-quality video cameras were all based on vacuum tubes for image capture until 1986, when the first practical cameras using charged-coupled-devices (CCDs) became available. It only took a couple a years for CCD cameras to almost completely replace tube cameras. The tube cameras were very sensitive, and were easily damaged if the imaging tubes were exposed to too much bright light. They also required frequent adjustment and calibration, were very power hungry, and required high light levels. The CCD cameras were more rugged, lighter, worked at lower light levels, used less power, and required much less maintenance. Although these cameras do use digital technology for image capture, they are still not considered to be fully digital cameras. In most early and many present CCD cameras, the image is ultimately decoded to analog and stored on an analog tape format. These videotape formats include 1-inch, ¾-inch, ¾ (or "U-matic")-SP, Betacam, Betacam SP, M-II, Video 8, Hi-8, VHS, Betamax, and others.

DIGITAL VIDEO CAMERAS AND DIGITAL VIDEOTAPE. In recent years, however, we have seen the evolution of cameras and tape formats to a fully digital state, where the signal stays digital all the way from the digital image sensor to the digital tape. At the high end, the most modern high definition video technology is all digital. For standard-definition digital video, Digital Betacam has become the most prominent format for distribution. Other high-end digital video formats include D-1, D-2, D-3, D-5, D-6, D-9, and others. The digital video (DV) format, however, is the first low-cost digital tape format. It has become pervasive and is rapidly taking over the camera market from the lowest-end consumer level all the way to some high definition video formats. Consumer cameras in the thousand-dollar range are being used to produce broadcast-quality programming and even feature films. Perhaps more than any other factor, the DV format is driving an explosion in digital video and making digital content creation with video a reality for a wide range of users.

There are various forms of DV with various data rates. The higher the data rate, the more digital information can be recorded for each frame of video. So, the higher the data rate, the higher resolution the picture. The most common format is the consumer DV format, which operates at 25 Mbps (megabits per second). There are also two professional formats, which operate at a 25 Mbps data rate: Sony's DVCAM and Panasonic's DVCPRO. Panasonic has also developed 50 Mbs and 100 Mbs versions of the DVCPRO format. The 100 Mbs format is called DVCPRO-HD and is a high-definition video format.

Sony has a high definition format called HDCAM. It is based on a ½-inch tape format and built around the Digital Betacam tape transport. HDCAM's 24-frame-per-second frame rate has captured the imagination of the film community and is becoming a popular alternative to 35 mm film.

POSTPRODUCTION

During the 1980s and early 1990s, complex and expensive linear editing bays in high-end postproduction facilities became the standard for broadcast-quality editing. Because of the expensive nature of these systems, a stratification evolved that separated the editing process into two parts, known as "offline" and "online" editing. *Offline editing* refers to the process of viewing the source material, trying out possible edits, and making decisions regarding how and where each shot will be used. *Online editing* refers to the process of finishing the video to a master tape, with all possible attention paid to achieving the highest quality finish.

The process of editing a video involves many decisions about what video clips to use, what order to put them in, and where they begin and end. Many other decisions are also made about the use of graphics, sound, effects, and other elements. Typically, the process involves a lot of experimenting with ideas. Many versions of an edit are often created before the final decision is made. The idea behind offline editing is that a lot of money can be saved by doing this work on a cheaper editing system, using exact copies of the original source tapes in a lower quality and cheaper format. Both the original high-quality master tapes and the low-resolution copies have identical SMPTE time code. (SMPTE time code is a system that applies a unique sequential number to each frame of the video on the tape. It provides an absolute reference back to the original source material from the copies.)

The main output from the offline editing session is not a finished program, but a computerized *edit decision list* (EDL) that provides a complete list of every edit to be made in the *online edit session*. The EDL is then transferred to the online editing system. That system, using the original camera masters and the best support equipment, is used to make a finished version and edited master tape of the show. An offline editing system may cost one-tenth or even less of what it costs to use the online system, making this a cost effective, but time consuming solution.

NONLINEAR VIDEO EDITING. In the late 1980s, a new kind of technology began to emerge, based on a new way to use computers in the editing process. It was called *nonlinear editing* (NLE). The first nonlinear video editing systems provided a better way to perform offline editing. In the new nonlinear process, the video material was first "digitized" to computer hard drives. Software-based editing programs were used to experiment with the shots and determine shot order and placement and other editing decisions. Because the video had been digitized to the computer's hard drives, it was available on a random-access basis: There was no searching through linear tapes to look for a shot. All the material was immediately available as files from the computer's hard drive. This saved a tremendous amount of time in the editing process. Like their tape-based predecessors, the nonlinear offline editing systems' output was an EDL based on SMPTE time code that could be loaded into an online linear tape-based finishing system for completion of the project at the highest possible quality.

This was a relatively effective way to work, and it is still the way most feature films are edited today, with a few variations. When offline editing a film-based project on a computer system, it is necessary to take a few more steps in telecine (transferring film images to videotape) to ensure that the EDL produced in the offline editing session can be traced back to the edge-code number on the film negative. Once the edit decisions are made in video, a negative-film cutter takes the list and conforms the film cut to it.

For video, however, the offline editing process has all but disappeared. This is because advances in the computer systems' processor speeds, hard drive systems, and software have made it possible to work at online quality in the computer system. The video can be digitized into the computer at the highest quality and completely finished in the computer system. Once completed, the only thing left to do is to run the show back out to a master tape format for storage and distribution. The expensive online tape-based editing systems are quickly becoming a thing of the past. There are a wide variety of nonlinear editing systems available from a number of manufacturers that vary significantly in features, functions, and price.

Currently, online editing at uncompressed high definition quality still requires a linear system, but standard-definition nonlinear editing can easily be performed on a variety of computer-based systems. The first viable nonlinear high definition editing systems are becoming available now. Nonlinear editing has many advantages over the linear process and will no doubt soon become the standard for high definition editing as well.

VISUAL EFFECTS. Generally speaking, visual effects are defined as creating, combining, or altering images in such a way as to synthesize a new image that didn't exist before. Visual effects are as varied as the imagination and can be accomplished in a wide variety of ways.

Before digital technology, most visual effects were accomplished optically by manipulating film in unusual ways, or by the use of props and optical illusions, such as miniaturization, motion effects, or optical compositing. Today there are sophisticated computer programs for all kinds of 2D and 3D image creation and manipulation. It's to the point where character animation is approaching a level of photo realism that makes it difficult to tell the difference between real and computer-generated actors. Realistic motion can be captured from real actors and applied to computer-generated models. Such details as wispy hair and fire can now be created realistically on a computer. There are entire movies being generated on computers. The trend will continue to blur the line between reality and the imagination. It may even give actors cause to be nervous.

Like all these other technologies, the acceleration is phenomenal in terms of capability and cost reduction. The use of blue- or green-screen technology, combined with various computer keying, compositing, and animation techniques can provide all kinds of virtual sets and elements that can be used to create almost any kind of effect or illusion.

The computer power and software technology necessary for the execution of large-scale effects used to require multiple, massive, mainframe computers and expensive custom designed software. Now, the same results can be attained on PCs with off-the-shelf software. There are dozens of companies that offer hundreds of programs that enable visual effects on desktop computers. Talent, however, is not available off-the-shelf; we can buy better pencils and pens than were available in Shakespeare's day, but how many pencil- or pen-owners can write as well?

AUDIO. Audio content creation and manipulation has been completely transformed by digital technology over the course of the past 20 years. In music recording, digital technology is everywhere, from the musical instruments themselves to the recording, editing, and mixing tools. The entire discipline of "sound design" has evolved around the digital audio toolset.

Today, most audio for film and video is created, edited, and finished on computer-based digital audio workstations. The systems are very powerful and make it possible to do almost anything you can imagine with sound. Digital sound libraries can contain thousands of samples of all kinds of sounds. Using a digital audio workstation, a user can edit and mix an unlimited number of sound effects into the soundtrack of any production.

The tools range from computer-based *digital audio workstations* (DAWs) to all kinds of outboard digital effects devices and digital mixing boards. There are digital tape machines and digital disk-based recording systems. There are also a number of new low-cost hardware-based DAWs that are like complete little digital recording studios, capable of multichannel recording and mixing in a single inexpensive unit.

Practically speaking, the audio capabilities of many NLEs are sufficient for the needs of audio post. Unless, there are serious audio problems to be solved, very large numbers of tracks, or sophisticated theater or surround mixes to be created, an NLE may be able to do the job.

THE TECHNOLOGY CHASE

The rapid evolution of technology has a huge effect on the way content gets created and even on what kinds of content gets created. This can either be a benefit or a curse to the creative process. Greater capabilities require greater restraint. It is evident that a lot of films have thrown in a lot of unnecessary or ill-advised effects shots just because it can be done. Entire genres, such as special-effects films, have evolved. This can be interpreted as business and technology getting together at the expense of creativity. Some Hollywood formulas figure that putting a lot of expensive-looking effects shots on the screen adds to the "production value" and visual wonder of a piece and somehow makes it worth more. This has led to an effects "arms race" in which the next big film to have a huge effects budget will automatically make huge box office numbers. Frequently, the story, the script, the acting, directing, or other elements do not measure up, however, resulting in a disappointing overall experience for the audience.

This kind of technological escalation takes place at all levels. Even relatively inexpensive software packages keep adding all kinds of effects capabilities, cheesy as they may be, in an effort to distinguish themselves or deliver the appearance of value.

The best tools are those that perform their core functions very well, with clean, intuitive user interfaces and controls—and a lack of unnecessary gingerbread. These are the most likely to help the creative user to accomplish his or her storytelling or communications goal with style and quality—which, in the end, is what good content creation is all about.

CHAPTER SIX

THE NONLINEAR AGE

Bob Turner

Editing is considered (at least by editors) as the most important craft in content creation. An editor can make a bad project look good and a well-written, directed, and photographed production look terrible. This chapter will take a look at a few of the aspects of the editing process that digital content creators should consider. These days, editing is done on digital nonlinear editing systems running on Macintosh, Windows, Linux, IRIX, and other computer systems; when all your content resides on computer storage, your creative options for cutting, pasting, and otherwise manipulating images expands tremendously. Such is today's nonlinear editing environment.

Let's start with some terminology. *Postproduction* (after production) is the act of assembling raw, created media. *Nonlinear,* although used to describe the process associated with digital video applications, actually existed long before linear editing existed. Film editing has always been "nonlinear" in nature. To explain, let's first take a look at what *linear* and *nonlinear* refer to in the creation of moving-image content.

Using *linear editing* in the simplest terms, editors had two videotape recorders (VTRs); they selected a clip (a defined portion of the video on the source tape) from the "source" VTR and rerecorded it on the "record" VTR in the order the editor wished. Of course it was actually a bit more complex than that with pre-roll synchronization of the two machines, and the involvement of one or more additional VTRs if you wanted to "transition" or "effect" from one source to another (fades, dissolves, page-turns, etc.). The important point is that you rerecorded one clip, followed by the next, and then the next.

Now, if you want to shorten the second clip of the ten assembled clips, you need to rerecord clip 3 over the portion of clip 2 you want to eliminate (or record a new clip 2 if trimming off the beginning of that clip). Once the clip is shortened you have to rerecord the remaining clips to eliminate the gap. Likewise, if you want to lengthen clip 2, then you need to rerecord the longer clip 2 and then all the rest of the clips to shift them forward. This process of expanding or contracting and readjusting what followed was known as "rippling the list."

I think you can easily understand how this time-consuming process might discourage people from wanting to try a lot of trimming and playing back after a series of edits have been made. Even by keeping a clean list of editing decisions (points where separate shots begin and end, and where they are placed in the program) in the computer and allowing the computer to autoassemble the program after the trim point, it frequently took two, three, or more minutes per edit (depending on whether tape changes were required) due to the shuttling of the tapes, synchronizing the pre-rolls, and more. With a simple change involving eight clips, it could take 15 minutes or more to reassemble the program with the trimmed change.

Film editing, on the other hand, has always been nonlinear. If you want to shorten the second clip, you slice out the portion you want removed, and splice the film back together at that point. All the other edits remain intact. If you want to lengthen that clip, you find the film segment you removed and you splice it back in. (The hard part is finding all of the film segments you want to assemble into the finished movie, especially the clip ends that you've already trimmed away.)

With digital nonlinear editing, the process is far simpler. Nothing gets rerecorded. The computer accesses the designated digital media (audio and video clips) and calls them up sequentially and instantaneously to be viewed on a monitor (and heard from loudspeakers). It is a nondestructive process: You never lose the ends, should you want to lengthen a shot. A list of editing decisions is generally presented graphically on a timeline. Each clip is defined on that timeline with a start and stop point for the media stored on the computer editor's hard disks. A predetermined point at which that "clip" should be accessed in the program also is defined, together with instructions on how it should be heard or displayed. This is the state of nonlinear editing technology today.

But *nonlinear* generally means much more than just random access to digitally stored media: You must be able to make adjustments and play back those adjustments easily. Today, nonlinear means more than timing adjustments on a horizontal timeline. You must also be able to make adjustments "vertically," or in any specific layer of a complex image composite, or any track of a complex audio mix, to meet the standards of what most professionals think of when referring to nonlinear editing technology.

OFFLINE VS. ONLINE

While we are on the subject of editing concepts, you may have heard the words *offline* and *online*. These words are the two most confusing words associated with editing. They have nothing to do with image quality or what type of computer you use, and their definitions are very different from those used in the computer industry. In the video industry, the word offline means a postproduction process wherein the primary result is a list of edit decisions most commonly know as an edit decision list (EDL). There may be a videotape created as a result of this process (mostly used for client-approval purposes), but the primary product is an EDL. In the video industry, the word online refers to the postproduction process wherein the primary product is a video program—usually an edited master recorded on videotape. There could easily be an EDL output (on a computer disk or printed out on paper), but the primary result is a video program.

The terms offline and online originated because of economics. Normally the much more laborious process of making offline edit decisions happens in a less expensive editing suite or on less expensive equipment. Signal quality is not necessarily monitored; the object is the arrangement of the clips with a focus on pace, rhythm, and how to most powerfully communicate the story. Many combinations can be tried and rejected in this process, requiring a major amount of time allotted for it, which explains why the cost of the equipment is of concern. Once the edit decisions are determined for the most part, the list is taken to an online editing system.

The online process generally focuses on signal quality, as well as on the technical and/or artistic processing of the images when those processes require expensive equipment (e.g., elaborate effects or sophisticated graphics composites). In film, the equivalent process for an offline edit is the creation of a workprint; the online editing process is equivalent to the negative cut.

Today, these concepts are evolving. It may be that workgroups of various creative professionals are all working with specific equipment and accessing media and data from a central Storage Area Network (SAN). Another possible evolution would be preselecting clips on location via a Personal Information Manager (PIM), like a Palm Pilot or WindowsCE-based PocketPC with logging software, perhaps connected to the camcorder via wireless or infrared technology. Alternative preselection choices include a browser and editor in an office, or even some of the clip-notation functions built into the latest camcorders. There are several systems that allow editors to create a program in one location while various other decision makers (directors, producers, clients, etc.) watch in another geographic location via the Internet or similar remote communications technology.

One manufacturer (Avid Technology) allows an editor to send an audio or video stream representation of the edited program to the manufacturer's server, where it automatically notifies the decision-makers (clients, producer, director, music composer, graphics designer, etc.) that the file is available. When a decision-maker is ready to view, he or she responds that they're ready to receive this stream, and an applet opens the stream for viewing. This applet offers the decision-maker an opportunity to make notes and attach those notes to specific frames. There is also a high degree of security with this system. Not even the server administrator can access the file. When each viewer finishes his or her comments, the file is sent back to the editor, with a track on the timeline where added comments (notes) reside and frame-location markers to indicate the notes and frames referred to.

As for tomorrow, there are standards in place where various image clips, audio clips, graphics, and other media become objects. Rather than combining them in the postproduction process, directions will be given as to how they will eventually be combined; the final marriage of those various media in a program might not occur until the final delivery or distribution point. The Motion Picture Experts Group (MPEG) is the leader in this effort.

METADATA AND ESSENCE MEDIA

Two more terms relevant to the nonlinear age of editing are the words *metadata* and *essence media*. Both are different types of digital data, and soon both will coexist in most digital media streams, be it camera-original footage, edited master programming on videotape, or media on a server. *Essence media* refers to actual media (video and cinema images, sounds, music, graphics, animations, text, etc.). Metadata refers to all the ancillary data that is associated with essence media. This could be anything from technical information such as sync and blanking data; library information data; editing decisions; audio levels, transitions, and equalizations; color adjustments; and even the geographical location where the images or sounds were created, and by whom they were created. This type of information can be contained in various portions of the signal stream or in separate data streams associated with the essence-media data streams.

Various standards organizations and manufacturer groups are defining how the two types of data will be organized, defined, combined, and used. These organizations include:

- The Society of Motion Picture and Television Engineers (SMPTE)
- The European Broadcast Union (EBU)
- The Advanced Authoring Format (AAF) Association
- The ProMPEG Forum
- The Media Asset Management Association (MAMA)
- The Motion Picture Experts Group (MPEG)

ADVANCED AUTHORING FORMAT (AAF)

The Advanced Authoring Format, or AAF interface, may be one of the most important of the standards from a postproduction point of view. Remember EDLs? Several different manufacturers developed their own structure or format for communicating these lists. The most popular were developed by CMX, Grass Valley Group, and Sony. Attempts were made to standardize on a single EDL, but that effort was not successful. The most popular of the various EDL formats was the CMX EDL, which is based on 40-plus-year-old technology! It assumes a single video track and a maximum of four audio tracks. Today 8, 24, or more video layers are common. Furthermore, with 5.1-channel sound-surround programming, limiting the EDL to four audio tracks is absurd. Nevertheless, there had been no universal EDL format standard developed to replace this antiquated standard until the development of the AAF. (Actually, AAF follows a previous attempt called the Open Media Framework [OMF]. This was partially successful, but was based on an Apple architecture that has been abandoned. Furthermore, it was ahead of its time and the acceptance by manufacturers was spotty at best.)

AAF is a subset of the overall metadata standard being defined by an SMPTE/EBU Metadata Taskforce, and has a subset of its own: Media eXchange Format (MXF), which is for less sophisticated production (e.g., news) and is focused on the needs of media servers and streaming. It may, in the future, also be useful in interactive programming.

AAF was developed for a rich exchange of postproduction metadata including multitrack EDLs, as well as exchanging effects compositing metadata and various sound and picture adjustments and enhancements. The architecture allows for both manufacturer-proprietary metadata and universally defined metadata. Although developed as a metadata standard and not an essence-media format, pressures from users and manufacturers—and perhaps an unwillingness by the owners of other essence-media formats to adapt AAF compatibility—have meant that the developing format now includes the possibility of an essence-media component. More importantly, several of the most popular digital audio and video signal, transport, and tape specifications now provide for a metadata component where AAF metadata can reside. The AAF Association Software Developers Kit (SDK) is an open-source standard available to any manufacturer for free. Membership is encouraged to provide input for future developments.

AAF provides for the possibility of creating a program on a low-cost compressed AAF-compliant editing system and having the AAF metadata sent to a online system such as that offered by Post Impressions.

COMPRESSION TRENDS IN POSTPRODUCTION

Digital film and video editing systems started out using Motion JPEG codecs (or compression–decompression algorithms) for their media. This was successful because most systems could offer a variety of compression options. Those options ranged from high compression with low-image quality (but with the ability to store many hours of moving images on a limited amount of storage), to low-compression, where the image quality was excellent (but the number of minutes per gigabyte of hard disk storage was low). This allowed the editor to start with a wide choice of low-resolution images and use an offline process to make editing decisions. As the editor deletes this low-resolution media from the hard disk storage, it goes through an automated "batch redigitize" process where selected clips are reloaded on the hard disk with a low-compression but higher image quality setting to finish the program.

The problem with Motion JPEG was the proprietary nature of the technology. It was not easily exchangeable, and in the early years manufacturers sold many systems based on the image quality available from their unique codecs. As digital video recording removed the need to digitize video at the workstation, the recorded digital codecs (e.g., DV and MPEG) replaced working with Motion JPEG codec images.

(Please note that the digital codec used in the editing process does not necessarily lead to using that codec for distribution purposes. Editing generally requires intraframe compression [each frame available to view forward, reverse, or still by itself]. Therefore, it is very common for a compressed program go through a secondary compression process or a transcoding process to a higher compression algorithm prior to distribution via streaming or optical media [e.g., DVDs].)

DIGITAL SIGNAL LOSS

Another compressed digital video issue that must be addressed is that just because it is digital does not mean that you can edit without signal loss. Let's use the DV videotape format as an example. You can store the DV "signal" natively on a hard disk, and with the appropriate editing software you can do a "cuts-only" edit without any loss of image quality. Unfortunately the moment you need to process the image (e.g., fade, dissolve, effect transition, superimposition, color correct, etc.), the signal goes through an uncompress–recompress process that can introduce artifacts. Multiple processes may degrade the image noticeably.

Furthermore, although video may look good with a certain degree of compression, graphics and animation sequences may not look as good with the same amount of compression. These two factors are the primary criteria that many content creators use when choosing a more expensive uncompressed video editing system. Such a system can uncompress camera-original compressed images, and all processing steps—including the compositing of those compressed video images with uncompressed graphics, text, and animation—occur in this uncompressed realm without any loss of image quality.

HIGHER RESOLUTIONS

You can read about SDTV and HDTV (both compressed and uncompressed) and even higher resolution digital cinema standards. With regard to postproduction, you should know that systems exist that allow editors to work with uncompressed, 2,000 × 2,000-pixel moving images and even 4,000 × 4,000-pixel (or greater) moving images. These images usually come from digital film transfers and allow the editor to select a portion of the overall image for display without unacceptable image resolution when displayed.

A major problem in this world of high-speed evolution is the selection of a standard format on which to "master" your efforts and archive your work with relative confidence that it will be able to be played back in the future. The only solution to fit the bill is 35 mm film, which has been in existence for more than 100 years. It is impossible to predict which digital imaging format will be around in 25 years—or even in five years. The expert view today is of a trend toward standardizing on 1,080-line, 24-frame, progressive-scan video (1080 24P) for mastering, but many arguments are being made promoting alternatives. Even the trend toward MPEG and DV compressed digital formats is countered by their rapid evolution. For example, in the DV realm there's now consumer DV, DVCPRO-25, DVCAM, DVCPRO-50, DVCPRO-50P, and DVCPRO-HD. Who knows what tomorrow will bring? The only answer is to use the highest quality that the budget will bear, and hope.

AN IMPORTANT TREND: CONSUMER GEAR

Consumer video technology has been one of the most important trends in professional video. It is important to note that the electronic signal recorded onto the videotape in an inexpensive consumer DV camcorder is identical to the signal recorded onto the videotape in the top-of-the line professional or broadcast-grade DVCAM or DVCPRO-25 camcorder. The difference in image quality is entirely the result of the differences in camera and lens technology, and not recording technology. As consumer camcorder image quality improves, and as consumer computer technology becomes faster and more powerful, the differences between consumer and professional equipment narrows. For many purposes the price difference grows harder to justify. This has already resulted in significant price drops for professional equipment.

The fact is, for simple editing with DV-format recorded images, there is no image quality difference between editing a cuts-only video program with professional editing software and with a consumer editing software package (e.g., iMovie-2 from Apple) that may come free with the computer. A low-cost compositing software package used by amateur film or video artists (Adobe After Effects, Artel RED, or Discreet Combustion) is technically capable of the highest quality work, and in fact is quite popular with many professionals as well.

The bottom line is that you can find lower-end consumer and prosumer equipment used for professional content creation. These professionals believe that talent, more than equipment, makes the difference, and because today's consumer equipment is of a higher quality than the professional equipment used just a decade earlier, they believe the delineation between professional and consumer gear has blurred.

More important, video editing is becoming as commonplace today as desktop publishing. An increasing number of grade-school children, executives, and home-video enthusiasts are familiar with editing concepts and techniques via such applications as PowerPoint and Apple's iMovie. Editing is no longer an arcane skill limited to a select fraternity of experts. In such a world, creative talent becomes the deciding factor in achieving success.

HOW DO YOU JUSTIFY THE COST OF HIGH-END NONLINEAR EQUIPMENT?

When asked about the high cost of nonlinear equipment, I ask a question in return, "Why does a carpenter purchase a $45 hammer, when a $7.50 hammer sold at a supermarket also drives nails?" The answer is that the professional carpenter appreciates the difference. The professional carpenter can feel how the $45 hammer is better balanced, which makes it healthier for him to use; the nails may be driven more efficiently; and the claw is designed to remove nails more easily. Using a hammer as much as a carpenter does, the difference in the cost over the period the hammer's life is negligible. In fact, if that carpenter can work faster, or if it protects him from downtime caused by a repetitive-motion injury, the investment can pay for itself in a very short time. More intangibly, as a professional, that carpenter takes pride in owning good tools; he makes a purchasing decision based, in part, on the value of the image that comes from owning professional-quality tools.

The justification for purchasing a much more expensive edit system can be clearly related to the story of the carpenter's hammer. Although a higher-end editing system may not offer higher-quality images for a specific type of product and source, there may be many more tools available for the editor to craft his or her work effectively.

Compare an Avid Media Composer to a consumer edit system. The Media Composer offers more than 30 ways to trim an edit and includes four different Trim Window modes; a consumer system offers two ways to trim an edit and no Trim Window mode. When you consider that a "trim" adds or removes frames to either end of a clip, you realize that this is a fairly simple thing to do. But each of those 30 processes and four trim windows serves a specific purpose. Moreover, there are layers of complexity to each of those modes (looping preview, specifying pre-roll and post-roll lengths, selecting which tracks are selected for trimming, out-of-sync indicators, amount-of-trim indicators, etc.). The decision to use one of the many processes available when trimming an edit may even be the simple choice of personal preference to the editor, who prefers a process that's intuitive instead of distracting.

The higher-end product typically has more processing power, so that rendering is quicker and effects are in real-time, which means greater efficiency. There may be more advanced tools for color correction, more sophisticated transitions, more complex compositing, tools to enhance sound or pictures, and the system may actually provide higher-quality images when the work is done. The editor may find it faster and easier to find the clip she is looking for. She may be able to clearly understand the organic pace and rhythm via the graphic user interface information. There may be more tactile control of the various processes for efficiency and effectiveness. (Examples of tactile control include drawing with a pen and tablet rather than with a mouse, or mixing in real-time with multiple faders rather than programming changes on a timeline—track-by-track and point-by-point for every adjustment.) The professional nonlinear editing system may offer the ability to work efficiently in workgroups, sharing media and metadata and efficiently using different software for various parts of the content-creation process without file incompatibilities—speed and ease-of-use solutions.

As with the hammer, when you amortize the cost of high-end nonlinear editing software versus a low-cost package over the lifetime of the system—and especially when you include essential professional peripheral equipment (computer monitors, video monitors, the computer itself, storage, waveform monitor and vectorscope, VCRs, audio mixer, audio sources, patching and routing, signal generator, power conditioning, etc.)—you find that it's far less expensive than it initially appeared. It may pay for itself with the efficiencies available, you may be able to charge a higher rate per hour, and it may bring in more business because of the prestige of the branding. Finally, as a professional, it may generate a sense of pleasure and pride.

MEDIA ASSET MANAGEMENT

Organizing content on hard disks, being able to find a specific clip quickly and easily, and being able to archive and library the media so that you can easily find and make changes to a program later are all important editing capabilities. Most editing systems can sort stored clips by a variety of parameters (e.g., alphanumerically by date, reel number, description, timecode, scene or take, character, or user-defined category). You should be able to sift your collection of clips (e.g., displaying only scene-1 clips or only clips from a graphics reel, or only clips where a certain character is visible) in a multitude of ways. You should be able to find a clip by one or more parameters. You should be able to condition your search request with terms like *and, or, not, nor, near*, and the like.

A common complaint when editing is that it takes more time to find clips than to edit them. Media-asset management software is designed to eliminate this problem and enable content creators to find specific clips quickly.

INTERACTIVE CONTENT

Interactive content adds a new twist to media-asset management. The editing is not as complex, but generally consists of many short, edited programs. In addition to managing the source media, system operations should facilitate managing the many edited segments as well.

Although it is common to find interactive-content editing as a separate operation, frequently on a separate application and used by a separate operator or workgroup, the tools available for interactive content are rapidly evolving. What once involved technical code-writing and occurred after the postproduction process today may occur in conjunction with the editing process and be far easier to do. For example, programming nodes can have one or more tracks on an edit-system timeline.

REPURPOSING CONTENT

Marketing departments are all creating slogans for reusing content: Anycasting (Sony); Create Once, Publish Anywhere (Avid);

Make, Manage, and Move Media (Avid again); Publish Anything, Anywhere, on Any Device (Adobe); Adobe Everywhere You Look (Adobe again); Create Once—Use Everywhere (Discreet); Multi-Mastering (Discreet again); No Limits (FAST Multimedia); New Media, New Formats, New Economies (Quantel); and Media Without Bounds (Grass Valley Group) are all examples of promoting the concept of distributing in a variety of formats while using the same essence media and metadata.

Repurposing content has recently become the buzz phrase, but the concept has been utilized and has evolved over decades. Twenty-five years ago corporate video communicators understood the cost-effectiveness of reusing previously shot or edited videotapes.

Using storage-area networks and the ability to use digital content in a variety of ways (e.g., cinema, video broadcasts, interactive games, and/or programming—including product promotions, music videos, Web pages, video-on-demand, VHS/DVD, etc.), producers are eager to maximize the profit potential for their content. Existing technology is making this much easier, with metadata descriptions, media-exchange standards for content reuse using a variety of software and computer platforms, better media asset management, and easy access to the original content on workgroup-capable media servers or archived nearline storage.

Of course, the ability to work with the highest quality sound and image files also facilitates repurposing the source footage afterwards; it is easier to go from cinema-quality images to streamed video than it is to go the other way around.

One concern in repurposing content is the need to change aspect ratios for cinematic distribution, wide-screen video delivery, standard 4:3 aspect-ratio television, and a variety of aspect ratios that can be used with Web page frames or interactive programming. Let's assume that the image quality allows for modifying aspect ratios. (If the image quality has to be enlarged and cropped to provide for a different aspect ratio, the image quality may degrade rapidly as the image is enlarged). If you start with a very-high-resolution image (e.g., widescreen digital cinema) you can crop the image with little or no noticeable loss in image quality.

One function required for adjusting aspect ratios is *pan and scan*, the ability to pan a wide-screen image in a 4:3 window. Generally producers want the time to remain constant for both normal and wide-aspect ratio programming. One problem is that a person entering a field of view does so sooner on a wide screen than on a centered 4:3 screen. You must select and then move the narrowed field of view to accommodate this. You may also need to make cuts in the 4:3 program to accommodate dialog between characters who may not be visible together in this narrow aspect ratio, but are visible on a widescreen aspect ratio.

CREATIVE RULES

There are rules on editing composition and changing planes of view. There are concepts taught concerning the selection of clips from high- and low-camera positions (and the implications in doing so), and the purposes of selecting different camera angles to incorporate into the program (from establishing wide angles to extreme close ups). The two most significant rules however, are that all these rules were created with the knowledge that they can be broken and that ultimately, creativity is what truly rules!

CHAPTER SEVEN

DIGITAL RECORDING

JOHN RICE

What do George Lucas, Spike Lee, Harry Shearer and Mike Figgis have in common? They have each created a feature film using digital recording. But, each of these filmmakers has used different, and essentially incompatible digital videotape formats.

The buzz that has become "digital" is confusing. Whereas Lucas garnered headlines in trade magazines and *The New York Times* for shooting *Episode II* of *Star Wars* with high definition 24P, Spike Lee has released two films—one shot using DVCAM, and another shot with consumer digital video (DV). Figgis used DVCAM for *Timecode* and Shearer shot in 720P. All digital is not alike.

There was a time, not all that long ago, when the idea of using digital videotape meant, unequivocally, the highest possible quality, and most often a high price tag. But as digital technology expands and excels, there are now myriad choices that fall under the banner of "digital." And we are not only talking about videotape here. Digital signals are being stored on servers for broadcast; on hard drives for editing; and will soon be available on disks for shooting editing and distribution for professional and broadcast signals for acquisition, postproduction, and distribution.

Digital videotape, once the exclusive domain of the high-end, high-budget production, has now become the mainstream, with over 15 formats or versions of formats currently available. But in that selection process there are an ever-increasing variety of applications and cost-points that make different formats more viable for different uses.

To say that there is a leading digital videotape format in today's production and postproduction market is virtually impossible. Some formats are succeeding on their technical strength and the marketing muscle of their makers. Others, although having good arguments for their benefits, have been caught at higher pricing levels than newer competitors. And it is safe to say that there are more iterations on the horizon as manufacturers can adapt formats with processors, tape pitch, speed, and various sampling rates to make their products more efficient and more cost effective.

There is no simple decision on what digital format is best. For some, quality is, and will always be the primary consideration. But quality does not in and of itself make for a successful format. (It happened in consumer video with Betamax vs.VHS, and in professional analog, and it will happen again in digital videotape formats.)

Price considerations must also be taken into account. And price goes to more than the cost of purchasing or renting a given camcorder or deck. Tape cost must be factored in. And maintenance will affect the long-term cost of any new equipment purchase.

The application of digital gear should also play heavily into any decision. Some formats have found their strength in postproduction or graphics, others in news, still others in either high-end or low-budget acquisition.

To try to address the best format for any given use, the first step should be to identify the application strength of the product. If the primary goal is to shoot in digital, then there are formats that provide quality and cost benefits, as well as weight and operational considerations that make them strong contenders to take into the field. Some of these also have viable applications in postproduction.

But other formats shine brighter in postproduction. They may be too big (in equipment weight or tape size) to make them reasonable for field work, but they may best serve in a linear postproduction environment. Some others may best serve specific types of postproduction work, such as graphics and film transfer, where quality is arguably the primary concern.

Another issue is one of compatibility and generational impact on the image. At the high end, uncompressed formats promise the highest quality over multiple passes in an edit or processing environment. When creating graphics, for example, a multigenerational uncompressed format may be the only choice. But more and more of the compressed formats are finding ways of maintaining high quality (if not quite as high as uncompressed) by maintaining the digital information as data during the passes.

That would eliminate concerns over concatenation (the issue of uncompressing and recompressing a signal with each editorial pass). Such a scenario gives strength to "closed" systems where all equipment operates in the same compression mode. But again, that may be unrealistic for any facility or operation that needs to integrate newer gear into an existing system.

Again, the choices are many, and there can be good and valid arguments for selecting almost any format included here.

Here we've broken down the videotape formats into two fundamental categories: standard definition (525/625 lines of resolution) and high definition (720 lines and above). Even with this breakdown, be careful. Many of the standard definition formats are capable of widescreen recording (16:9 aspect ratio) although the signal remains in the realm of current NTSC and PAL signals. To be honest, any format can record a 16:9 aspect ratio if the wide image is anamorphically squeezed via a lens.

High definition formats will record higher resolution images. Most of the initial offerings operate using 1,080 lines interlaced. But these formats and their VTRs are already promising to record the broad range of high definition signals that are being promised, including 720 progressive, 480 progressive (arguably not high definition), and even 1,080 progressive. And, high definition has now introduced different frame rates. 24P is a progressive mode format that records 24-frames-per-second. It is being adopted by more and more filmmakers and professionals who find 24P to be more the equivalent of film (which also shoots at 24-frames-per-second). And, through the use of variable recording time machines, or computer-aided postproduction systems, some HD systems now offer variable speed recording (again, something once the exclusive domain of film).

The standardization process has already confirmed eight different digital videotape formats (from D1 to D10, skipping D4 and 8). Some of the formats listed will likely fade as they are usurped by better or cheaper variations. And there is always a chance the new formats will emerge sooner than later.

There was also a time when videotape formats could be easily categorized by their application—postproduction, graphics, news, corporate. But many of today's digital formats can run across applications and, in fact, are marketed as providing "complete end-to-end solutions." It is not uncommon to shoot, edit, and distribute on the same format tape.

One of the benefits of staying within a single format is the signal integrity. Virtually every digital format has an incompatibility with every other format. In some cases there is compatibility in decks, allowing a VTR to play back tapes of a variety of formats. But, it is important to remember that the information recorded on these tapes (and hard drives and servers) is data—1s and 0s—and those need to be translated into the video, audio, and other information such as timecode to be used in a production environment. Then add in the issue of compression. No matter how you look at it, a compressed signal is missing some data that created the original image. Robust and sophisticated algorithms will create an image that is close to the original, but it is coming from less data. Most of the time, this is invisible to all but the incredibly well-trained technical eye. But compressing and decompressing the data over multiple passes will certainly generate artifacts that are visible to the untrained (read: viewer) eye.

One answer is to work in an uncompressed world—and that is often the choice in high-end production and graphics applications. Another answer is to work in an environment where the data remains data and all steps along the production process are using the same type of signal, staying with a DV or MPEG data system, for example.

PRODUCT FAMILIES

It is not uncommon to have a new product announced as "the next generation of...." In the case of VTRs, this sense of lineage can be truer than imagined. It is no mistake that Panasonic's D3 and D5 formats are built on the same type of transports and recording/playback heads as found in MII (analog) and even in their consumer VTRs. Similarly, there is more than just a name that connects Betacam and Betacam SP with Digital Betacam and Betacam SX (and even HDCAM). Look at the cassettes. Nearly identical in design, the differences are in the tape formulation inside the cassette and the color of the plastic shell.

This "generational" connection means more than a company's ability to produce VTRs and tapes at lower costs (although those manufacturing economies do have a bearing on the cost of the hardware and tape stock in the marketplace). It has also led to a new kind of compatibility in playback.

Panasonic's D5 and HD-D5 decks will play back D3 tapes. And Sony's latest digital offering, the top-of-the-line MPEG IMX decks, can play back all of Sony's $^1/_2$-inch videotape standard definition formats from good old Betacam on up. Take it a step further, and Sony's HDCAM decks (with an optional board) can play back standard definition and high definition $^1/_2$-inch videotapes that cover the full range of Sony's product line—well Sony's $^1/_2$-inch product line.

DIGITAL VIDEO (DV)

But not all families are so "genetically" tied. In some cases, there is a common ancestor, but division amongst the progeny. Hence the legacy of DV. Originally established as a consumer digital video format, and agreed upon by 17 companies worldwide, digital video (DV) was intended to be the next wave of home video technology for recording, playback of movies, and other uses.

Shortly after all the DV standards were accepted by all these companies (many of whom were and are competitors in the broadcast and consumer marketplace), there emerged two incompatible iterations of DV to provide professional features such as timecode. From the Panasonic camp came DVCPRO, which led to DVCPRO 50 and DVCPRO 100 (or HD). From the Sony side, came DVCAM.

These two professional, essentially incompatible formats, along with the growing use of consumer DV for professional applications, have created a market necessity for all three versions of the format to "play nice." Which they can do, because the basic compression and digital signal on all the formats is identical. The differences are in tape speed and secondary variations in the format for other data and operational elements.

So, in today's VTR marketplace, most DVCPRO decks will play DV and DVCAM, and most DVCAM decks will play DV and DVCPRO.

PICKING YOUR FORMAT

There is no clear and obvious argument for using one digital format over another. Some are market leaders in units sold—which is a good test of a format's acceptance. But some users never need to venture outside of their own walls and select a different format for technical reasons.

If we could tell you which would slip away and what is yet to come, we would. But we can't. The choice, I'm afraid to tell you, is yours. What follows is a brief description of the formats available today (see also Table 7.1).

STANDARD DEFINITION FORMATS

D-1. The first digital videotape format to come to the broadcast and video market was D-1, developed by Sony Corp. in 1986. At the time, it offered the first capabilities for multigenerational editing without the degradation associated with analog tape passes. The D-1 format uses 19 mm (or $^3/_4$-inch) metal particle videocassette tapes with a maximum recording time of 94 minutes.

The format records component video signals with a resolution of approximately 460 lines of horizontal resolution. The advantage of component recording is that chrominance and luminance signals remain separate. In this manner, the signals maintain their maximum bandwidth and thereby maintain a higher quality.

Unlike future digital formats, D-1 does not compress the video signal. Although this maintains the highest quality, it also affects the cost of machinery and tape stock. Early testing of D-1 showed that nearly 100 passes (playback and rerecording of an image) could be accomplished before seeing any noticeable difference or degradation in the signal.

The initial, and still probably the most widely used application, for D-1 is in graphic production where multiple recording passes for layering and effects are required. The introduction of such capabilities opened up new levels of creative and technical capabilities for such work.

D-1 has always been limited in its application, primarily because of the high cost of equipment and tape stock. More recent format introductions have proved to be more cost-effective for postproduction and graphics applications.

D-2. Shortly after the introduction of D-1, Ampex Corp. unveiled a composite digital format called D-2. Composite recording, where chrominance and luminance information are combined, allowed for lower cost decks. The composite process, however, does not maintain image quality at the level of component recording.

The D-2 format was quickly adopted by Sony, which began marketing its own decks under a licensing agreement with Ampex. D-2 also uses 19 mm (or ¾-inch) metal particle videotape cassettes but they are not compatible with D-1.

D-2 recording provides a resolution of approximately 450 (horizontal) lines. Maximum recording time on a cassette is 208 minutes. D-2 decks were used to a great extent as a replacement for 1-inch open reel analog decks for postproduction recording.

Although the composite signal does not allow for the same depth of generational passes as D-1, this format was more broadly adopted in postproduction because of its digital quality. A key feature for the format and its decks is "pre-read," allowing for the signal to be read from the tape, passed through a switcher or edit control system, and recorded back on the identical tape. What minimal degradation that might be introduced by multiple passes is still well within generally acceptable standards.

D-2 also found applications in program playback, particularly when configured into robotic cart systems.

D-3. A comparable, but incompatible format to D-2 was developed by Matsushita in 1991 and was standardized by SMPTE as the D-3 format. Using composite digital signals and metal-particle tape, D-3 also provides horizontal resolution of around 450 lines. Matsushita, which is marketed in the United States under the Panasonic brand, was a strong manufacturer of ½-inch tape formats, including consumer VHS. This in no small part contributed to D-3's use of ½-inch videotape cassettes. The format is capable of recording up to 245 minutes on a single tape.

Although D-3 was worthy competition for D-2 in edit configurations (except, perhaps, in the Hollywood postproduction community), the strength of the format came from the ½-inch tape, which allowed for a broader line of equipment—including the first digital videocassette camcorder.

Although D-3 did not make its mark on the Betacam SP camcorder market like some had expected, it did move Panasonic Matsushita into a much stronger market position with broadcasters and production facilities: this would prove to be very important as subsequent digital formats came to the marketplace.

D-4. There is no SMPTE D-4 format. The explanation most often offered is that the Japanese word for "death" is very similar to "D4."

D-5. In 1993, Panasonic made its entry into the high-end world of video recorders with the introduction of the D-5 format, an uncompressed, component (10-bit) digital videotape format using 4:2:2 8-bit sampling. Initially offered as a studio format to compete with D-1, D-5 provided strong competition to Sony's high-end, uncompressed format and provided the additional benefit of playback and record compatibility with D-3.

Using $^1/_2$-inch metal-particle tape, it can record up to 124 minutes on a single cassette.

D-5 was initially accepted as a replacement for D-1 because of lower cost. It has taken a reasonable hold in the high-end postproduction, graphics, and film-transfer markets. The compatibility with D-3 may have been responsible for some of its early strength, but Panasonic also let it be known early on that D-5 would contain a direct migration path to high definition recording. (See HD D5, described later.)

D-5 was introduced around the same time as Digital Betacam, setting off fierce positioning over the issue of compression. Targeted for high-end film transfer, graphics, and compositing work, D-5 demonstrated its strength in the postproduction environment.

DVCPRO (D-7). With the development of the small-sized, digital video (DV or DVC) format for consumer use, manufacturers looked to expand the fundamentals of the format for professional and broadcast applications. Matsushita Panasonic's answer became DVCPRO, which was standardized by SMPTE as D-7.

Using the same sampling rate and compression as DV, DVCPRO operates with 4:1:1 quantization and a compression ratio of 5:1 at 25 Mpbs. It is important to note that DV, DVCPRO, and Sony's DVCAM all record using the same fundamental quantization and compression scheme. Differences in the format are limited to tape speed (and therefore recording time) and features such as timecode—as well as equipment configurations based on the markets using the formats.

DVCPRO tape speed runs at twice the speed of DV tape, allowing for recording of the control track and timecode that is necessary for professional applications. DVCPRO is also differentiated from DV and DVCAM in its use of metal-particle tape (as opposed to metal-evaporated for the other formats). Metal-particle is a more robust formulation that reduces dropouts often associated with multiple passes of a piece of tape—for example when editing.

The DVCPRO cassette is a 6 mm tape, with cassette sizes up to 123 minutes. DVCPRO decks can play back all DV-based formats (DV, DVCAM, etc.).

All three DV-based formats record component digital signals and provide horizontal resolution of just under 500 lines, putting them in a league with D-1, D-2, and D-3 for quality, and making them superior to analog formats.

Another key feature of the DVCPRO format is 4× playback (later introduced for DVCAM as well). This is implemented for input into computer-based nonlinear systems utilizing DVCPRO cards for direct input of data. Although the format does offer a "laptop" cut-only linear edit system, the primary applications of DVCPRO have been for news and acquisition.

DVCPRO 50. The second generation of the DVCPRO family answered many concerns voiced over the first manifestation. DVCPRO, the original, sampled at 4:1:1, raised concerns over applications such as chroma key. By modifying the format for 4:2:2 sampling at a data rate of 50 Mbps, chroma resolution is enhanced and the format performs on a level comparable to other 4:2:2 formats, such as JVC's Digital-S (D-9).

Faster tape speed and a lower compression ratio (3.3:1 compared to DVCPRO's 5:1) account for better image quality. But the speed also reduces recording lengths for cassettes. Most decks, camcorders, and cameras can record both DVCPRO50 and DVCPRO signals.

Although DVCPRO has found its niche in acquisition and news, DVCPRO 50 is much better positioned as an all-around format, including postproduction applications. The format's related equipment also allows for 16:9 as well as 4:3 aspect ratio recordings.

The DVCPRO50 format can also be configured for progressive recording, with a horizontal resolution of 700 lines in 16:9 format.

DVCAM. Sony's professional adaptation of the consumer DV format is named DVCAM. It operates at a higher tape speed than DV, but uses the same component digital signal and sampling rate (4:1:1 in NTSC, 4:2:0 for PAL). DVCAM uses metal-evaporated (ME) tape comparable to DV, and at a lesser cost than metal-particle (MP). DVCAM tape moves at a higher speed than DV, but slower than DVCPRO.

Product configurations for DVCAM equipment include emphasis on FireWire (IEEE-1394) for integration into computer-based edit systems. With higher-end digital formats from Sony (Digital Betacam, Betacam SX, etc.), the positioning of DVCAM in the marketplace is oriented much more toward midrange corporate and small broadcast production. But DVCAM has found marketplaces in everything from corporate and event videography, to documentary and even some experimental and feature film production.

DCT. Digital Component Technology (DCT) is Ampex's offering into the high-end digital video arena (not to be confused with discrete cosine transfer as used in the Digital Betacam format). DCT technology is targeted almost exclusively to postproduction and film transfer applications.

Ampex DCT records compressed (2:1) digital component signals and boasts error correction, as opposed to other formats' error concealment processes. Decks are switchable from 525 to 625 and can record more than three hours on a single, 19 mm (3/4-inch) tape.

DCT may be more in line with data recording machines (a current strength of Ampex's offerings) than in comparison with digital videotape decks. Processes of error correction put it more in this realm, although the products do offer error concealment as well. Interestingly, Ampex has never strongly referred to its machines as VTRs, preferring to call them "tape transports." It may be more semantic than technical, but it is an indication of the company's position on the gear.

DIGITAL BETACAM. Introduced by Sony in 1993, Digital Betacam is the digital successor to the popular Betacam SP line. Using cassettes that resemble Betacam SP, Digital Betacam offers backwards compatibility—thus allowing Betacam SP tapes to be played back, but not recorded on Digital Betacam decks.

The initial product offerings for Digital Betacam looked to replace Betacam SP in acquisition and postproduction environments. More recent products have highlighted the format's capabilities to record widescreen 16:9 imagery for digital television (DTV) and to emulate popular film stocks.

One strength of the Digital Betacam format has been the familiarity of the marketplace with Betacam SP.

Although Digital Betacam uses component digital signals, it compresses the digital signal by a factor of 2:1, using discrete cosine transfer (DCT). The format records component digital at a sampling rate of 4:2:2, 8- or 10-bit.

One advantage of DCT over other compression systems is that each frame is compressed individually (intraframe), allowing for cleaner and more precise decompression—especially in editing scenarios. Other interframe compression schemes actually compress data from a number of frames simultaneously. This hampers efforts to accomplish frame-accurate editing without decompressing and recompressing with each edit pass.

Digital Betacam also uses metal-particle tape and provides horizontal resolution of just under 500 lines. Many Digital Betacam decks are backwards compatible, allowing analog Betacam and Betacam SP tapes to be played.

BETACAM SX. Sony introduced Betacam SX in 1996; SX uses MPEG-2 compression at a compression ratio of 10:1, and sampling at 4:2:2.

Initial offerings of the Betacam SX line were oriented toward recording and playback, especially for broadcast, but later equipment announcements broadened the market for acquisition (news and production) and even a field-edit configuration. Today, Betacam SX is found primarily in broadcast applications for news gathering and program playback.

The use of MPEG-2 compression requires all production and postproduction systems to operate within an MPEG-2 environment to maintain image quality.

D-8. The SMPTE committee members responsible for naming formats felt that, to avoid confusion, no format would be given the "D-8" name, because it was too close to Tascam's "DA-88" audio tape format, which is used extensively for archiving and transporting audio and video data.

DIGITAL-S (D-9). Digital-S is JVC's first offering in the digital videotape arena; it has been standardized by SMPTE as the

D-9 format. Recording component digital signals and using $\frac{1}{2}$-inch tape similar to S-VHS, the format uses 4:2:2 sampling, which leads to image quality comparable to Digital Betacam and is judged by many to be superior to DV-based formats. Horizontal resolution is in line with the DV formats and Digital Betacam and Betacam SX, at just under 500 lines.

Digital-S also provides *preread*, a unique feature in digital formats. Preread allows for a single deck to playback and record simultaneously, making it possible to configure A/B roll systems using only two decks. Price points of Digital S equipment put it in line with Betacam SP gear.

Although originally positioned as a digital upgrade for existing S-VHS users (corporate, industrial, small broadcast), Digital-S has found acceptance in the broadcast community as well. The full product line provides camcorders, record and edit decks, and equipment for integration into nonlinear edit packages.

MPEG IMX (D-10). The MPEG IMX format is based on the MPEG-2 standard for digital audio and video. Introduced in 1994, MPEG-2 became an important standard in many nonlinear editing systems and hard-drive and server recording units. The IMX standard is a 4:2:2 digital component video signal. The I-frame–only specification allows for frame-accurate insert and assemble editing. MPEG IMX can support 30, 40, and 50 Mbps formats for both input and output, allowing recording and playback units to support a variety of tapes—and output their signals in their original format or as MPEG IMX signals. MPEG, which provides for the development of fully integrated systems, can be found in recorder/players as well as hard drive recorders and servers. Prototype MPEG IMX camcorders have also been demonstrated at industry shows and events.

DV. In the early 1990s, a number of manufacturers cooperated to create a new consumer format to replace VHS. The result was Digital Video Cassette (DV or DVC). Shortly after development of the format, it was supported by 17 different manufacturers that planned to offer DV and related products.

To date, however, the idea of a new consumer format has not taken off, in the face of competition from disk-based playback formats like DVD.

The professional community took notice of the format's 4:1:1 digital signal and soon Panasonic/Matsushita and Sony unveiled their upgraded version of the DV format for professional applications (see DVCAM and DVCPRO).

But DV's digital signal, ability to record widescreen images, and upward compatibility with both DVCAM and DVCPRO, have made the format a popular acquisition tool for broadcast and higher-end productions. Because the compression and digitization processes of all DV-based formats are the same, the only perceivable difference in image quality among the three is usually associated with the optics of the camera. The digitization is identical.

Although there is little if any activity in developing complete production systems based exclusively on DV, the small, lightweight cameras have found a role in news, corporate, and documentary production.

HIGH DEFINITION DIGITAL FORMATS

HDCAM. Sony was the first company to offer a high definition digital videotape recorder, the HDD-1000. It is a full-bandwidth, uncompressed recorder using 1-inch open reel tape. Although it is still available in the marketplace, equipment cost and tape cost ($1,300 for a one-hour reel) have limited its use.

Sony's current foray into high definition production and postproduction equipment comes under the banner of HDCAM. Initial offerings of studio and record decks have been joined by camcorders and additional production gear including HD/SD nonlinear editing systems, such as Sony's XPRI. Further expansion of the product line includes VTRs that can play back Betacam SX and Digital Betacam, in some cases allowing for upconversion of Digital Betacam signals to high definition (1080/60I).

HDCAM records using 3:1:1 quantization (10-bit) at a compression ratio of just under 4.5:1. (If you consider the difference between 4:2:2 and 3:1:1, the compression works out to about 7:1.)

Originally released to record 1,080-line, 60-frame interlace (60I) images, HDCAM can also record and play back 24-frame progressive 1,080 HD images.

D-6. D-6 was the first of the high definition formats, developed by Toshiba and BTS. Using 19 mm ($^3/_4$-inch) cassettes, it records uncompressed HD signals; maximum recording time is 64 minutes on a single cassette. At present, Philips appears to be the only active marketer of the product.

The Voodoo Media Recorder is Philips' digital VTR designed for high-end, postproduction applications. This is a re-engineered HDTV derivative of the D6 recorder, jointly developed by Philips and Toshiba, and now supports a variety of DTV formats. It is a cartridge-based, uncompressed, HDTV tape machine offering 12 channels of digital AES audio. The basic system will record 4:2:2 video uncompressed in 1080I/60, 1080P/24, 1080P/25, 1080P/48SF, 1080P/50SF, with cross play possible between 1080P 24/25/48SF and 50SF. Although the basic system provides for 4:2:2 recording, there are already future options under development to provide 4:4:4:4 (RGBK) recording using lossless compression and a data interface to provide data recording up to 128Mb/s. All of this is achieved on the same basic transport deck.

HD D-5. Panasonic expanded its D-5 product line to high definition with the addition of a 4:1 compression add-on that allows D-5 machines to record high definition signals. Originally the processor was a stand-alone unit to be integrated with a standard D-5 deck, but later models incorporated the processor and VTR into a single unit. Whether internal or as an external unit, the processor still allows D-5 VTRs to record and play back the uncompressed D-5 signal (in both 720P and 1080I line resolutions). D-3 playback is an option on most of the latter decks.

DVCPROHD (DVCPRO100). The most recent iteration of the DVCPRO format is DVCPROHD. Originally called DVCPRO100, it uses essentially the same tape transports and mechanics of the other DVCPRO formats, although it records at 100 Mpbs tape speeds—four times those of DVCPRO and twice the speed of DVCPRO50. The first offerings of DVCPROHD equipment operate in the 1080I standard for high definition, but Panasonic has made a strong commitment to 720P and is now marketing camcorders and decks that record in the 720/60P format.

Initial product offerings for DVCPROHD to reach market are camcorders and a studio deck, targeted for acquisition and postproduction applications.

This has been a discussion of *only* videotape formats. Increasingly, hard drive units and servers are being called upon to fill rolls traditionally held only by VTRs, including program playback, live recording (for slow motion playback, for example) and, of course, as sources for editing systems.

In the acquisition arena, hard-drive–based cameras have been demonstrated since 1995, and with the standardization of the MPEG-IMX (D-10) format, new models have been demonstrated in prototype in recent years. Also, looming on the horizon are disk-based cameras, which promise the benefits of high quality recording on small, CD-sized disks.

An oft-repeated complaint in the hardware marketplace is that there are too many different options from which to choose. In recent years, the options have dramatically increased and there is no indication that there will be any reduction in offerings in the future. The marketplace is often the deciding factor in which format and hardware will succeed and which will fall away to be future museum pieces. But in today's technological world, with its ever-increasing needs from high-end broadcast to Web-casting, it appears that this variety of choices is, in fact, allowing users to select the particular technology and its related hardware offerings that best match needs and budgets.

There is a lot to choose from. Pick wisely.

TABLE 7.1 Digital Formats

FORMAT	SIGNAL/ COMPRESSION	SAMPLING	TAPE SIZE	TAPE TYPE	RESOLUTION	EQUIPMENT	PRIMARY APPLICATIONS	COMPATIBILITY WITH OTHER FORMATS
D-1	Component digital/ Uncompressed	4:2:2 8 bit	3/4-inch (19 mm)	Metal-particle	460 lines	Recorders	Post, graphics, film transfer	None
D-2	Composite digital/ Uncompressed	8 bit	3/4-inch (19 mm)	Metal-particle	450 lines	Recorders	Post	None
D-3	Composite digital/ Uncompressed	8 bit	1/2-inch (12.65 mm)	Metal-particle	450 lines	Recorders, camcorders	Post, acquisition	D-5 decks will play back and record D-3
Digital Betacam	Component digital/2:1	4:2:2 8/10 bit	1/2-inch (12.65 mm)	Metal-particle	500 lines	Recorders, camcorders, field-edit package	Acquisition, Post	Decks will play Betacam SX, Betacam (analog) and Betacam SP (analog)
D-5	Component digital/ Uncompressed	4:2:2 8 bit	1/2-inch (12.65 mm)	Metal-particle	500 lines	Recorders, camcorders	Post, acquisition, graphics, film transfer	Switchable decks record and play back D-3 (see also HD/D-5)

continued on next page

TABLE 7.1 Digital Formats (continued)

FORMAT	SIGNAL/ COMPRESSION	SAMPLING	TAPE SIZE	TAPE TYPE	RESOLUTION	EQUIPMENT	PRIMARY APPLICATIONS	COMPATIBILITY WITH OTHER FORMATS
DVCPRO (D-7)	Component digital/5:1	4:1:1	1/4-inch (6 mm)	Metal-particle	500 lines	Camcorders, recorders, field-edit package	Acquisition (news), post	Decks play back DV and DVCAM
DVCPRO 50	Component digital/3.3:1	4:1:1	1/4-inch (6 mm)	Metal-particle	700 lines	Camcorders, recorders, field-edit package	Acquisition, post	Decks play back DV and DVCAM
DVCAM	Component digital/5:1	4:1:1	1/4-inch (6 mm)	Metal-evaporated	500 lines	Camcorders, recorders	Acquisition, post (especially as source for computer-based edit systems)	Decks play back DV and DVCPRO
DCT	Component digital/2:1	4:2:2 8 bit	3/4-inch (19 mm)	Metal-particle	500 lines	Recorders	Film transfer, post, graphics	
Betacam SX	Component digital/10:1	4:2:2	1/2-inch (12.65 mm)	Metal-particle	500 lines	Recorders, camcorders, field-edit package	Post, acquisition	

continued on next page

Table 7.1 Digital Formats (continued)

Format	Signal/ Compression	Sampling	Tape Size	Tape Type	Resolution	Equipment	Primary Applications	Compatibility with Other Formats
Digital S (D-9)	Component digital/3.3:1	4:2:2	1/2-inch (12.65 mm)	Metal-particle	500 lines	Camcorders, recorders	Acquisition, post	Decks play back S-VHS (analog)
DV	Component digital/5:1	4:1:1	1/4-inch (6 mm)	Metal-evaporated	500 lines	Camcorders, recorders (especially for use in computer-based edit systems)	Acquisition	
MPEG-IMX	Component digital	4:2:2P@ML	1/2-inch (12.65 mm)	Metal-particle	500 lines	Camcorder, studio and field recorders	Acquisition and post, storage	Decks play back Betacam, Betacam SP, Betacam SX, Digital Betacam
HDCAM	Component digital/4.5:1	3:1:1 10 bit	1/2-inch (12.65 mm)	Metal-particle	1080 lines (interlace), 720 lines (progressive), 30-frame (60I), and 24P	Camcorder, studio and field recorders	Acquisition and post	Decks play back Betacam, Betacam SP, Betacam SX, Digital Betacam, MPEG-IMX
D-6	Uncompressed		3/4-inch (19 mm)	Metal-particle	1080 lines (interlace)	Recorders	Post, graphics, film transfer	

continued on next page

Table 7.1 Digital Formats (continued)

Format	Signal/ Compression	Sampling	Tape Size	Tape Type	Resolution	Equipment	Primary Applications	Compatibility with Other Formats
HD D-5	Component digital/4:1	4:2:2 8 bit	1/2-inch (19 mm)	Metal-particle	1080 lines (interlace), 720 lines (progressive)	Recorders	Post, graphics, film transfer	Decks play back D-3
DVCPROHD DVCPRO100	Component digital/3.3:1	4:2:2	1/4-inch (19 mm)	Metal-particle	1080 lines (interlace), 720 lines (progressive)	Recorders, camcorders	Acquisition, post	

CHAPTER EIGHT

CGI AND DIGITAL CONTENT CREATION

BRIAN MCKERNAN
AND RANDY CATES

Computer graphics, digital by nature, are at the heart of content creation. The uses of computer-graphic imaging (CGI) are many, and can be used to produce almost anything the mind can conceive.

Of all the tools of digital content creation, CGI is the most versatile and pervasive. Versatile, in the sense that making pictures with a computer can include everything from simple Flash animations on the Web to photorealistic movie dinosaurs. Pervasive, in the sense that CGI software applications and those for video and audio editing often reside on the same computers. Digital content can be passed back and forth among these different applications for manipulation with a wide variety of creative software tools. In such a computer environment, all the image content exists as digital data (zeros and ones); in that sense it's all "CGI."

The term *CGI* is most often applied to the creation of photorealistic 3D objects (rampaging dinosaurs, *Toy Story* characters) or entire environments (alien *Star Wars* landscapes). Such high-profile uses of CGI technology grab attention, but other

CGI applications may be less obvious. These include character generation, for typing words over video or film footage; paint, used to retouch still or moving images; and Flash or Photoshop animation, named after the software itself, often used for simple Web cartoons and compositing, in which live-action digital video (or film images transferred to digital video) are seamlessly combined with computer-generated imagery. Even today's classic-style TV and movie cartoons are usually created using CGI, with animators drawing on computer screens instead of the clear plastic sheets ("cels") traditionally used.

Because computer technology is integral to CGI, it gets more powerful and less expensive over time. Moore's Law, named after Gordon E. Moore (physicist, co-founder, and chairman of Intel Corp.), describes the tendency for computer microprocessor chip performance to continually improve. Mass production and competition, meanwhile, tend to drive down the cost of these chips. This is good news for CGI—and for every other part of digital content creation.

The sophistication, complexity, and resolution (amount of picture detail) of different CGI applications depends on many factors, chief among them being the software and hardware power of the computers used to make computer-generated images. The CGI tools in a Hollywood studio are typically far more powerful than those of, say, a Web-page design firm. The two applications have different quality and performance demands. Equally crucial is artistic talent, without which all the technology in the world is useless.

CGI USES

Although CGI is integral to a wide variety of digital content—movies, TV, corporate video, DVD menus, and the Web—the technology originated and evolved in the research labs of universities, governments, and private industries. Such organizations had the resources and the need to develop CGI to accomplish a variety of imaging tasks. CGI is used for design and modeling in the aerospace and automotive industries, for medical imaging, and for complex scientific visualization. Architects, for example, can use visualization software to create videos showing what it would be like to walk through a proposed building; changes can be made with a few mouse clicks before any actual construction commences. Computer-aided design/computer-aided engineering (CAD/CAM) software is routinely used to design everything from airplane and automotive parts to plastic detergent bottles. There are even milling machines (3D printers) that can carve a plastic or metal object according to the dimensions modeled in a three-dimensional CGI program. Advanced scientific theories have been advanced with CGI; examples include IBM researcher Benoit Mandelbrot's research into fractal geometry. His Mandelbrot set includes a CGI pattern that has been described as one of the most remarkable discoveries in the history of mathematics—but because of its complexity, it could never have been drawn or visualized without CGI tools.

As with every other area of digital content creation, CGI technology (hardware and software) continues to advance. Some experts in this area are predicting photorealistic CGI actors in a few years that will be indistinguishable from real human beings. Many fine books are available on the numerous aspects of CGI production. This chapter's purpose is to provide an overview of what CGI is, how it has evolved, and its role in digital content creation.

PIXELS AND PROGRESS

It is generally agreed that the field of CGI began in 1962 with a graduate student at the Massachusetts Institute of Technology named Ivan Sutherland. His doctoral thesis (titled "Sketchpad: A Man–Machine Graphics Communication System") included a way for people who weren't programmers to "draw" simple shapes on a picture tube connected to a computer. Prior to Sketchpad, "computer graphics" entailed writing lines of programming code that dictated paper print-outs of crude patterns of X's and O's (Snoopy on the roof of his doghouse was a popular test image).

Sketchpad made history by enabling anyone to use a light pen and a row of buttons to create basic images. According to the late Robert Rivlin's book *The Algorithmic Image* (Microsoft Press, 1986), "except for the addition of color and a few minor details concerning how the graphics processing is accomplished, the 1963 version of Sketchpad has remained virtually unchanged in 95 percent of the graphics programs available today, including those that run on home computers."

In 1964 Sutherland teamed up with Dr. David Evans at the University of Utah to develop the first academic computer graphics department. CGI advanced through the years as a part of computer science, evolving in the labs of universities, corporations, and governments. A long series of innovations continually improved the technology, including: the display processor unit, for converting computer commands and relaying them to the picture tube; the frame buffer, a form of digital memory that improved upon that principle; and the storage-refresh raster display that made computer-graphic screens practical and affordable. These screens use cathode-ray tubes (CRTs) that divide images into picture elements, or *pixels*, the basic unit of computer graphics. Digital, random-access memory (RAM) determines the number of pixels on the screen, their location, color, and other parameters. Each pixel is assigned a particular number of memory bits in a process known as *bit-mapping*. The more bits, the greater each pixel's potential color combinations. *Bit depth* refers to the number of these bits. The greater the bit depth, the more color levels possible. In a 24-bit system, the computer provides for 8 bits each of red, green, and blue. It is from these basic colors that all other hues are created in CGI. A 30-bit system provides for 10 bits per basic color; a 36-bit system, 12. More color means more shading, detail, and—in the hands of a skilled artist—realism. Hollywood filmmakers, for example, crave as much bit-depth as possible in the CGI systems they use so that their images have the beauty and realism of movie film.

Hand-in-hand with CGI technology improvements came ever-increasing advances in writing the programming code—software—necessary to make graphic objects and present them with three-dimensional perspective. This long process of discovery went from inventing ways of drawing wire-frame "skeletons" of objects, to developing algorithms for what's known as polygon rendering to give them surfaces, to animating them in ways the human eye finds appealing.

As with every field of endeavor, CGI researchers build upon past accomplishments and the knowledge they gain to continually refine their technology in an ever-evolving process. Drawing upon mathematics, physics, and other sciences that measure and describe the physical world in which we live, the science of CGI continually seeks to improve the ways that images can be created. A major goal is to visually achieve the "look and feel" of matter as it responds to the universe's natural laws. Such knowledge is incorporated into CGI software, which is constantly being revised and improved.

In the 1970s, a computer science Ph.D. at Xerox's Palo Alto Research Center (PARC) named Richard Shoup developed not only one of the first digital frame buffers but also the first sophisticated video painting software. NASA used the technology to create TV news animations of its *Pioneer* space missions. Known as SuperPaint, it was used to create the first computer graphics ever broadcast. In 1979, Shoup left Xerox PARC to found Aurora Systems, an early manufacturer of digital video paint and animation systems.

Alvy Ray Smith, another Xerox PARC researcher, made significant software-design contributions to Shoup's paint program. Smith went on to work at the New York Institute of Technology and NASA's Jet Propulsion Laboratory, two hotbeds of CGI research in the 1970s. In 1980 Smith and fellow CGI pioneer Ed Catmull were hired by *Star Wars* producer George Lucas to run what later became Pixar, a company formed to advance the art and science of CGI in motion-picture production. That company later went on to produce *Toy Story* and other movies that propelled CGI into mainstream entertainment and advanced the capabilities of the technology.

While these and many other pioneers worked to improve the ways in which computers could make pictures, they were also finding applications for the use of such pictures, and writing the necessary software to generate them. At the same time, computer technology itself was undergoing a revolution. Processing power continued to rise and costs dropped—Moore's Law at work.

The year 1974 saw the birth of the Association of Computing Machinery (ACM)'s Special Interest Group on Computer Graphics (SIGGRAPH), which went on to play an instrumental role in providing support for this new science. SIGGRAPH also organized conferences and provided a valuable information-sharing environment. Today the annual SIGGRAPH conference and exhibition is CGI's most important event and a reference point for the technology's progress.

TELEVISION

The elaborate graphics and moving text one sees on TV and cable shows these days is a relatively recent CGI development. Less than 25 years ago, TV graphics consisted of hand-made art cards, which would typically be set on an easel and shot by a studio camera. News graphics such as local, regional, and national maps were created as base images that could be altered with clear plastic overlays for repeated use. Art departments at television stations (and elsewhere) used rub-on text, cut paper, photos, and illustrations to add a level of professionalism to their broadcasts. These methods were crude by today's standards, time-consuming to create, and not very forgiving if rapid changes were required.

Computers that could generate text onscreen began taking over from TV-station art cards in the late 1970s with the invention of the Chyron. Then, in 1983 a British company known as Quantel introduced a graphics system specifically for generating graphics for television. It didn't take long for Quantel's Paintbox to became the standard by which all other 2D-computer graphics systems were judged. Type setting, airbrushes, paint tools, drop shadows, beveled edges, matte tools, the ability to create masks, animation features, and electronic storage to save the fruits of all these CGI labor were all part of Paintbox, not to mention a graphical interface that was designed to be user friendly to artists.

Working at video resolution (72 dpi, or dots per inch) the original Paintbox not only had the ability to create still imagery, it could grab frames from video, layoff (record) animations created in the system to videotape, and capture anything put under a video camera. Not only was the Paintbox capable, it was as fast as the user could work. This revolutionized the way in which television graphics were generated, and that had a huge impact on how news graphics and graphic animations were created.

The Paintbox became the main graphics computer program for television production all over the world. Production facilities used it to generate other forms of content, as well, including corporate projects, trade show animations, and interactive touch-screen kiosks.

Then in the mid 1980s a company called Abekas was formed based on technology developed ten years earlier by Ampex, the company that invented the first practical videotape recorder(in 1956). Ampex had developed the ESS1, which was the first digital recording device that could write to multiple computer disks simultaneously. Abekas was founded by the people who developed this technology; they called their new device an A62 digital disk recorder. The A62 digital disk recorder revolutionized the layering of video; it used RAM chips and computer disk drives to store videographics digitally prior to their output to tape. Layering became much easier, and graphics designers were given new creative latitude.

More complex CGI to create photorealistic 3D objects, meanwhile, required computers more powerful than the IBM PC-style engine of the original Paintbox. But Moore's Law was ever-present, providing ever-improved processing power for comparably less cost. In the past ten years the computers necessary to create all forms of CGI have evolved from large mainframes, to large workstations, to custom-built PC/ATs, to off-the-shelf (but increasingly powerful) Macintosh and Windows computers optimized for video. The high end of CGI technology is still a place where millions of dollars are easily spent, and although high-end capabilities invariably gravitate down toward affordable systems, the high end is continually developing its own special capabilities. Computers based on UNIX operating systems, such as those manufactured by SGI and Sun Microsystems, tend to dominate here. Nevertheless, Moore's Law continues to rule, and ever faster and more powerful microprocessors, ever greater storage-system capacities, and ever better software have made Macintosh and Windows computer serious competitors to larger workstations. CGI technologies today run the gamut from turn-key systems that package software with a particular brand of workstation to shrink-wrapped programs intended for personal computers.

In addition to a computer, monitor, keyboard, and some form of data storage, nearly all CGI systems also employ a mouse, graphics tablet, and electronic pen. The tablets usually work by means of an embedded wire grid, which senses the location of the pen and relays it to the computer; the "drawing" is displayed on a computer monitor. CGI systems typically use some sort of graphical user interface, usually running under the UNIX, Macintosh, or Windows operating systems. (Linux is a public-domain operating system with more than 20 million users that runs efficiently on a wide variety of machines, including both Windows PCs and Apple Macintoshes. Its popularity is growing rapidly.)

THREE-DIMENSIONAL IMAGES

When used in connection with CGI, the term "3D" usually refers not to 3D (stereo-optical) movies or images, but to creating two-dimensional images that have three-dimensional realism. When rotated or otherwise manipulated, these 3D CGI objects give the appearance of having realistic mass and dimension (think of the characters in the *Toy Story* movies, which looked like actual plastic objects—albeit intelligent ones).

Stereo-optical 3D, which typically requires viewers to wear special polarized eyeglasses, can be created with 3D CGI, but 3D movies don't necessarily require; it these kinds of movies have been made since the 1950s (e.g., *House of Wax*).

Three-dimensional CGI programs create objects with x, y, and z coordinates to correspond to what their real position would be in three-dimensional space. They begin as wire-frame objects composed of *vectors*, or lines of specific length connected to other lines. The more numerous and complex these lines get, the more variety and detail there can be in the polygons they enclose. Creating objects with the kinds of random, multiple curves found in nature requires greater detail than even complex polygons can describe. Drawing from the mathematics of irregular surfaces, CGI developers have turned to an exotic variety of surface elements—known as *patches*—to build their models. *Metaballs*, *nurbs*, and *Bezier curves* are among the geometric "building blocks" that enable CGI designers to create objects of sophisticated complexity, including objects that are organic in appearance.

Animators use their keyboard, mouse, and pen-and-tablet interface tools to tell the computer the shape of these mathematical models, some of which may be chosen from a library of CGI shapes already created, like clip art. Other means of modeling 3D objects include caliper-style input devices to plot surfaces, or even laser scanners for inputting the dimensions of everything from an automobile to a living human being.

Once all the contours of the object have been defined by whatever form of modeling is used (constructive, procedural, and solids modeling are three different approaches), animation of the wire-frame object can begin. Animators typically storyboard the action of the scene they intend to create for effective planning. Motion paths can be determined and camera positions (the point of view that viewers of the finished animation will have) and movements specified. *Key frames* are specific, periodic points at which certain actions will occur; the computer can then interpolate and generate the *in-between* frames between key frames. As software improves, more and more of these functions become easier for the CGI animator to use, with the computer sometimes automating parts of different tasks.

When the animation process is complete, the object or objects being animated can be given a surface via *rendering*, which assigns whatever color, shading, and other surface attributes the animator desires (metallic, matte, textured, etc.). *Mapping* refers to the computationally intensive process by which the computer calculates and draws in the skin or surface of the object(s) in the animation. It's not as if the computer is taking over from the human designer; the human designer has instructed the computer on what the animation should look like. The human designer has determined where the light source is coming from in the scene, what the surface should look like, what the motion path should be, the colors, the quality of the camera's lens, etc. The computer then relies on its rendering program to carry out these directions, mathematically calculating and drawing the finished action for each successive frame of the animation. The final animation can be output to videotape, motion-picture film, or other storage media, such as a digital data recorder.

WORKGROUPS

Realistic 3D represents the high-end of CGI. Most CGI content-creation tasks are more mundane, and include making graphical, animated menus for DVDs, simple cartoons for the Web, or retouching portions of television commercials. The beauty of CGI is its capacity for unlimited image manipulation, assuming you've got the right software, a powerful enough computer, and sufficient creative talent. Given the rising power of even off-the-shelf Windows and Macintosh computers, small boutique studios have proliferated in recent years to serve a variety of clients—and not just in major cities. Operated by content-creation entrepreneurs, these boutiques often have multiple CGI-based software applications and those for other video-production tasks (such as editing picture and sound) running on the same computers (as mentioned earlier). Avid Technology's Digital Studio (DS) is a good example of this kind of CGI–video production system, which is based on a Windows computer.

Higher-end studios, such as those working on Hollywood movies or network TV shows, may opt for specialized CGI talent—3D modeling, digital character animation, and the like. The logic here is: Why have an editor do graphics when you can have a graphics designer do it faster and better? In such environments, workgroups speed efficiency.

Workgroups use Ethernet or other high-speed computer networking systems to link editing, CGI, and other specific-function computers in such a way that different people can work on the same content simutaneously. In a workgroup, editing, graphics compositing, 3D animation, paint, and audio talent can all access the same content and work in a timely manner. Workgroup editing enables users to expedite the adjustments between, say, a paint workstation, an animation workstation, and an edit system. The ability to log into large commercial *render servers* for compute-intensive operations is increasingly attractive. Extremely photorealistic CGI requires much computer-processing power, but if you can access other computers via a network, "borrow" their compute power, and distribute the task over multiple machines, it's a workflow advantage.

Such industry expansion requires management, both in terms of the networks themselves and the CGI, video, and other data they carry. The stakes are high and there are a host of relatively new companies vying to define a new industry segment known as *media asset management*. Such systems, as the name implies, are used to catalog, search, retrieve, and distribute large databases of all kinds of media. Without question, the eventual impact of this technology will fundamentally affect the way creative CGI and other content-creation talents work and get paid in the future.

VIRTUAL SETS AND ACTORS

Everyone is familiar with the use of *chroma-keys* in TV, the most familiar being an announcer superimposed over a weather map. A new CGI-based technology know as the *virtual set*, however, expands on chroma-keying by adding the dimension of movement. Virtual sets can place actors in a synthetic CGI environment (realistic or otherwise) and allow them to walk around in it in believable fashion. In reality, however, the entire set is a photorealistic 3D CGI model. This tend to be high-end technology, and it is being promoted as a way to turn any blue- (or green-) screen stage into any kind of environment you can create with CGI—from the Old West to outer space. In any case, using a virtual set should be cheaper than building a real one (assuming you use it properly).

Virtual set installations start with a CGI model of the set, created using such 3D software as Alias Wavefront, Kinetix, or Avid Softimage. That model is then stored in a powerful computer (such as an SGI Onyx) capable of rendering it in real time. A moveable video camera (or cameras) captures images of the actors' performance on the blue-screen set. Using either motion sensors, a motion-control camera "head," or pattern-recognition technology, the camera's position is fed into the computer, which then draws (renders) the background as it should be seen from that particular point of view. The actor(s) and the CGI are then combined, and a virtual environment is created—hopefully one that convinces viewers that there's no trick involved.

CGI objects can also be introduced into the scene; with proper rehearsal and blocking they can be made to look as if the actor is interacting with them (a computer-generated bird, for instance, flying through the scene). Camera movement is what makes a virtual set seem real; the computer generates a background image that corresponds to what the camera's (audience's) point of view should look like from any given position. The computer must be powerful enough to redraw the scene in real time as the actor(s) move within the set.

Although several companies sell virtual-set systems, this is still a technology in development, and virtual sets tend to be expensive and balky to operate. But as the technology matures it's entirely possible that Hollywood carpenters may have something to fear from CGI artists.

Virtual actors may scare Hollywood actors even more. As each new CGI software revision further advances humankind's ability to synthesize images, we come one step closer to being able to create photorealistic images that will be indistinguishable from real human beings. Each year, the SIGGRAPH conference's film and video show (the "electronic theatre") spotlights the best CGI clips; synthetic humans are inevitably among the featured attractions. Each year these images are that much more convincing.

A decade ago critics contended that CGI actors could never display sufficiently convincing facial expressions or physical movement to pass for real or even earn an audience's empathy. In 1995 Disney released Pixar's *Toy Story*; directed by CGI pioneer John Lasseter, this Oscar-winning first feature-length CGI movie touched audiences with appealing performances by its all computer-generated cast (albeit with voices supplied by Tom Hanks, Tim Allen, and other leading human talents).

Total realism is still to be achieved, but it comes closer every year. As it does, software engineers learn more and more about the complexity of human movement and facial expression. Humans are conditioned from birth to "read" faces, which is why anything less than total realism is so noticeable; movies such as *Final Fantasy* are bringing it ever closer.

Assuming that totally convincing virtual actors (also called "synthespians," for "synthetic thespians") can be created, their advantages will be many: They don't age, don't get sick or need vacations, aren't potential subjects of personal scandal, and they don't demand raises. A synthespian can be crafted to please a selected demographic profile and combine all the best traits of the most beloved movie actors. Digital compositing technology's cut-and-paste capabilities have already brought forth synthespians from the past—sort of. Deceased performers such as John Wayne, Groucho Marx, Fred Astaire, Jackie Gleason, and others have been digitally lifted from various film performances and composited thanks to CGI software (and creative talent) into new live-action commercials hawking everything from vacuum cleaners to beer.

Can the age of the synthespian be far behind? Perhaps it's just a matter of time before Moore's Law brings forth computers so advanced that Hollywood sound stages will be sold off for valuable real estate and major studios will operate in modest offices located wherever creative CGI talent wishes to reside. Or maybe by that time computers will be writing scripts as well.

For the present, however, one type of synthetic actor technology, called *motion capture*, is seeing increasing use. Motion-capture systems use a human performer—usually a professional dancer or mime—to direct the movements of a CGI character, which can either be cartoonlike or photorealistic. Using a motion-capture system, the movements of a person—wired at multiple points to a computer—translate to corresponding movements in a CGI character. The movie *Shrek* made extensive use of this technology.

ANIMATION FOR THE INTERNET

Although in existence for some time, the Internet had seen limited growth until the mid-1990s. The rest, as they say, is history. Today the creation of animation and graphics content for the Web has become a huge market for CGI tools. Originally limited in visual imagery because of slow access times, the first Web pages were simply information in the form of text. As the potential of the Internet became better realized, faster connection methods were developed and pages became more visually stimulating. Software aiding in the development of Internet content has become popular and almost all existing graphics applications are being upgraded to be capable of outputting Web pages. These formats condense the information of bit-mapped images to create extremely small files while doing a great job at maintaining the original image. Smaller files are easier to transmit on the Web's generally limited bandwidth.

As one might expect, animation was soon desired for the Internet. The public immediately saw that computer screens look like "televisions," and so any Web site that could deliver TV-style graphics was that much more desirable. Animation for the Internet today is best achieved by using *vector-based graphics*. Vector images are smaller, because of the fewer points in space needed to define an image. These control vertices are then filled with flat areas of color. The other option for graphic animation is *bit-mapped imaging*. Bit maps are more painterly and loaded with color information. Bit-mapped images are saved much the same way as graphics for CD-ROM, in an indexed or compressed fashion. Bitmapped images, although saved in a JPEG, GIF, or other condensed form, still don't reduce as well as vector graphics, but they are needed for photographs and other still images not handled by vectors.

The method of choice for small file-sized animation uses compressed audio and vector-based imagery. With this approach a very large amount of animation and sound can be output at an amazingly small size, making downloads much faster. MacroMedia Flash or Adobe LiveMotion are two popular programs that animate content for the Web; there are a number of software packages for both PCs and Macintoshes that aid in Web-site production. In addition, some 3D applications output animations as vector images, allowing 3D for the Internet. Digital compositing programs such as After Effects now support .swf (the Flash format), allowing anything animated to be placed on a Web site.

Another means of enhancing the Internet experience is with *streaming video*. For this approach, video shot specifically for the Internet is converted to a number of various frame rates, sizes, and compression qualities and saved on a server. The user, depending on connection speed, selects the option best suited for him, thus allowing all Internet users some level of the same visual experience. The quality you see is proportional to your connection speed. Today you can log onto a news Web site and see variations of news stories previously only available by television. In the not-so-distant future computers may replace televisions for personal viewing.

CHOICES

As with every other area of digital content creation, consideration of which CGI software or system is most appropriate depends on what it will be used for. For content creators, the best course of action in choosing CGI tools is to clearly understand the tasks you wish to perform, obtain as much information as possible (talking to people who do similar work is a good place to start), and investigating all the products on the market. The Web is a rich research tool. In a technology arena as dynamic as this one, it's not impossible that a lower-priced system will outperform a more expensive one. Compromises in speed can yield savings; a less expensive system may offer all that its more costly counterpart provides—if you're willing to tolerate getting your work done more slowly. Whatever you choose, rest assured that CGI technology continues to offer more functionality at ever-improving price–performance ratios. Moore's Law is on your side.

CGI's most compelling content-creation aspect is its role as a tool that can be used to make just about any image the mind can conceive. It can be used to make a simple Flash animation that can download from the Web over a limited-bandwidth connection. It can modify live-action digital video or film imagery in unlimited ways. It can even be used to simulate reality itself. Any definition of digital content creation should start with CGI.

REFERENCES

1. *The Algorithmic Image*, Robert Rivlin (Microsoft Press, 1986).
2. *Computer Graphics Applications*, E. Kenneth Hoffman and John Teeple (Wadsworth, 1990).
3. *Desktop Computer Animation*, Gregory MacNichol (Focal Press, 1992).
4. *How Did They Do It? Computer Illusion in Film & TV*, Christopher W. Baker (Alpha Books, 1994).

CHAPTER NINE

AUDIO: MORE THAN HALF THE PICTURE

TIM TULLY

"*Content*"...what does the word mean? The most high-profile definition is pictures, video, and "the movies." Well, yes. But there's another element as well. One that, for many years, got last consideration and the smallest part of the content creator's budget.

That content element is sound—audio—the part most people hardly think about unless it gets done so badly it knocks your suspension of disbelief down to earth. An early colloquial term for cinema was, as a matter of fact, "the talkies," the miracle being not that pictures could move, but that they could talk.

How important is sound to the experience of watching digital content? Renowned French director Jean Luc Goddard is purported to have claimed that "sound is 60 percent of cinema." This may strike some as inappropriate, but try this experiment. Go out and rent the scariest, most shocking horror flick you can find, pop it in your DVD player, and start it up. Now turn off the sound. Are you as scared as you would be with the sound on? I'll bet you're not.

Even better, if you want to alleviate the outrageous insult to your intelligence most TV commercials constitute, forget turning off the video monitor and mute the audio instead. You'll find it amazingly easy to turn away from the screen, have a conversation with the person next to you, and experience a welling sense of liberation. Part of this is due to one of the fundamental differences between sight and hearing. Sound spreads and permeates, grabs you from every angle, and hits you at an emotional level. The visuals just interrupt your eye rhythms.

SILENT TREATMENT

Although this chapter cites the use of audio in television and movies to make its point, these examples are applicable to all content that combines sound and picture. Consider the following story....

In the late 1980s, television broadcasts began including stereo sound. At the time, I happened to be touring the Los Angeles facilities of Fairlight International, manufacturer of one of the premiere digital music and audio production systems of the day. At that time, *Miami Vice* was one of the top television shows. One of its most powerful elements was the hot, electronic, jazz–fusion score that saturated its soundtrack, composed and produced by the talented Jan Hammer. What made the show's use of sound revolutionary were Hammer's methods: He produced a new score for each weekly episode, on the fly, entirely with the Fairlight. The device was a music synthesizer of magnificent proportions combined with a hard disk audio recording and playback system, all in one box. Hammer created, played, and mixed while watching the edited tape of the episode. In the process, Hammer was helping redefine the way soundtracks were put together.

Hammer had been in Los Angeles appearing at a trade show for the company. He'd apparently done so only with the understanding that Fairlight would provide him with a duplicate of his personal audio system so that he could pursue his demanding composing schedule after hours. On the day I visited the facility, Hammer had just left for home, but the system he'd used was still set up. It consisted of a six-foot-square video screen, a Fairlight, and some synchronizing gear for locking Hammer's sounds to the video of the next week's *Miami Vice* episode.

Needless to say, I was thrilled at this glimpse of gleaming cutting edge audio-for-video production, and convinced the man giving me the tour to fire up the system. We actually never got the Fairlight to play back Hammer's work—we guessed he'd taken all his files with him—but the video played back just fine.

And as soon as the video came up, it hit me like a burst from a TEC 9. We watched the two *Miami Vice* stars act out what was doubtlessly the kind of tense, dramatic scene that made the show a hit. But what we actually saw on that screen had not a note of music behind it. What should have been pregnant pauses in the dialog were vast empty tracts. Potentially trenchant give-and-take was banal and flat. They looked silly up there. We were looking at a final edit, but it felt like raw footage. I'd known that Hammer's driving music was important, but I'd never had it demonstrated so vividly.

Notwithstanding the power of sound in the movies, between the time when "the talkies" became "the movies" and the mid-1970s, sound played second fiddle to picture, for reasons that are a matter of endless debate. But then, in 1977, George Lucas' *Star Wars* was released. The film presented the popular imagination with an audio phantasmagoria that set a new standard for the way movies were designed and perceived. This is not to say that movie audio was dormant in the intervening years; but in the same way that *Star Wars* set new directions visually and thematically, the sound of the film became the benchmark against which all subsequent films were compared.

Interestingly, the watershed sound track for *Star Wars* predated digital production. It did, however, institute the demand for more and better audio effects. In ways that could never have happened using traditional analog audio, that demand was met by the digital audio technology that emerged well after the first *Star Wars*.

DIGITAL AUDIO

Historically, the technology that enabled the acceptance of digital audio arrived about ten years before digital moving images became technically feasible. In the mid-1980s, companies such as Digidesign and Sonic Solutions were developing the hardware and software foundations of digital audio: computer cards that converted sound to digital data and back again, and the software needed to control the recording, playback, and editing of digital audio. By this time as well, musicians and composers who used electronic musical instruments—synthesizers—were benefiting from the development of MIDI—the Musical Instrument Digital Interface. The MIDI specification laid out the basis for designing hardware and software that enabled computer control of electronic instruments. By the mid-1980s, a number of software developers were already offering sophisticated computer programs—MIDI sequencers—for desktop computers; these would give composers digital control over enough MIDI instruments to simulate entire orchestras. In addition, pioneering programs such as Blank Software's Alchemy gave movie sound designers more, better, and faster ways to create never-before-heard sounds for film. By the end of the 1980s, at least some of the music in most Hollywood films and television was created using MIDI, and the dialog and effects had gone through at least some digital process.

Two aspects of the digital audio phenomenon became foundations for the development of digital video. The demands put on computers by MIDI and digital audio production were different from and greater than standard home and office applications had ever levied. Hard disks had to be larger, faster, and able to read and write data continuously and without pauses or gaps. Data communication via serial ports, SCSI, and other devices likewise had to be faster and more stable than previously needed.

In addition, the graphical interface of MIDI sequencing software and audio editing software introduced and refined the visual metaphor common today in video and animation application programs. The scrolling timeline onto which media events are placed, arranged, and edited was later adopted by video editing programs such as the Avid Media Composer digital nonlinear video editing system, which was quickly adopted by Hollywood.

Avid is one of the earliest examples of digital video and audio technology in film. Although it did not immediately deliver benefits to the audience, the low cost and visual flexibility offered by Avid nonlinear editing quickly replaced traditional editing techniques. The Avid made a compressed digital video copy of the film or video footage needing editing, and it stored this data on hard drives. Like a word-processing program, the Avid could be used to edit video quickly, then generate an edit decision list (EDL) specifying which clips go where. As a result, the final film edit became much simpler and less costly. Interestingly, the Avid system comprises elements from different manufacturers, including an audio system built on Digidesign hardware and software. A large part of the acceptance of the Avid system in the film industry was because its audio format was already compatible with Digidesign Pro Tools, the system used by most sound designers, editors, and composers. Convergence began to happen.

SOUND AND EMOTIONS

So what's special about digital audio in content that most people think of as visual? At first, the answer to that question lay more in the creation process than in the final quality of the product. Digital audio data provides all the benefits of any other kind of digital data—words in a word processor or calculations in a spreadsheet, for example. Like any digital data, that representing audio is more portable, malleable, and copiable than its analog counterparts, and can be transmitted, edited, and reproduced much more easily and effectively than one on analog tape. After all, digital data is just a series of zeroes and ones; manipulating that data is essentially a data-management task.

For years, audio has been the last thing addressed in the process of filmmaking. This means the sound people are inevitably caught between the film's expected release date and the inevitable delays created by all the rest of the production. Still, changes are invariably called for by frantic, demanding directors, producers, censors, and so forth, and these changes must be made on the fly. A certain number of such edits are possible with an analog track, but digital presents an enormous leap of flexibility. With a digital audio editor, not only do the standard computer techniques of cut, copy, and paste apply, but sounds can be merged, pitched, lengthened, shortened, convolved, compressed, reversed, and otherwise altered in an almost infinite number of ways. It took little time for the harassed composers, sound designers, and dialog editors to see how digital flexibility could make their lives easier.

But human ambition being what it is, the advantage of just making life easier wasn't enough. The technology created a new and enormous sonic palette. Sound designers had means to create sounds that were not only more realistic, but also super-realistic. Sound could not only represent reality, but create a more-than-real sound track that could create an impact on the audience every bit as profound as the visuals. In fact, as good sound people know, the impact of music and other audio works in a vastly different, and more deeply emotional way on the audience than does visual images. And this is where the role of digital audio began to have a direct impact on the audience.

WHAT DOES IT MEAN TO ME?

Today, the benefits of digital audio to the listener are multifold. Outside the realm of audio for film and video, the convenience and endurance of the audio CD is now axiomatic, lovers of old vinyl LPs notwithstanding. In the future, DVD audio, recorded at the new 32-bit, 96-kHz specs, will add to that a sound quality that will best all previous recording technologies (yes, the truth can now be told; CD audio is cold and brittle, inferior to the best analog recordings). Another digital technology, the MP3 compression codec, has added to that portability, both via the tiny, rerecordable MP3 players that hold scads of songs within a tiny box, and by taking advantage of the home computer for storage, transmission, and editing.

The sound that accompanies movies, video, and other digital content derives the most from these technologies. We have a wealth of technologies competing to provide the most compelling movie audio: THX, Dolby Surround, SDDS, and DTS all offer crystal-clear surround-sound that all by itself makes watching and listening to a movie a more immersive experience.

As wide as screens have become, and as brilliant as the visual images are, watching a movie with traditional audio is still an I'm-here-in-my-seat-and-it's-up-there-on-screen proposition. But the moment the audio starts coming from rear speakers, and the low frequencies emerge from a big subwoofer, the viewer is immediately placed in the middle of the action, and the world changes. Not only does that motorcycle zooming onto the screen roar right past your ear, or that helicopter fly right over your head, but the small sounds—crickets, footfalls, even just a breeze or a bit of reverb—make you no longer separate, but part of the action.

The overall effect here is subtle, even if specific effects are not. If you're watching *Saving Private Ryan*, you not only hear what you see, but you also hear a barrage of battlefield horrors that you never see: they're behind you and off to either side. They bring you closer than ever to feeling the presence of the battlefield all around you. In a film as well crafted as *Saving Private Ryan*, this is not just a gimmick, but a powerful device for delivering, with yet further impact, one of the dominant messages of the film: the sheer horror of war. I recently watched a well-restored DVD version of *The Sands of Iwo Jima*, the classic 1949 John Wayne World War II movie. There's nothing to take away from this well-made film, but watching *Saving Private Ryan* the next day drove home the vastly different nature of the two works. *The Sands of Iwo Jima* seems like theater; you can almost see the curtains and wires. (Let me stress that I'm in no way derogating that film. It would be the height of arrogance to think of it as "quaint" because it came from another era; in its way, this film is no less valid than Shakespeare.)

Saving Private Ryan is another kind of experience, regardless of the fact that you sit in a theater to watch it. It's shot on film with cameras, actors, and all the rest of the common elements. But whereas any 1949 film rejected graphic violence, 1998's *Saving Private Ryan* specifically intends to put the reality of battle right to you, in every way possible. To that end the unnerving and frightening surround audio is every bit as vital as the graphic visuals, and makes the film a qualitatively different experience.

Nor is surround sound the only technique in the digital audio toolbox. At the risk of sounding like a fan of war films, a great example of the power digital technology gives a sound track is the 1999 Roland Emmerich film, *The Patriot*. Shortly after the film came out, I interviewed Per Halberg, supervising sound editor. Halberg was acutely aware of the sonic achievements of *Saving Private Ryan*, and talked about a very specific detail of his work on the later film. What I'd noticed in *The Patriot* sound track was the sounds of the individual bullets—flintlock rifle balls, in fact. The whiz as they flew sounded more than real, yet not so much different as to be hokey. I'd thought it was just enough to enhance the impact of the battle scenes. Halberg told me he'd gone into the desert with people who shoot antique firearms as a hobby, and recorded everything they had, from single shots to barrages. His intent was to develop a palette of whizzes ranging from musket balls and cannonballs to bullets from modern weapons with much higher muzzle velocities. To create the movie sounds, he layered up to three or more whizzes together in many different ways, making consciously ultrarealistic sounds. His combinations yielded sounds with noticeably different qualities that Halberg eventually used as leitmotifs, signaling a specific character or kind of event. He also spoke of mixing a certain scene where the main character is rescuing his son. The scene had no music, and no bullet sounds in the front speakers, only in the back speakers. This effect not only accentuated what sound effects the audience did hear, but also perfectly separated the rescue scene from the main battle.

Again, this metaphoric use of the audio adds dimension and depth to the film.

SOUND OF *EVIL*

Of all the stories coming out of the digital sound design world, probably one of the most interesting is that of the 1998 release of the re-edited, 1957 Orson Welles film *Touch of Evil*. Having been absent from the final editing of the film (in 1957), Welles was furious at the result and sent Universal Studios a 58-page memo and nine pages of "sound notes," detailing how he wanted the film to be re-edited. Universal implemented very few of Welles' suggestions, aborting the director's vision, and ultimately, his career.

Seen today, the Welles memo is a blueprint for differentiating the 1957 movie sound philosophy from today's. Instead of having Henry Mancini's orchestral score dominate the film, Welles wanted certain scenes in the film—which takes place in a dark and spooky Mexican border town—set to music emanating from on-screen cantinas, radios, and jukeboxes.

"To get the effect we're looking for," Welles' memo insisted, "it is absolutely vital that this music be played through a cheap horn in the alley outside the sound building. After this is recorded, it can be then loused up even further by the basic process of rerecording with a tinny exterior horn."

This illustrates an aesthetic unheard of at that time. Welles called for the "above music (to be) realistic, in the sense that it is literally playing during the action." He continued: "....this music will be referred to as 'background music,' as distinguished from 'underscoring,' a term which will be used to designate that part of the music which accommodates dramatic action and which does not come from radios, night clubs, orchestras, or jukeboxes. In other words, the usual dramatic music (used) in a picture.

"This underscoring, as will be seen, is to be most sparingly used, and should never give a busy, elaborate, orchestrated effect," Welles wrote. "What we want is musical color rather than movement; sustained washes of sound rather than...melodramatic or operatic scoring."

Welles was way ahead of his time. The distinctions he makes here developed in the film world at large only years later. Indeed, when multi–Oscar-winner Walter Murch was recruited to re-edit the 1998 version to conform to Welles' memo, Murch was stunned to read that document describing techniques Murch believed he had developed for the film *American Graffiti* nearly 20 years later.

Ultimately, Murch and his team used a boatload of digital audio gear to re-do the entire track, making it as close to Welles' wishes as possible. Perhaps the crowning achievement of the new version is the opening sequence. The sequence is a three-minute tracking shot that sets up the plot, characters, and flavor of the rest of the story. In the original release, not only did credits roll over this entire scene, but the only audible sound was Henry Mancini's proto-*Peter Gunn* music.

Murch replaced the Mancini "underscoring," to use Welles' terms, with the kind of "background music" Welles would have preferred. After removing the Mancini music from the sequence, Murch discovered an effects track on the opening shot that had never been heard before, buried by Mancini's score. It was a full track, with traffic, footsteps, goats, and more. By eliminating the underscoring and boosting the level of the effects track, Murch enhanced the realism of the sequence. Then, going to Welles' sound notes, Murch saw the director had called for "a complex montage of source cues." Murch constructed this source music using the Mancini sound track and the *Touch of Evil* CD that Mancini had released in the 1980s, including source music absent from the original film.

The result is wonderful. The rolling credits and Mancini underscoring are replaced by a mix of multisource rock, jazz, and Latin sounds, fading in and out, along with dialog central to the plot and a barrage of sound effects, creating an intense and realistic introduction that foreshadows in many ways the drama to come.

The rest of the film continues to combine source and underscore in ways that enhance the drama, giving the re-edit a quality quite different from the original.

Digital audio is not just a technology, and in itself, not just a tool. It's a fundamentally different entity that is helping us understand the many and profound differences between the experience of hearing and the experience of seeing. In the hands of artists such as Murch and others, these differences have provided—and will continue to provide—wonderful new ways for audiences to experience moving-image content in such a way that they hear, feel, and even see it with more impact and relevancy.

CHAPTER TEN

THE CASE FOR PROFESSIONAL COMMUNICATORS IN A DIGITAL AGE

CAMERON SANDERS

Visual communication is a human characteristic; creating this kind of content has become more important than ever before. Two historic trends collide in our time. The first is the long-term trend, accelerated in the Digital Age, of ever more inexpensive, easy-to-use and universally available communications tools—the "democratization" of communication, which reduces the role of the craftsperson in media content creation. The second is the increasing integration of previously distinct media into a new, multifaceted, visually dominated medium—a "convergence" that challenges traditional ways of communicating.

Both of these trends were widely predicted. Little attention, however, has been paid to the inevitable result: A fast-paced, media-saturated world in which more people attempt to communicate more messages more widely than ever—and yet with fewer messages actually breaking through the clutter.

In this era in which talk is cheap and understanding not necessarily abundant, there's an unprecedented opportunity to elevate the public perception of the content-creation or communications expert to that of an indispensable professional.

AN AGE OF INCREASINGLY DYSFUNCTIONAL COMMUNICATION

The traditional mailbox is as full of junk mail as ever, even as home and office e-mail inboxes burst with pitches. There are more magazine titles than ever, not to mention the e-zines that pop up on every subject. New cable and digital television channels proliferate—as do DVDs—and streaming video is emerging as an Internet staple. Satellite-delivered radio is taking its place beside terrestrial broadcasters, delivering sound—wanted or not—to every corner of our environment. Mass-audience advertising seems ubiquitous, supplemented now by tightly targeted appeals, some reflecting every left click of a mouse. Yet, the ability to reach an audience appears diminished.

Total time spent with media grows. On a typical day in a typical American home, the TV is on for seven hours and 53 minutes, up nearly an hour from a decade ago.[1] Internet use grows at a dramatic pace; 66.9 percent of Americans now surf the Web[2] and it's estimated 75 percent will by 2005.[3] The average "netizen" spent more than 17 hours and 49 minutes per month online in March 2001. Each session lasted about 31 minutes, but on average just 55 seconds was spent viewing any particular page, scarcely enough time to glean a message.[4]

In the workplace, every form of media is now employed to train, inform, and motivate. The memo still reigns supreme, but is supplemented by video, internal publications, PowerPoint presentations, and computer-based training. The vision of a "paperless office" is long forgotten. Paper consumption actually increased 40 percent as e-mail was deployed in the average office. It seems employees print-out longer missives to read and file.[5] And, it's estimated that paper consumption will increase by an additional 50 percent in the next ten to fifteen years, primarily because of increased computerization.[6]

All this is true, and yet it often feels as if the whole world is talking and no one is listening. In a very real sense, that's precisely what is happening. At the dawn of the digital content creation age, communication is increasingly dysfunctional. To understand why, it's important to review how we view the process of communication and compare that with how we actually learn.

A "LITERAL" VIEW OF COMMUNICATIONS

When we communicate, we focus on what we want to say. We try to find just the right words. It seems logical. We assume that if we write or say something, our audience will focus on our message and ignore everything else. And we assume our message will be interpreted exactly as it was meant. In short, we think we can dictate a message. It's as if we think communication just happens. But it doesn't.

The intense focus our society puts on words and—in particular—the written word, can be traced to the Sumerians 5,000 years ago. They assigned meaning to some 2,000 simplified drawings of objects, creating a visual alphabet. For the first time, written communication wasn't an original drawing limited by the skill of the artist and subject to widely varying interpretation. The Phoenicians, 2,000 years later, took a step further, creating a much smaller and more manageable set of symbols, representing the basic sounds of language. The written word was born.

It was hardly inevitable that thought would be communicated through such abstract symbols. Humankind had communicated over distance and time using visual imagery. But before film and electronic imaging, words proved the most efficient way to send messages. It was the reproducibility and transportability of words—not superior effectiveness—that led to the de-emphasizing of images as the primary means of communication.

Johannes Gutenberg's completion of his moveable type "42-line" Bible in 1456 not only served as a critical milestone in the democratization of communication, it cemented the concept of the word as primary in intelligent thought. Images were rare and subordinate to text. In the centuries that followed wisdom not only arrived in written form, wisdom was seen as virtually indistinguishable from text. Even the primary definition of literate, from the Latin word for letter (*littera*), is "an educated person."[7] Scores of generations were taught that educated people communicate—even think—in words. Primitive societies, we were told, used drawings and icons. Modern societies use text.

Today we still often associate text with wisdom and images with ignorance. But we also often hint we're not so sure that's how it should be or really is. In the Disney classic *Beauty and the Beast*, Gaston challenges Belle's love of reading with "How can you read this, there are no pictures?" She retorts, "Well, some people use their imaginations." Yet, the close-up shows her pointing to a picture of a castle to make a point. We often portray television as shallower than the written word, yet 43 percent of us spend more than an hour watching television news every day. Just 26 percent spend as much time with newspapers. And we tell researchers we find local television news more believable.[8]

IN THE MIND'S EYE

This view of the word as preeminent is at odds with millennia of thought. Artistotle said, "There can be no words without images." Plato gave primacy to sight over all the other senses, saying, "Of all the senses, trust only the sense of sight." While intellectually extolling the virtue of text over image, Western culture has clearly retained a special place for images. Many of today's most commonly used phrases suggest we agree with Aristotle and Plato that we perceive primarily through sight: "I see your point," "Let me see," "Seeing eye to eye," "Love at first sight," "See what I mean," and, of course, "In the mind's eye," just to name a few.

We are, after all, a visual species. Yes, a scent, a taste, a sound, even the texture of an object can evoke powerful memories, but it is sight that adds definition. Think back to the last time you sensed a particular perfume or smelled a favorite food. What images came to mind?

Other creatures, of course, perceive the world differently. Bats and dolphins "see" predominately in sound. Many mammals "see" through scent. Some creatures see nothing at all. Even those that do see generally see things differently. The compound eye of a fly, for example, enables detection of the slightest motion, but lacks the depth of field and range of our vision. Along with less than 6 percent of the animal kingdom, we perceive the world through what can be described as "camera-style" eyes: A series of rapid-fire "snapshots" make up our view of the world. And that makes all the difference when it comes to how we perceive.[9]

For most of the last 500 years, the printed word was the dominant mass medium. There was little or no visual communication. Yet, the word itself was most often used as a means of delivering images from place to place. Writers turned pictures and objects into words. Readers interpreted those words and turned them back into rough approximation of the original image in their imagination. It's no wonder, then, that as images became more commonplace, words began to seem less potent.

A picture, it is often said, is worth 100,000 words. We repeat this truism because we understand quicker, more completely, when we see it for ourselves rather than having it described to us. Given the opportunity, we strain to see, rather than read or hear. Ultimately, we trust what we see over any other sense. Seeing is believing.

IMAGES SPEAK LOUDER THAN WORDS

The siren song of a word-dominant view of communication is this: If we can just choose precisely the right words, we can guarantee that our message will be received and understood exactly as intended. Yet experience tells us that even when we choose our words with utmost care we can be misunderstood or ignored. That's because the words we focus so much attention on have a remarkably small impression on the audience, compared to other factors.

Successful speakers and salespeople understand this. In one-on-one communications, the importance of nonverbal cues has been widely noted. Speech coaches and management consultants routinely cite the 1967 conclusions of psychologist Albert Mehrabian. He found that just 7 percent of what the listener understood was verbal content. Put another way, 93 percent of the information received wasn't words! Thirty-eight percent of the information received was gleaned from vocal tones, inflections, and accents; 55 percent from visual cues, ranging from facial expressions to physical appearance and context.[10]

Different people understand these cues differently. A good many popular books have been written about the different ways in which men and women interpret the same cues. Similarly, different cultures imbue mannerisms, gestures, and images with different meanings. Even audiences composed of differing income levels and age ranges interpret what they sense differently. The more we learn, the more we discover that perceptions vary much more than we assume.

Effective communication, then, isn't a "brain dump." We really can't "tell" anyone anything. A message is sent only if we are successful in evoking understanding in the "mind's eye" of another. In short, the listener must "see" what is being said and relate it to a past experience.

In relationship counseling and employee communications seminars, experts seek to help us bridge this gap between what a communicator says and what the listener "sees," through what is called *active listening*. The listener is taught to try to envision what is being said, relate it to an experience in his or her life, and repeat the experience back to the speaker until both speaker and listener reach a shared understanding. Although an effective means of solving the communications gap in one-on-one encounters, this approach is hardly a solution in public speaking and mass communications. Communicators in these environments have one chance to evoke images that resonate.

Clearly, if 93 percent of one-on-one communication is nonverbal, we must assume that this is true in other forms of communication. Can anyone doubt that images are critical in film or video?

THE RISE OF MASS VISUAL COMMUNICATION

Over the last 150 years, mass visual communication emerged alongside print, then began shoving it aside and reaching a dominance that couldn't have been imagined earlier.

It began with the increased use of the woodcut depictions and reality photography first used on a large scale in publishing during the American Civil War. Then came the motion picture, silent at first, yet astonishingly effective in its ability to communicate—so effective, in fact that Harry Warner of Warner Brothers was quoted as saying in 1927, "Who the hell wants to hear actors talk?"

The dominance of the picture in magazines such as *Look* and *Life* made clear the value of the image, increasingly taking more space in publications of record. Then came television, putting visual communications front, center, and predominant in our popular culture. In our time it is clear that every medium of communication has been altered by the dominance of television, from magazines and newspapers to—yes—public speaking and everyday speech.

The newly rediscovered primacy of sight over other means of perception became clearest in politics. John Kennedy's confident, tanned, and relaxed appearance in the first 1960 Presidential debate, alongside the sweating and seemingly nervous Richard Nixon, is thought to have given Kennedy the edge on television. Yet, on radio, listeners gave Nixon the edge. Ronald Reagan's handlers boasted that it didn't matter what reporters said as long as the image on the television screen portrayed him as "Presidential." George Bush's handlers even placed him in a flag factory. In 1988 Michael Dukakis learned the lessons of image after posing infamously in a tank and after being on the receiving end of ads critical of his Massachusetts Prison Furlough program's release of convicted murderer Willy Horton. In politics the phrase "perception is reality" is here to stay. And, there's no doubt sight is king when it comes to communicating perception.

Canadian educator Marshall McLuhan, once called the "Oracle of the Electronic Age," understood this phenomenon. It was McLuhan who posited that "what" we say is of little importance, only "how" we choose to deliver it.[11] The title of his best-known work, *The Medium is the Massage* (sic),[12] an apparent word play on the then-new "mass age" and his dictum "the medium is the message," dramatically makes his point that the visual nature and mass reach of television so overshadows the word as to make virtually irrelevant anything we say on TV. And McLuhan went further, arguing we perceive our world in the form of our most dominant mass medium (television) saying, "we shape our tools and they in turn shape us." Taken to its logical conclusion, the image had emerged as the dominant element in all means of communication, from one-on-one conversation to mass communication.

The emergence of the Internet in the last decade hardly suggests a less visually dominated future. Streaming video, extensive graphics, and animation are already swamping text in importance. Though vast libraries of text lie behind flashy home pages, the audience must know what it wants and "pull" that information out of the archives. If communicators wish to "push" a message, they typically do so in images.

In our time, images simply overpower words, relegating them to an important—but clearly context-setting or supporting role—when used in combination with images. When we merely "talk at" our audiences they tend to fail to "see" our point. And, there's clearly less time to deliver a message. Audiences not only want to see the point, they want you to get to it right now. In such an environment, the most effective means of reaching an audience is often through a well-chosen, fast-moving image accompanied by a few well-chosen words.

WHAT YOU SEE IS NOT NECESSARILY WHAT YOU GET

Even imageless media, such as novels and radio, can be described as "visual" at their core. An effective writer or speaker readily evokes images within us because we humans are hard-wired to demand them. Think about the last good novel you picked up. What images did you "see?" In the absence of pictures, well-chosen words can effectively paint images. But words that fail to create images fall flat. It's clear that if an image is not conjured up or provided, the message cannot be received. The word, then, can best be seen as a means of conjuring up images where none existed, creating context between them, and adding definition to them.

This human insistence on the existence of an image both enhances and complicates communication. When an image is lacking, we simply create one. When I was host and business editor of the public radio program *Marketplace*, I was occasionally told I didn't look at all like listeners had imagined. Taking cues from the sound of my voice, my vocal mannerisms, and the words I chose, listeners envisioned profiles. Many were, I'm sure, disappointed in my actual appearance! Although some imagined me older, others imagined me younger; some imagined a button-down business executive-type, others an ex-university radical; some imagined a full head of hair, others a bald dome.

Where did those images come from? And why were their imaginings so different? The answers lie in some curious characteristics of our sense of vision and thus our ability to interpret what a communicator attempts to show us. Studies of visual understanding show that humans tend to modify the images they see to match their expectations. When an "impossible" two-dimensional image is presented to us, we imagine it as three-dimensional and possible, for example. When test subjects were shown a man entering a barbershop, then leaving, many insisted they had seen him getting a hair cut when they had not. Prejudice colors perception, too, as shown in a study in which black and white men were shown together. Many insisted the black man was holding a razor and looked "threatening" although a white man held the weapon.

To communicate effectively we must understand these characteristics and expectations or the images we send may be perceived quite differently than intended. It's clear that just as words are an imperfect delivery mechanism, images—although arguably much more effective—are imperfect as well.

The biggest challenge in this visual age is the lack of a visual "dictionary." Just as writers early in the age of the written word had to seek out commonly understood words and phrases to communicate effectively, today's communicators must identify commonly understood images. Researchers have begun to catalog some of the many hundreds of thousands of nonverbal cues we use. And, yes, we do know some of the emotions certain images generally conjure up. But there remains so much we don't fully know. Perhaps someday there will be comprehensive visual reference guides, but until then, we must explore and learn from one another. Intuition and trial and error are our only tools.

A CRAFT MAKES WAY FOR A PROFESSION

So, in our visual age, effective communication is clearly a very complicated business. Yet, paradoxically, many have recently come to believe that communication content creation is—or may soon become—a do-it-yourself activity.

This so-called "democratization" of communications isn't what it seems. It is true that the invention of the offset printer, the photocopier, the word processor, mass-delivered e-mail, and Internet authoring software has turned written communication into a pursuit that can be engaged in inexpensively and easily. Similarly, affordable digital video camcorders and PC-based editing and graphics software are increasingly democratizing the means of visual content creation. And, yes, it is true that communicators are no longer forced to consult with printers, artists, and producers to create even the most basic of projects merely because of the expense or complexity of communications equipment. But while the democratization of communication has put tools in many more hands, it hasn't offered insight into how to use them. Without this understanding, the present state of increasingly dysfunctional communication was inevitable.

A new communications profession is now emerging, driven precisely by the need to break through the noise of the digital age. This may seem counterintuitive to some communications experts still struggling with TV-addicted, camcorder-toting clients who think they know as much about content creation as the experts they retain, but the increasing failure of companies and individuals to get the results they demand offers an unprecedented opportunity to position the communications professional as more indispensable than ever.

Perceptions are slowly changing. Once seen primarily as craftspeople toiling with ink, paintbrush, typewriter, or bulky camera gear, today's content-creation practitioners are transforming themselves into professional communications consultants as well, armed with sophisticated strategies and reams of data. And it isn't a stretch. By far the most important skill set the communications professional brings to clients isn't the ability to execute a project, it is the ability to define the intended result, divine the message, and discern the best way to connect with the audience.

The communications professional has already emerged in politics as a critical player, and businesses and individuals are following. There's an increased understanding that just as an architect is necessary to create a safe, practical, and attractive structure, and that an attorney is needed to discern and effectively argue the finer points of the law, a communications professional is needed to turn content into an effective message that resonates and gets results. The day is coming when corporations will recognize the communication professional's rightful place in senior management as chief communications officers beside chief financial officers and chief information officers.

TEN STEPS TO EFFECTIVE COMMUNICATION

For the professional content creator and communicator, success in today's environment requires educating clients and superiors about the power of the image in today's world and the importance of focusing on the audience—not just on the words spoken.

Communications has always been a collaborative process. In today's world you'll need your clients and bosses to be full partners in the process. It also requires managing their expectations. To do this, professional communicators are finding it essential to map out a series of well-thought-out methodical steps to keep projects on target. Here are ten steps used by many successful communicators today:

1. *Don't jump to conclusions.* The message you assume you'll want to deliver is seldom the one that should be delivered. In the parlance of the everyday, it's important to think before you talk. In years of communications consulting, I

am constantly struck by how often a client (or superior) is absolutely certain he or she knows what the message is and precisely how it should be delivered, before having given a thought to what he or she really wants the audience to do—or even who the audience is.

2. *Establish a realistic, measurable goal.* What exactly should the audience do immediately after reading, listening, or viewing the message? If the audience doesn't act immediately, it's unlikely to act at all. Identify specific, measurable goals that are achievable. At all costs, avoid trying to accomplish too much, or you'll accomplish little or nothing.
3. *Define the type of communication.* Should the project train, inform, motivate, entertain, or sell? Be careful. It isn't as easy as it seems. Atlanta scriptwriter Grant Williams created a "TIMES" grid, based on these five categories. He asks his clients to choose category by category which type of communication is closer to the mark; giving the nod, ultimately, to the category receiving the most marks. Using this approach, I've discovered a remarkably high number of clients conclude their original assumption was wrong.
4. *Identify your target audience as narrowly as possible.* Because different groups perceive images differently it's critical to focus on just that portion of the audience that must be convinced or persuaded, rather than on wasting time preaching to the choir.
5. *Understand the target audience.* The professional communicator must know everything there is to know about what the audience perceives: every demographic, every dislike, every shared experience. This does not mean asking a representative group what they want to see or hear. If that worked, public television would draw more viewers than sitcoms and exploitive reality programming. Understanding the audience means understanding what they perceive and do, not what they say they like and will do.
6. *Where will the audience hear the message?* Will it be heard in their home? If so, will the viewer be alone or with the

family? What will the effect of the co-audience have? Will it be heard in the car? Will it be viewed in a group? If received in a group, will it be supervised or among a group of skeptical viewers? Consider the environment closely. If it interferes in any way, the message will be lost.

7. *Define the message at the latest possible opportunity.* If you are to effectively connect, the message will be necessarily altered as you gain a full understanding of what you want the audience to do, what type of communication is to be employed, and the characteristics of the audience targeted to receive it.
8. *Decide carefully on a delivery medium for the message.* All those involved should remain open to using a different medium if indicated. Too often messages that should have been delivered on audiotape or CD during a commute or computer-based training session became ineffective videos just because the client or communications practitioner always assumed it would be that medium.
9. *Stay on message.* Make sure every image, every word, every sound is consistent with the communication's goals and the audience's perceptions.
10. *Evaluate the results.* In step two you established realistic, measurable results. How'd you do? Do it as a team, production team and client, corporate department and boss. Don't let others evaluate the effectiveness of a project on a "gut" feel or feedback from colleagues or family. After all, the project wasn't designed to reach them!

SHOW ME!

The good old days of "tell 'em what you're going to tell 'em, tell 'em, and then tell 'em you told 'em" are clearly over. A more apt dictum for the digital age is "Don't just tell 'em. Show 'em!"

It's never been a better time for professional content creators and communicators. Everything is visual now. We're in demand. And for the first time we're being viewed as professionals critical to the bottom line, not just glorified A/V guys or artsy folks playing with electronic toys.

REFERENCES

1. Nielsen Media Research 2001 survey, from Frank Magid Morning Report.
2. "Surveying the Digital Future" UCLA, cited on cnn.com, October 2000.
3. The Gartner Group Inc., cnn.com October 2, 2000.
4. Nielsen/Net Ratings, Inc., March 2001.
5. PriceWaterhouseCoopers, *New York Times* Business Section, April 21, 2001.
6. 2000 Canadian Forest Products Association report.
7. *Webster's Third New International Dictionary of the English Language, Unabridged* Merriam-Webster, 1986.
8. Pew Research Center's biennial survey of the national news audience, 2000.
9. Thanks to Daniel Chandler, lecturer in media theory, University of Wales, Aberystwyth.
10. Mehrabian, A., *Silent Messages: Implicit Communication of Emotions and Attitudes*, Second edition, Wadsworth Publishing, 1981.
11. Griffin, E., *A First Look at Communication Theory*, McGraw-Hill, 1991.
12. McLuhan, M. and Q. Fiore, *The Medium is the Massage*, Bantam, 1967.

CHAPTER ELEVEN

THE PRACTITIONER'S VIEW

GEORGE AVGERAKIS

Throughout the history of humankind, there were only two times when all known media were in the control of one practitioner. The first was in pre-history, when tribes bestowed upon one individual—the shaman—the role of painter, storyteller, singer, choreographer, and even physician. The second time is right now, when a creative individual—using affordable digital content-creation tools—is capable of competitively executing entire projects within a given business category.

I am such a worker. This is my story.

From the ancient days of the shaman until recently, society has continuously divided labor into ever more specific categories. Specialization is a process of civilization, where large groups of people—cities, corporations, churches—are organized for the common good. Smaller organizations, on the other hand—nomadic tribes, family-owned companies—demand that each member be a generalist.

The choice between specialization or generalization is often determined by the opportunities offered any given individual. I began my media career when large corporations offered workers the best opportunities. Today, that is no longer the case.

To understand this vast transition from specialization to generalization, from civilization to economic nomadism, from large to small enterprise, you have to look back a bit. A good place to pick up the trail is at the advent of the Industrial Revolution.

From the time of the Industrial Revolution (beginning in Britain about 1750) until about 1975, all professions—and the profession of what today we might describe as "media content creator" in particular—could be divided neatly by purpose (such as printer, scribe, crier, etc.) and by class (lord, master, apprentice, journeyman, etc.).

Although this period represented, economically, an improvement in efficiency over the feudal production of all goods, including media, Industrial Age technology dictated that the high cost of production tools like printing presses, typesetting machines, audiovisual recording and broadcast devices, and distribution channels, presented nearly impenetrable barriers to entry. Although media content creators enjoyed a distinctive increase in status during this period, their employers—those who owned the means of production and distribution—could exploit the creators of media to a significant degree.

Computers had been available to the public at affordable prices as early as the late 1960s. Few people took notice, however, because practical applications for computers were few and computers were complex. Gradually, however, certain individuals exploited the computer, making it the core technology of many useful tools.

The first of these tools was the *word processor*, a replacement for the typewriter. Word processors, like those produced by Wang, offered the advantages of typing text into a storage device, after which the text could easily be edited and printed. Word processors were, however, expensive, and only the wealthiest companies could afford them for secretarial staffs.

A device similar to the word processor was introduced in 1970 by the ancestor to today's Chyron Corporation; it allowed text to be typed over a video image. This *character generator* meant that to superimpose words onto a video image you no longer had to create type on paper, shoot it with a camera, and then use a special-effects generator to mix it with a second video image. Character generators and special-effects generators were, however, expensive and only the wealthiest of broadcast companies could afford them for their studios. These and other video-production devices functioned as discrete standalone units requiring separate operators.

In 1975 I noticed a trend in the printing industry and decided to capitalize on it. This trend was caused by the invention of *phototypesetting*, and it reminded me of how all the dogs in a neighborhood are said to begin barking just prior to an earthquake. Something big was going to happen. Consider: In 1975, if you wanted to create a simple brochure, five categories of media artist were required: copywriter, graphic designer, typesetter, engraver, and printer. The text, created by the copywriter on a typewriter, was submitted to a typesetter. The typesetter retyped the text into a hot type machine that created metal rows of text, from which a single black and white impression (*camera-ready* art) was made. The engraver took the camera-ready art and any photography to retouch and perfected them using photographic techniques. The final page was then photographed and chemically impressed on the printing plates, which were then used to create the final brochure.

The process required a hot-type machine (over $100,000*); a highly skilled (often union), laborer billing over $150* per hour; photography; engraving machines that cost over $250,000*; engravers billed at over $250* per hour; and million-dollar scanners—yielding a prepress creation cost of about $1,200 per page.

In the 1970s, a digital revolution began in printing. Color Electronic Prepress (CEPS) systems from companies like Scitex and Chromacom brought digital control to the engraver using machines costing as much as $1.2 million.* Soon after, companies like Compugraphic and Mergenthaler introduced machines that allowed a minimally skilled operator to create camera-ready type in minutes. Employing a process that projected images of type fonts onto photosensitive paper under the control of a computer, these machines cost about $150,000*

*The formula used to convert 1970s dollars to present value is times three.

and could be operated for a lot less than what a union typographer would charge.

Most of these machines were purchased by the same companies that had previously invested in conventional prepress machines. The companies, owning a new means of production, realized that they had a window of opportunity to outsmart the unions. Unions had monopolized the typographer and engraver labor pool and consequently controlled labor costs. If the companies could find low cost, nonunion labor to operate the new computerized machines, they could drastically lower their labor costs.

I knew a lot of underemployed artists in 1975. Most of them were working freelance, hand-to-mouth for less than $15* per hour. Splitting the difference between what a union typographer made and what my artist friends were making, I set an arbitrary price at $60* per hour. Then, I went to Compugraphic with a proposition. If they trained five of my artists to operate their new machines, I would have those artists train others and put all of them to work as temps in Compugraphic's client facilities.

I was in business. My profit margin allowed me to invest in training more artists. My artists were happy making double the money, and while I had a full-time job as an artist myself, I was making enough money to buy a Brooklyn brownstone before I was 30.

Although the average price of a page of camera-ready text dropped to less than $500 per page, the new technology, *phototypesetting*, did little to improve the life of the media artist. Exploiters, like the old companies that owned the machines, and new guys like me, got most of the bucks and kept most of the creative control.

But even the phototypesetting business bit the dust. Just a few years later, the software program WordStar enabled desktop computers, connected to high-quality printers, to create camera-ready art for about $60* per page. In 1988 the raster-image processor (RIP) propelled the development of device-independent prepress software such as Quark and Adobe Photoshop. Eventually the expensive CEPS machines were replaced by Macintosh computers and art directors themselves took advantage of the full digitization of printing technology.

Now media artists, investing just $15,000 in equipment and software, could write, typeset, and print a brochure all from one workstation. Finally, the two remaining barriers to efficient media creation had fallen, at least for the medium of text printing. The means of production were affordable and one set of equipment could do the whole job from start to finish. I sold my temp agency out to my partners and moved on.

If you stop and examine the history of computers and software, you can see this story repeated for every media type. Any reasonably intelligent tech-watcher in 1980 could see the progression written clearly on the wall like a blank check. By this time, I had put my profits into real estate, so that when the domain of business math got sucked into the computer, I resisted the urge to buy any hardware and continued to do my spreadsheets with pencil and paper.

Visicalc, the first spreadsheet program, expanded the usefulness of the computers that had already been purchased by graphic artists to facilitate typesetting. Although this advancement might not seem at first a benefit to the artist, one only had to consider how the typical media maker obtained the cash to finance his or her work. Every job begins with a creative proposal to a sponsor. That proposal usually includes a typewritten description of the work and a budget of costs. Prior to the computer, the average time required to create a proposal was three full days.

Using Visicalc, and later, Lotus 123 and Excel, the media artist could quickly produce budgets and proposals. By archiving various proposals and budgets in the computer's digital memory, the producer could "boilerplate" proposals, further saving time and effort, and finally reducing the time to create a proposal to a couple of hours.

The next medium to become computerized was still photography, a significant development. The invention of such high-quality drawing programs as Adobe Illustrator, and picture-retouching programs like Adobe Photoshop, along with other prepress tools such as PageMaker, allowed the digital graphic artist complete prepress control of images.

In the span of a year or two, the entire print business was fundamentally overhauled. Typographic houses, engraving shops, stat centers, pen and ink artists, retouchers, and most of the people and companies that flourished for generations in small niche segments of the print-production business were put out of business. If they were smart, they got computers. If they got computers, they learned how to do the work that niche vendors had held to themselves. Most important was the trend toward generalization of the workforce. Artists formed new companies that were smaller, based on desktop-computer technology that required fewer people. These new companies were also highly profitable.

These new studios tended to work on a project basis rather than specialize. Instead of taking on one specific function in a project's "food chain," they went for the entire project and performed all the functions. Eventually, the term, "project studio," came to be used to designate such firms.

By the early 1980s, project studios were clearly developing in print and it seemed inevitable that the concept would one day take hold in more complex media types: audio, and later video. Looking further ahead, clever folks could see that once the media content-creation business had been retooled to the project-studio model, the news would spread—because the media was spreading it—and other industries—such as power production, banking, construction, medicine, retail, fashion, and automotive design, would take their turns.

A second industrial revolution was beginning. This revolution fulfilled some of Karl Marx's theories about class warfare, but the battle was taking place at the intersections of silicon, not asphalt. This revolution, based on technology, was replacing earlier ones based on politics and violence. And this revolution was beginning to fulfill the dreams of enlightened laborers. It was placing the means of production into the hands of the worker, and it promised to offer the same advantage to all labor categories because it depended on information—not capital—as the principal ingredient of value.

At about this time, I got bored with real estate and decided to take a job at a large corporation to revamp their in-house audiovisual facility. The job allowed me an opportunity to test a lot of my project-studio concepts in real life. Instead of buying new cameras, I hired freelancers who owned them and kept track of the annual expenditures with a computer. Instead of building a studio, I shot all my projects on location, using freelance lighting crews that owned their own lights.

Prior to being hired, my employers had brought in a consultant to submit an equipment configuration. Their detailed estimate, based on traditional assumptions, came in at about half a million dollars.** That was a lot of money. But the argument in those days was that if you could amortize the investment in equipment in three years or less the equipment was a good investment. In other words, my employers figured that I would be able to produce $15,000** worth of production in three years. To break even.

Don't laugh; prior to the 1980s some electronic gear actually lasted longer than three years before becoming obsolete! Now, of course, a computer is obsolete before it's delivered.

In short order I proved to my employers that I could produce $900,000** worth of video per year in facilities costing less than $160,000.** Done deal? Nope. No corporation moves as quick as a video producer. The corporation went into its usual nickel-and-dime routine and I was out of there. In less than two years I had worked out a plan to set up my own shop with less than a $60,000** investment, and I had that much in my new company's first contract. My formula for amortization was more aggressive. I was looking to amortize my equipment not over three years or even one year. I wanted to pay for my equipment with profits from the first job!

The 1980s introduced the commercial availability of relatively low-cost personal computers and a growing library of software. For the first time, since prehistory, media-content creators could ply their trade on equipment that they owned themselves. And in many cases, the investment in essential equipment could be recouped quickly, maybe not on the first job, but certainly after three or four.

If you were a savvy video producer in the 1980s you could see the revolution coming in a lot of ways. Even if you weren't familiar with the project-studio trend in print production, you could still extrapolate from other technological developments.

**The formula used to convert 1980s dollars to present value is times two.

Small-format video began to replace more expensive film shooting and processing in TV news. Audio technologies such as Musical Instrument Digital Interface (MIDI) changed the way music was produced and recorded and allowed for the growth of audio project studios, which began competing with traditional sound-recording companies, even putting some of them out of business. Typically the owners of these project studios had worked in their larger, specialized antecedents, and could now apply their expertise and creativity toward businesses they could afford to own.

Technological developments in video also surged ahead, all of which took advantage of the accelerating improvements in computer technologies. Video editing is a good example. In the 1980s, video editing was extremely expensive compared to film editing, whereas video shooting was extremely cheap compared to film shooting. Although a simple "two-machine" edit could be used to create a "cuts-only" production, competition with the sophisticated effects of film editing (which had the advantage of an 80-year head start in such areas as optical printing technology) drove video producers to demand electronic editing systems that were capable of video dissolves, wipes, titles, and other special effects.

To execute such a sophisticated video edit, several playback sources—videotape playback machines, character generators, special effects generators—had to be synchronized mechanically and electronically to a video tape recorder (VTR). This required a lot of expensive equipment, which could only be owned and maintained by large, well-financed companies. These multimillion dollar *online* facilities charged huge rates of $500 to $1,000** per hour, including an operator. (Film editing was going for about the same price per day!)

Back then, the term "online" editing was derived from the unique capabilities of these media factories, which used sophisticated computers to control and synchronize all the various source and record machines. An edit was said to be "online" when more than one video source was employed in the edit. In an online session, the computer would read SMPTE code (an invisible pulse code developed by the Society of Motion Picture and Television Engineers) that was embedded on a source tape, enabling the identification of every frame of tape and the synchronization of the machines.

In an effort to reduce the cost of video editing, the concept of "offline" editing was conceived. Offline editing used just two, less expensive, professional VTRs, such as the U-Matic videocassette format from Sony. Although these decks were not, at the time, computer-controllable, offline editors could create simple, cuts-only programs by playing a tape on one deck, while copying the selected scenes, sequentially, on another deck. The resulting *offline edit master* would allow a client to view a rudimentary rough edit of the production, allowing some level of control, change, and approval prior to the producer spending big bucks for the online, or final edit.

The producer began the offline process by taking his original tapes and copying them to a less expensive tape format, such as the then-new VHS cassettes. During the copy process, the SMPTE code on the original tapes was superimposed as a visual numerical readout over the picture material. Using the SMPTE (pronounced "SIM-tee") code numbers, the editor could assemble a list of "in-cues" and "out-cues" for each scene. These cues would then be entered into the computer at the online facility and the original tapes would then be used to completely reassemble the show with all the online effects.

The advent of offline editing, employing equipment that could be purchased for under $30,000,** allowed some media artists to acquire the means to production, even before the computer revolution improved things even further. Consequently, numerous offline editing houses, owned by the editors themselves, appeared on the market.

When I decided to go on my own, I used my lunch breaks to begin calling nearby corporations and offering my services as a freelancer. I got little response. But when I started calling on behalf of "Avekta Productions," a name I had registered a few years earlier, the response was significantly different. First off, my calls were now directed to marketing executives instead of the in-house video department. Secondly, it was assumed that my firm consisted of several people and that it was housed in some facility.

Within just a few weeks of starting, I had an appointment with a large pharmaceutical firm, which led, only a month later, to what would become a four-year contract to document a newly defined malady—Alzheimer's Disease.

With enough production on contract to last several months, I quit the full-time job, bought two Sony U-Matic decks, a video monitor, and a cuts-only controller, and I set up shop in two spare rooms of my Brooklyn brownstone. I was in business. Because I very carefully executed every offline edit, taking careful notes of the time code so that my online sessions were as short as possible, I was also profitable.

A year or so later, a new computer came on the market, called the Via Video. Costing $45,000 in 1983 dollars (several times more than the downpayment on my brownstone) it was easily the most expensive investment my wife and I had made. But the computer was capable of displaying a whopping 16 colors on the screen at one time, and—by fading the colors on and off through a simple program—we could actually create animation for video!

At this time, computer animation was only available to TV commercial producers using devices costing over five times as much as our machine. Although not as sophisticated as the high-end systems, our machine allowed us to offer animation to the corporate community at a price the market could bear and which, we hoped, would be highly profitable. We were the first company in New York City to sell animation to the corporate video market. The computer paid for itself in one year and generated tons of additional leads for work that did not actually require animation.

But now I was doing two jobs. I was a video producer and an animator. I took a moment to make an important realization. If each medium was going to be absorbed by the computer, the various vocations within the media profession were going to merge. No longer would there be video editors, sound editors, and Chyron operators. Soon, there would be one guy on one computer doing all three jobs. Maybe, I thought, even tasks as diverse as art director, video producer, and music producer would merge. Wow....

The term *multimedia* in those days referred to large-scale slide shows, run on dozens of computer-synchronized, 35 mm slide projectors. In the 1980s the companies making these shows started to flounder. Competitors were offering computer-generated slides that could be designed on desktops and developed at centralized service bureaus. Later, even these would be replaced by LCD screens running sophisticated shows directly from computers. Video projectors, running computer-generated images that were recorded directly to tape, soon reclaimed the name "multimedia" and redefined it. Banks were left to reclaim everything else.

Step by step, each of the media was converting into a form that could be created, manipulated, and distributed at low cost. Using equipment whose price could be amortized by the artist after only one or two assignments, media content creators could easily learn their crafts, find apprenticeships, form companies, and become independent, wholly owned operations.

As the big slide houses crumbled, my clients and potential clients were asking me about using the new video projectors to mount "slide shows."

"Why do shows with still pictures?" I asked, "why not project real video?"

"It's too expensive," was the immediate response.

"Not compared to programming a dozen slide projectors," I responded. I found myself winning creative accounts that had previously been the exclusive territory of multimillion-dollar firms with five-year exclusive contracts at IBM and Pfizer. These companies didn't exist anymore and their creative people were scrambling around trying to learn how to do what I had already been doing for five years. But then, I was no longer a video producer. Now I was making video, animation, and staged presentations. And I was not the only one doing this. What were we all becoming? Digital content creators.

By 1991 the remodeling of the media business with computer technology was in full swing. A Topeka, Kansas company called NewTek introduced a peripheral card called the Video Toaster for Amiga personal computers. This card took a $2,500 (all prices from here on are unadjusted) Amiga computer and made it into a full production studio. The Toaster featured four camera or tape inputs and a switcher with hundreds of cool transition effects. It didn't kill the market for traditional, stand-alone video production switchers and the technical directors who operated them, but it may have caused some worry for a few of them. What it did do was democratize expensive technology by making it affordable and accessible. *Wayne's World*-style basement production was born.

The Toaster also featured a character generator (a source of concern for operators of single-purpose character-generator devices), a video color enhancer, and a 3D animation program (goodbye Via Video). How much did it cost? I wrote a check on the spot when I saw it at a trade show, but an attractive redhead working for NewTek handed it back to me. "No, no," she said. "The Toaster board and software is fifteen hundred dollars, not fifteen thousand."

Suddenly, in 1991, everyone could afford to become a full-service production company, assuming you had talent and good business sense. Suddenly, the whole gang in the edit/production room was merged into one person—the "desktop media producer." The room itself—here comes that term "project studio" again—became an ever-changing, plug-and-playground as new devices were invented to quickly replace several that came before.

Simultaneous with this technological boom, there occurred an economic bust in the world economy and a deep recession hit all industries. Corporations began to strip out their middle-management category, the residence of most full-time media producers. These polished pros, no strangers to job-hopping in a lucrative media market, soon found their resumés bouncing back with depressing regularity. Rather than don paper hats and learn the rapid preparation of retail food, many decided to invest in this new "desktop" computer-based project-studio technology. A flood of new, small, one- or two-person companies appeared, further fueling the research and development of still better desktop media tools. Manufacturers such as JVC, Panasonic, and Sony, meanwhile, kept coming out with better and better quality professional video cameras and cassette recorders with ever-lower price tags. Cameras with built-in recorders—*camcorders*—began appearing, with digital recording technology increasingly employed. Camcorder prices declined as quality and features rose. Digital interfaces with

Toasters and other desktop computers made for increasingly efficient and better-quality productions. More often than not the cameraman became the editor once he had returned from the shoot and sat down in his studio at the editing computer. Apple Computer, meanwhile, introduced its video-and-audio friendly Macintosh and third-party companies such as Avid Technology and Digidesign turned it into a powerful video- and audio-editing platform. Larger media facilities started to flounder, feeling the pinch as much as the corporate behemoths. Many corporate media studios and large postproduction factories went on the auction block, making the acquisition of traditional hardware available at ten cents on the dollar. A lot of this equipment found its way into the small project studios. I couldn't afford a Betacam camera or VTR in the late 1980s, but I had my pick in the late 1990s.

Although equipment was cheap, jobs were hard to get. Rather than specialize, project studio owners diversified, exploring every opportunity to win new clients and retain those they already had. Our own firm employed the animation capabilities of the Toaster, and in our spare time produced a series of generic animations that could be employed to present any company's logo. We created a tape with 20 of these animations and called it "McLogo." Out it went to 50 medium-size corporations and in a month we had a half-dozen orders at $2,000 each. It paid the rent.

We also converted our small conference room into a shooting stage and sent out letters to hundreds of companies offering "Video Memo," a service by which a CEO could tape a one-hour presentation and have it distributed on VHS tape to his workforce in 48 hours. At a flat rate of $2,500, we created a steady stream of managers who found our service a palatable way to preach the gospel of downsizing and cutbacks. Video Memos got us through the recession and when the budgets came back in the mid-1990s, we were one of the few firms still floating with a reputation that dated back before the debacle.

This is a lesson well learned: A generalized, small company, such as a digital media content-creation studio, can survive economic downturns better than a specialized, large company. As I write this, we have enjoyed almost ten years of economic prosperity. The temptations to grow a project studio into a medium-size company are many. Resist them.

Your friends will ask you how many people you have working for you, with the implication that more people means more success. Do not be fooled. I believe that doubling the payroll rarely, if ever, doubles the "take home." But it does triple the risk. Use freelance labor until it's impossible not to.

Your bankers will offer you cash for new-equipment purchases, real estate acquisition, and "cash flow." Do not be fooled. Loans are easy to pay off in good times, but have a funny way of becoming voracious sharks in bad times. I know. I've attended the auctions and tasted the blood of over-leveraged media firms. It tastes good.

Your family will want you to tap out some of your company's cash for the good life. That Porsche looks good and a vacation in Cancun beckons. Buy the Chevy and make a documentary on Mexican culture. Stash your cash and extend the amount of time your company can survive without a single contract coming in the door. If you've got it up to six months, you're barely solvent.

Now, after more than 20 years in the media business, I can say it's a good living. You're challenged every day and never stop learning new ways to apply your skills. You meet and serve fascinating people who often have pivotal roles in the global society. If you're a talented generalist, you will eventually become proficient in all of the media of your profession, and your project studio will flourish into a well-run, profitable machine. Here's a typical workweek from a recent "To Do" list. *Monday:* Sales calls until noon. Draw a storyboard for an animation to be produced in high-definition video. Organize translators to convert the script of a planetarium narration into German, Italian, Spanish, Japanese, Portuguese, and French. Late afternoon meeting for new business at a clothing designer. Hire ten crew people for convention work next month.

Tuesday: Edit video for a documentary about new filmmakers using digital video for theatrical distribution. Review HD storyboard with client. Debug Cold Fusion programming and write user help screens for a Web site catalog in development. Dub location footage to burn-in VHS for client review. Sales calls.

Wednesday: Write second-draft script for a medical education video on the history of human fertility enhancement. Design Flash e-mail notice for launch of Web site. Write proposal for clothing designer. Arrange studio for CEO media training session next week. Sales calls.

Thursday: Start supervising HD animation from approved storyboard. Schedule recording studio, limousine service, and refreshments for celebrity voice-over talent on planetarium show. Test new product (a VCR that records video direct to a CD) for magazine review. Cast actors for trade-show booth. Review programming changes to Web site catalog. Sales calls.

Friday: Sales calls. Receive final rewrite directions from client on fertility video script. Finish editing on digital filmmakers documentary. Send casting tapes to client. Check 3D objects and scenes from HD animation. Design and write copy for a CD jewelcase on investing in coffee options, send with artwork to CD duplicator. Take rewrite of fertility script home for weekend work.

Notice that every day, a little time is spent making sales calls. Regardless of the media, regardless of the changes in technology or the fluctuations of the economy, this part of the job never stops. Stop calling, start dying. It's the price we pay to do the work we enjoy.

Notice also that video is only part of what's being done. Using basically the same computer-based desktop tools, every job involves creating some form of digital data that eventually takes the form of a videotape, audiotape, Web file, or print on paper. What kind of work is this really? It's digital content creation.

So do you want a career in creating this business? If you're smart, you'll learn as much as you can about the trend of merging media. It's not over yet by a long shot. Before the next decade is over, digital content creators will be deeply involved in theatrical motion pictures, broadcasting, interactive media, probably some form of 3D presentational reality, possibly some new medium we haven't yet dreamed of, and of course print—in English and—most likely—several other languages. You'll distribute over the air, over wire and fiber, and by packaged media (CD-ROMs, VHS tapes, DVDs). You'll be serving both the internal and external communications needs of your clients, working on recruitment, training, benefits, motivation, reorganization, and—when the market turns down—outplacement. You'll be working to develop new products, testing consumer response, creating advertising, and marketing media. And, yes, when you least expect it, the opportunity may arise where you get to make a movie or a national spot or even something you never dreamed of, such as directing an astronaut on the Space Shuttle or an actor you idolized as a kid.

You will do all of these things because each of the industry categories in their turn will undergo the same technological revolution that we are experiencing today in the media. Media makers are the first to enjoy this revolution, but the benefits of this age will spread. Soon, the revolution will spread from the media to all other forms of production and other business categories, eventually changing our methods of work, our social customs, our language, medicine, genetics, religion, and even our perception of reality itself.

We, as digital-media content creators are really lucky to be surfing on the first wave, but most probably it will be grandchildren yet unborn who will coast to shore on the foam of this one. You, as a student or producer of digital media, have a front-row seat with two views; you are the first to see your work change and you will be the first to see—and document—your world change—forever.

Here you are at the second time in history where all media are once again in the control of creative individuals. But what, exactly, will this word "media" describe. Will it include, as in ancient times, the shaman's wide aegis of wisdom, medicine, religion, art, and entertainment?

I would say, yes. In fact, it already does.

CHAPTER TWELVE

OPENING ONLINE DOORS FOR RICH MEDIA

JON LELAND

It's a reflection of the rapid changes and the intense innovation taking place in the online world that the term "dynamic media" was even invented. Some call it "rich media," others call it "online multimedia," and others simply talk about video and animation on the Web.

No matter what you call this kind of digital content (and I use most of these terms interchangeably), because of the Web's added dimension of interactivity, "moving pictures" on the Web represent a whole new breed of communication medium.

This chapter will help you to better understand how to create dynamic media for the Internet, including:

- Big-picture trends that have created and will continue to shape video and animation on the Web
- Bottom-line variables of bandwidth and their significant impact on your delivery options
- The five major steps in the process of creating, authoring, encoding, hosting, and managing your dynamic media creations
- What works and what does not work in this new media world

MOVING BEYOND TEXT AND IMAGES

It's always useful to understand the media environment within which you are creating digital content. Of course, entire books have been written on this subject alone, and for that reason I limit this section to three "Digital MegaTrends" that are certain to continue to transform the digital media environment within which you and I create and deliver our digital content.

DIGITAL MEGATREND #1: CPU POWER IS INCREASING EXPONENTIALLY

PC aficionados call this "Moore's Law." First observed by Intel founder Gordon Moore, the astounding high-tech fact is that the power of PCs doubles about every 18 months, and these exponential power increases continue to be available at roughly the same price. This has been happening already for about 20 years!

What this means is that the processing power of your current-generation PC is beyond even the most expensive supercomputers of ten years ago, and reasonably priced personal computers will continue to grow astoundingly more and more powerful.

DIGITAL MEGATREND #2: BANDWIDTH IS INCREASING EXPONENTIALLY

This law is newer and less proven, but high-tech visionary, journalist, and author George Gilder says that the available bandwidth (explained in more detail later) is doubling every six to nine months. Whether or not this proves to be precisely true, there's no question that available bandwidth is increasing very, very rapidly; this too is very good news for digital content creators. Think of conventional commerce before and after the roads-and-highway infrastructure was built. Don't forget, the information highway is still under construction, but it won't be long before the superhighway is built—complete with an on-ramp right into your living room or office.

DIGITAL MEGATREND #3: THE NETWORK IS GETTING MUCH MORE VALUABLE

3Com founder and Ethernet inventor Robert Metcalf gets credit for observing that the "value" or "power" of a network increases in proportion to the square of the number of nodes on that network. Or as Netscape founder and Web browser inventor Marc Andreesen has said, "Every new node, every new server, every new user expands the possibilities for everyone else who's already there." In other words, as more and more people across the entire planet get themselves online, the Web becomes a more valuable and more powerful medium. This ongoing increase in the value of the Net is not just linear; it's exponential. In other words, a network of 500 people is not 10 times more valuable than a network of 50 people. It's 50 times 50, or 2,500 times more valuable—at least according to Metcalf.

These are very exciting times for digital content creators because of these trends (and because of other factors as well). You can count on it that your personal computer will continue to get more powerful, bandwidth will continue to become more abundant, and the Net will continue to offer more and more kinds of value from more and more participants. So, as you continue to learn and grow as a content creator, the environment will continue to become more and more valuable and viable.

OVERCOMING BANDWIDTH CHALLENGES

Here's an essential technical concept. It's crucial for every digital content creator to understand "bandwidth." Although it can sound to new producers like a difficult technical term, it's really quite simple. Bandwidth refers to the size of the data pipe that is being used, and online multimedia requires a fat (or fast) pipe. A 14.4 Kbps dial-up modem provides a very thin, narrow, and slow pipe; thus, only a mere trickle of data. On the other hand, a broadband connection like a cable modem or a DSL connection provides a fat pipe that is a much more comfortable channel for dynamic digital data and its associated large file sizes.

As a digital content creator, it is very important to keep the limitations of your audience's connection speed—or bandwidth—in mind. If you know that you can count on your audience having a fat pipe (perhaps because they are all in educational institutions or large corporations), then you are lucky. More commonly, digital producers either need to create multiple files for users with different bandwidths, or at minimum, create a file for low-bandwidth viewers and another one for high-bandwidth viewers. Alternately, content creators must limit the data intensity of their productions by reducing image quality to accommodate dial-up viewers. Sometimes you must do all of the above.

The technology with which digital content is modified to enable it to be delivered more efficiently and with more economical use of bandwidth is called *compression*. As you will see in the next section, digital media files that are delivered on the Net are encoded or compressed by using an assortment of *codecs*, or compression–decompression algorithms. In professional circles, compression itself has become something of a specialty; but for many content creators, the use of basic compression software tools (including Media 100's Cleaner EZ or the more professional Cleaner 5, Real Producer, and Microsoft Windows Media Encoder) make the process relatively simple, although still time-consuming.

The essential thing to understand is that there is a wide assortment of compression options available depending on how you deliver your content (DVD, CD-ROM, Internet, etc.) and in what format you want to deliver it (RealPlayer, Windows Media Player, QuickTime, Flash). Because this chapter focuses on the online delivery of dynamic media, we'll explore these options and your available choices in the following sections.

FIVE STEPS TO PRODUCING, PROCESSING, AND MANAGING DIGITAL MEDIA

CREATE AND CAPTURE

Many people are familiar with this process. You start with a digital camera, you shoot the picture, and then you transfer it to your hard drive before you can do anything with it. The same holds true for the digital video that is most commonly captured in the DV format and transferred over a FireWire (a.k.a. i.Link or IEEE 1394) port. With animation, you usually storyboard your work in the design phase and then render a digital file. In all of these cases, you are creating a *digital asset*, and you are *capturing* the raw material of digital content creation.

Edit and Author

If all you want to do is to e-mail a movie of your daughter to her grandparents, then there may be nothing to this second step. If, however, you are producing a more interactive piece of digital content, this is a critical step. The most important feature that differentiates Web delivery from more conventional media is its interactivity. Not only can your digital creation appear within an interactive Web page, it may also contain hyperlinks of its own.

Media 100 calls this kind of technology "EventStream" and they were the first to offer it via their Cleaner encoding products. Event marks that allow you to jump to chapter headings or embedded Web-site URLs that allow you to click on a movie and jump to another Web page will become more and more commonplace through a wide variety of formats (see Delivery Platforms, in the next section).

Furthermore, there will be more and more searchability of video clips and other media. So, if for example, you are a training organization and you want your potential customers to be able to find your latest program, it will help to have what is called *meta information* or *meta data* embedded in your digital media files so that they will be more accessible to Web-style searches.

Bottom line: There is more to preparing a piece of digital content than merely producing the asset. You also frequently need to edit the content for easy accessibility and author the interactive functions that you want incorporated into it. If you want viewer interaction with what you have created, then you should plan to incorporate these added dimensions prior to encoding and, to facilitate this process, have already made your delivery option and media platform decisions.

ENCODING: A STRATEGIC DECISION

The good news is that there is a full menu of delivery and encoding options, and the bad news is the same. For better or worse, the online multimedia world is still evolving and there are no real standards. Because it's valuable to have a clear perspective on your choices, I'm going to describe the two most important variables that you will have to choose from.

The first is delivery options. There are several significantly different technological approaches to dynamic-media delivery, and they are not necessarily platform-dependent. For example, the Apple QuickTime multimedia platform is capable of delivery via all three of the delivery options described here.

DECIDING ON A DELIVERY OPTION. There are three types of dynamic media delivery on the Web: downloads, progressive downloads, and streaming.

Downloads are the most basic. The file is transferred in its entirety to the user's hard drive, and she can play it from there. The advantage of this approach is that the user has complete control over the file, including the ability to forward it to others (which can raise copyright issues.) The disadvantage is that the user cannot view the file until it is completely downloaded, so the "instant gratification" factor is severely diminished.

Progressive downloads allow the user to watch the clip while it is being downloaded. QuickTime has the most successful implementation of this capability with its FastStart feature, where the timeline bar at the bottom of the video frame fills with black as the file is downloaded and played. When there is enough data downloaded to play the file or when there is a fast enough connection to continue to download while playing, the file automatically begins to play. Macromedia Flash also uses a kind of progressive download to play files or parts of files while others are being downloaded in the background.

Streaming, on the other hand, is the Web's only real-time dynamic media experience. It is used, for example, for live Webcasts, whereas neither downloads nor progressive downloads can be applied to live events. Of course, streaming can also be used for on-demand or archived material that is served from a pre-existing Web page. But unlike downloads and progressive downloads, which can be delivered from conventional Web servers, because it is more technologically sophisticated, streaming delivers significant performance advantages. It also requires a special streaming server computer with special software, because there is a high level of technical "conversation" that takes place between the streaming server and the corresponding media player. This conversation or technical communication enables more media frames to be delivered in less time and for a longer duration, although there also may be more compromises in image quality. By using a cache or buffer, when it works correctly, streaming allows for improved user instant gratification. And in today's world of diminishing attention spans, that's a good thing.

PICKING A MEDIA DELIVERY PLATFORM. Online multimedia delivery platforms are highly competitive and rapidly evolving technologies. In the consumer market, RealNetworks and its RealPlayer is the most popular. In the corporate environment, Microsoft's Windows Media player is increasingly competitive, and many creative professionals prefer QuickTime. I also like Macromedia's Flash although it does not offer a streaming option.

These platforms are sometimes referred to as *multimedia architectures*. They are software formats, each of which produces its own file type. This is reflected in the file name extension. In the same way that a JPEG graphics file has a .jpg extension and a GIF graphic has a .gif extension, a QuickTime movie usually has a .mov extension and a RealMedia file has a .rm extension. These extensions represent the media platform and, in almost all cases, files that are encoded for one platform cannot be played or streamed by the media players of another platform.

Please do not be confused. In terms of delivery, these media platforms are cross-platform (or perhaps we should say "cross-operating system"). In other words, they exist independently of computer operating systems. Thus, Windows Media files can be played on Macs, and QuickTime files can be played on Windows (although sometimes with less dependability).

There are too many variables to make a single recommendation for all situations, so I will have to limit myself to a few insights about each of these competitors.

RealNetworks is a true leader and is highly committed to the online and digital-content markets. They not only have the number-one player in the market, they also are taking a number of initiatives, including an online music service with three of the five major record labels (as of press time), Real Jukebox, and a searchable online MLB baseball highlight service. Their specialty is streaming, and they do that very well within a very sophisticated technical architecture.

Windows Media has been playing catch-up, but in typical Microsoft fashion, it is doing it well. The quality of images produced by its recent codecs is excellent, and many people (especially corporate users) are comfortable with the Microsoft name. Microsoft has recently recognized that it needs to be a major player in dynamic media; thus it has been investing considerable funds in the research and development of its media technologies, and the quality continues to improve. Its biggest weakness may be continued incompatibilities on the Mac platform (which it claims it will overcome soon).

Apple's QuickTime, in my opinion, has perhaps the most elegant multimedia architecture, and does the best job of progressive downloads. On the other hand, the performance of its streaming option has been inconsistent at best and has not received popular support. So, I'd go elsewhere for a streaming solution. Because QuickTime also does the best job of incorporating multiple media types, however, it is very robust if you want to combine video with interactivity, Flash vector graphics, text tracks in multiple languages, or that kind of thing.

Macromedia Flash is my personal dark-horse candidate, but it doesn't do streaming at all. Flash, however, is exceptional in its ability to use vector graphics to give maximum screen impact at minimum bandwidths, and it also has a very sophisticated programming language (ActiveScript) that makes it extremely capable when incorporating complex forms of interactivity. Flash has also recently gained the ability to incorporate video clips in its native (.swf) format via a tool called FLIX from Wildform.com.

Obviously, the choices are yours. I recommend that you experiment and view various productions in various formats to see what you like best. Then learn your way around your chosen media platform. Or alternately, if you have the resources, you may want to offer viewers alternatives. For example, encode for multiple bandwidths of both the Real and Windows Media players, and offer a progressive download via QuickTime and/or Flash as well.

Host and Distribute

Web distribution is different from e-mail distribution. Unless you have prior permission, don't attach large files to e-mails; rather, post them on your Web server or link to them on a streaming server.

If you are offering downloads or progressive downloads, any Web server will work; but to do streaming right, you need a server computer with a special streaming server installed on it.

You can either set this server up on your own (which requires considerable technical expertise) or you can use a hosting service. For example, I use HostPro.com, which offers a shared ten-stream Real server with its $40/month UNIX server package. This kind of hosting service provider then becomes your Web ISP. You then upload your encoded files to it so that your users can access your files on the Web.

Even easier alternatives are application service provider (ASP) services like Popcast.com. These are very easy-to-use services with features that are controlled through a Web-based interface. These features allow you to upload and link to streaming files on the ASP's servers without any detailed knowledge of Web development. The downside is that you will have very limited options about how your video (or other forms of dynamic media) will appear on your Web site.

Managing Your Digital Assets

Finally, take good care of your digital media assets. Although a whole subindustry has grown up around commercial software that can help large entities organize their digital files (see emotion.com, for an example of a digital media asset management software company), on a smaller scale, basic common sense is still the key skill that you need.

Here's what to do: Keep your project files well organized and ALWAYS remember to back-up. A little bit of preplanning goes a long way. Also, develop a system for naming your files so that you can find them more easily. For instance, designate that all files for a particular project should begin with a very short three-letter project name or project code. Then, each time you create a new file, carefully place it within an overall project folder. Within the project folder, create subfolders for the various kinds of assets (like Photoshop files, audio files, text documents, etc.). In this way, if you save everything to its appropriate file, finding things will be much easier. The more organized you are as you create, the easier your project will be to go back to if you want to reuse assets or make changes.

In terms of back-ups, the ease with which we can all "burn" CD-ROMs now is a wonderful blessing. When you finish a project, if the files aren't too big, why not burn a CD immediately with all of the source, as well as finished assets, on it for easy future access? If your files are too big for CDs, then you must make other plans for back-ups (for example, with tape back-up drives), otherwise you may find yourself in an impossible position in the future when Murphy's Law inevitably kicks in.

SUCCESSFUL DYNAMIC MEDIA: WHAT WORKS AND WHAT DOES NOT

Here are a few basic guidelines to remember. If you think that channel-surfing TV watchers have a short attention span, Web surfers are worse. So, shorter is always better, especially if you are hoping to catch and hold someone's attention as they browse randomly through your Web site. Be clear and concise as well as creative.

Be bandwidth-friendly. If you are creating something for a broad audience, especially outside of educational and corporate institutions, remember that most people are still using dial-up connections. Sure, there is plenty of additional bandwidth coming down the pipe in the future, but for now, for broad audiences, be bandwidth-conscious. Your audiences will appreciate that.

"Eye candy" and unfocused communications do not work with dynamic media on the Web. I'm continually amazed at how little common sense some digital content creators display.

Eye candy is a broadcast term for animation for its own sake. Like sugar, it has no real nutritional value. Likewise, no one likes to hear incoherent babble; just because you can post something on the Web doesn't mean people will want to watch or interact with it.

For these reasons, it's always best to keep your communication basics in mind. Focus on your objectives. Make sure that you are clear about the purpose of the content you are creating and the audience for which it is intended. By hitting the center of the bull's eye of your target, you will deliver satisfied viewers and make the Web a better place for dynamic media.

I'm wishing happy and successful digital content creation to you all.

CHAPTER THIRTEEN

RICH MEDIA AND LIVE STREAMING MEDIA

AL KOVALICK

Unless you're Rip Van Winkle, it's very likely that you have been affected by Web-streaming media. The Web is being transformed from a text and graphics medium into one that includes integrated audio/video content. Because streaming media is the key to the future of Web-based communications, this chapter concisely reviews the basics of streaming media technology that content creators need to know.

WHAT IS STREAMING MEDIA?

Did you watch the evening news on TV last night? If you did, you experienced live streaming media. In the context of a traditional broadcast, a TV signal received and displayed constitutes *streaming media*. The concept is not new. What is new is how streaming media is deployed across the Internet/intranet framework. Streaming involves sending multimedia content from a media server to a client over a network such as the Internet. A streaming server emits the program as data packets and feeds them into the network. At the receiving end, the client reassembles the packets and the program is played as it is received. A series of these related packets is called a *stream*.

In a way, the Internet offers broadcasters and others a "new antenna" for reaching millions of untapped viewers. It's unlikely that streaming media will replace traditional television—but it will augment it. Many new and exciting streaming-media applications will be introduced over the next few years. In addition, stations may reap new revenue from T-commerce–related enhancements. Of course, streaming media isn't just a broadcast-related technology. Opportunities are exploding in almost every area of video media viewing.

Today, Webcasting is still in its infancy. (In this chapter, "Webcasting" and "streaming" are treated synonymously. A Webcast is an event, like a broadcast. A "stream" is the fashion in which the Web-encoded content reaches the viewing clients.) Some speak of Webcasting with contempt: "It's a toy with small, jerky images," they'll say. With the advent of broadband networking to the home, however, quality content is starting to flow. Various research sources have estimated that between 20 and 30 percent of U.S. homes will have broadband Internet access (of greater than 384 Kbps) in the next few years. This will change everything.

Content may be king, but bandwidth and reach are the king's best friends. In the corporate intranet environment, broadband access is readily available, and many businesses are using Webcasting for communications every day. Don't bet against bandwidth. As for content, the better it is going into a Webcast, the better it will look coming out. (See a later section "Eight Steps to Optimum Live Streaming Media," for suggestions on improving the A/V [audio/video] quality of Webcasts.)

STREAMING MEDIA CONCEPTS

Let's go way back in time (in Internet terms), to about four years ago when the Web was young. How was media distributed to end users (their PCs, that is)? Simple, just download the requested content as a file, decode it, and view it on the client station. If you have ever done this you know it's a painfully slow way to consume media. Sure, it has a place, but users often want to view the media without waiting for the download to complete. In addition, users desire immediate control over the position in the media (fast-forward, rewind, pause...) Enter streaming media.

Figure 13.1 gives a general model for this discussion. On the left side, media is captured, edited, or created. These are usually traditional solutions with the exception that the final display resolution (i.e., small video windows) should be accounted for. Once the A/V content is ready for distribution, it needs to be "encoded" for the Web. What does this mean?

ENCODING

The trick is to compress the original A/V content so that it may be streamed to users over narrow- or broadband pipes. The choice of encoding methods can be bewildering. Microsoft (Windows Media Technology[1]), Real Networks (RealVideo, RealAudio), Apple Computer (QuickTime 4-based[2]), and others offer proprietary solutions for encoding, streaming, and

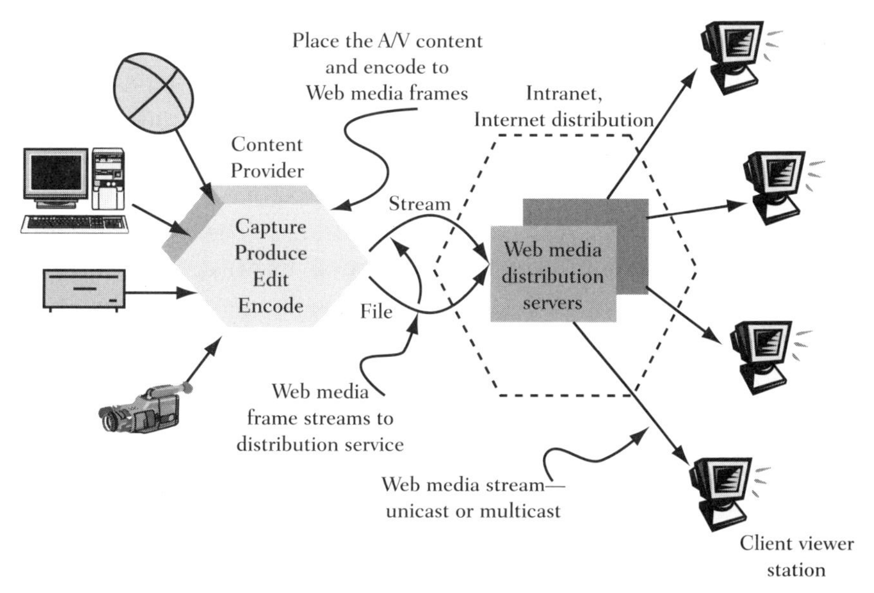

FIGURE 13.1 A general model for Webcasting.

client viewing. In addition, some vendors are starting to offer products that are 100 percent pure MPEG-4 compliant. Typical video encode rates range from 20 to 300 Kbps, with 1 Mbps and above on the horizon. Audio and video encoders are quite distinct in function.

The final encoded material is optionally multiplexed with text, graphics, and scripts and then placed in a file structure that is *streamable*. That is, the file has time stamps and other features that make it easy to stream over the fussy Internet and then decode on the client side.

Encoding is a content-creation art that should account for scalability across different encode rates, tolerance to packet loss and bandwidth fluctuations in the Web, good A/V quality at the lowest rates, costing of the encoding/streaming solution, stream control, and other aspects. The jury is still out as to the "best" solution. Pinnacle Systems, a company offering products for streaming, supports multiple encoding schemes in its streaming solutions.

STREAMING

Once the media is ready to be streamed to users, a Web media server is usually needed to perform the streaming function. It meters out the streams to the end users. Web media servers are also proprietary and strongly linked to the encoding method. Most contemporary servers are software-based and run on standard NT/W2K, UNIX, or Linux platforms. Anywhere from a few to thousands of streams may be served per platform. With a distributed server, millions of streams may be served. As you might expect, the content is streamed over Internet Protocol (IP)-based networks.

Where does the Web media server get its streamable content? There are two chief ways. One is for the content provider to file-transfer the Web-encoded material into the server. This is typically done for nonlive, VOD programming. For live content, the provider sends a single master stream to the Web server, and it distributes it as requested.

We often hear of the so-called "last-mile" problem regarding the available bandwidth to viewing clients. There is also a first-mile problem related to the connection bandwidth and quality of service between the content provider and the Web media server. This path needs to be very stable and provide low packet loss, because if the main feed to the server is faulty, then *all* viewers receive poor reception. Don't ignore this link when planning a Webcast.

An intranet-based server is a captive, private device and is usually managed by an IT department. Access and provisioning for these servers is normally a straightforward task if you have the rights. At Pinnacle Systems, we routinely use intranet Web media servers to broadcast company meetings to employees worldwide.

On the other hand, Internet-based media servers are available for public hire. An Applications Service Provider (ASP) offers to sell clients the right to stream using their Web-connected media servers. For a price, the ASP will usually guarantee to:

- Serve a maximum number of streams (e.g., 250 streams max per Webcast)
- Serve at specified rates (e.g., 50- and 200-Kbps encode rates)
- Serve with selected formats (e.g., Windows Media)
- Webcast time window (e.g., 2–4 p.m. on August 24, 2002)
- Store the content for future access (e.g., 1 GB of storage)

One of the challenges in using publicly provided servers is the provisioning step. How does one set up the Webcast? How is provisioning of the features accomplished? Who do I call when there is a problem? Companies are devising various solutions for such questions; Pinnacle Systems offers CastConnect, which is the result of a coordinated effort with leading ASPs to make sure that their customers' Web-casting experience is as easy as possible. By accessing Pinnacle Systems' CastConnect.com portal, users are able to register, pay, schedule, and provision their Webcast from one to thousands of viewers. CastConnect.com gives users the means to provision Webcasts anywhere, anytime, without weeks of planning. The goal is to take the hassle out of Webcasting—to focus on the Webcast without worrying about provisioning or stream load-balancing.

CLIENT VIEWER

Here's where the streams are consumed. The major streaming-media product vendors have each designed PC client-viewing environments. Besides the obvious A/V decoding and displaying, the viewing frames have browsing and stream-control buttons. No doubt, the interactive control of the streams is a major enhancement over traditional broadcast TV. In addition, the streams often have embedded text or graphics, producing a truly rich viewer experience. The Web we encounter today is the worst it will ever be (let's hope). As connection rates improve, so will the overall viewing experience.

STREAMING MEDIA DELIVERY OPTIONS

Streaming media is finding applications in every imaginable area. One way to categorize the delivery landscape is shown in Figure 13.2. The choice of delivery mode is made, to an extent, based on the type of application that must be supported. For example, if video conferencing is the goal, then the streaming-delivery infrastructure should support it.

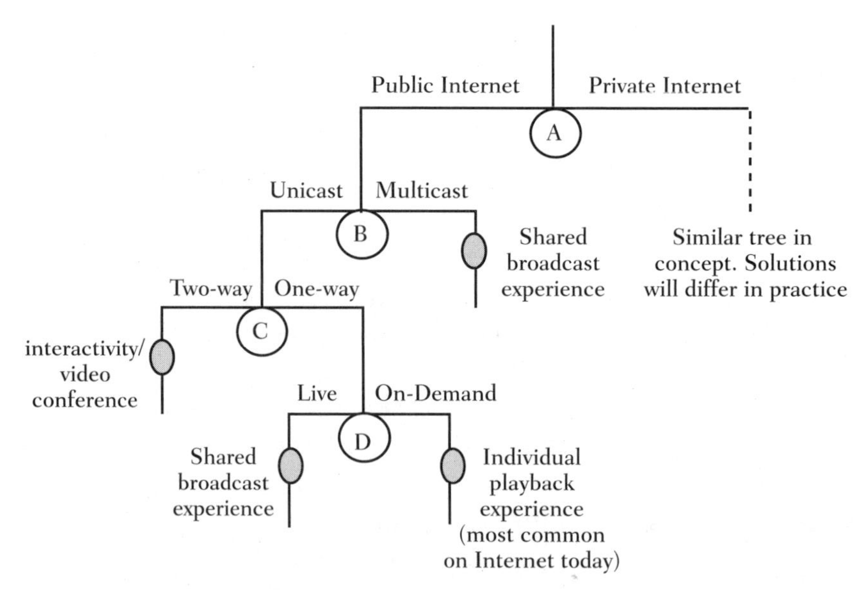

FIGURE 13.2 Streaming media distribution methods.

What are the choices that a Webcaster has? The Figure 13.2 tree shows four distinct branches—A, B, C, and D. Let's examine each choice.

THE "A" CHOICE

Of course, there are streaming applications within both private intranets and the (public) Internet. Whether you use one or the other depends to an extent on who and where your audience is. If your audience is within the boundary of a company's private network, then use that network. It will be easier to provision, secure, and control your entire Webcast environment.

THE "B" CHOICE

A TV station transmits in the *multicast* (one-to-many) mode. All viewers receive the same, undifferentiated content. Over the Web, however, *unicast* (one-to-one) is the most prevalent method. Each end-user gets differentiated content. Unicast is actually an easier paradigm to support for IP-based networks. IP routers and network equipment must have special protocols enabled to support multicast.

So, why use multicast? Multicast saves network resources when many end users receive the same signal. A unicast equivalent in the TV broadcast world would require a separate RF channel per end viewer! That would be crazy and a huge waste of precious bandwidth. Multicast is more common in private intranets than in public Internets. Some Internet ISPs do, however, support it. It will be many years, if ever, before the great bulk of the Internet supports general-purpose multicasting. So, "use it if you can" is a practical motto.

THE "C" CHOICE

All Web-based communication is inherently two-way (click and you shall receive). So why the choice at this branch? Actually, this choice is for two-way A/V, or at the least two-way audio only. The most obvious candidate for two-way communications is video conferencing.

One problem with Web-streaming media is delivery *latency*. Streaming has built-in buffering delays in the Web media server and at the client viewing stations. If you have familiarity with streaming viewers, no doubt you've seen the message, BUFFERING STREAM, PLEASE WAIT. A typical latency from the time a new stream is requested until it is played out is at least 10 seconds for most deployments. This delay may be mitigated, but it is usually present.

You may be wondering, "Why build in the latency?" If a received streaming data packet is detected as bad (or lost), the underlying protocols have time to retrieve it again from the server prior to it being needed by A/V decoding and display logic. It's a trick that improves viewing quality but at the expense of adding viewing delay.

THE "D" CHOICE

This one is easy to explain. Some content is live and some is prestored on the Web media server for access on demand. If the stream is live, all viewers receive the same program. An on-demand experience allows for fast-forward, rewind, and pause control of the stream.

STREAMING MEDIA PRODUCTION CHAINS

Figure 13.3 is an expanded version of Figure 13.1; it shows four methods that produce content that is Web-cast ready. This is not meant to be an exhaustive list but it does show some representative product solutions that Pinnacle Systems has announced.

CONTENT PRODUCTION WITH EDIT STATIONS—*PUBLISH TO THE WEB*

Here's the place to create original content (or repurpose content) for the Web. Once an edit is complete, users may want to encode the finished program to Web media within the edit environment. Typically, there is little demand for real-time encoding. A non–realtime encode process is adequate. Some encoders

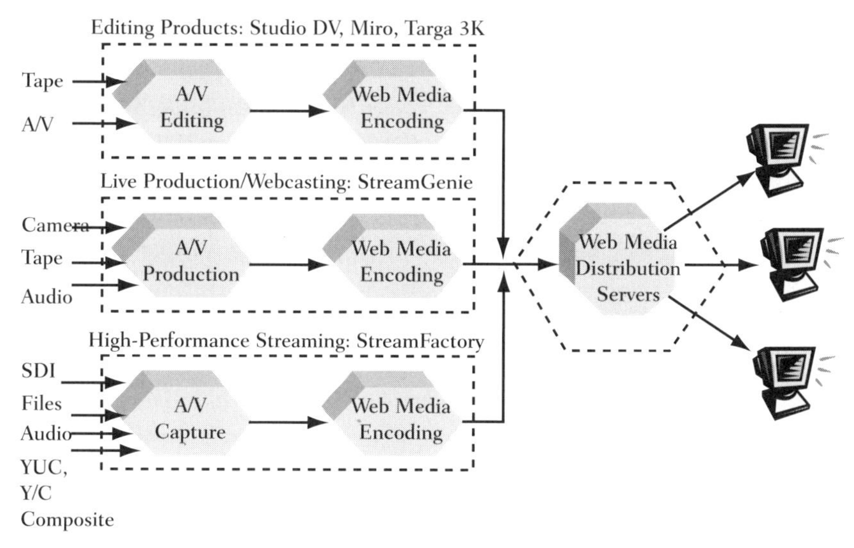

FIGURE 13.3 An example of streaming media chains.

provide for a batch mode, where many files are encoded to one or more Web-media formats. Once the content has been encoded, it's ready to be transferred to the Web media server of choice. This operation is usually a simple file-upload process. Most of Pinnacle's content creation (editing) products, such as Studio DV and DV500 products, offer Web-media encoding.

Pinnacle's StreamGenie (and StreamGenie Presenter) is a truly portable, total solution for real-time, multicamera Webcasting. A product such as this is great for corporate presentations, conferences, news, distance learning and education, event videography, and live e-commerce. The idea is to break the complexity barrier to Webcasting by allowing users to connect cameras and microphones, and then use the on-screen GUI to produce a Webcast.

Because I'm plugging Pinnacle Systems' products, let me describe one more. We've taken a lot of time to make Webcasting simple for content creators, and we believe that the products mentioned are establishing the paradigm for this form of communication. The StreamFactory family of encoders is designed to provide everything necessary to broadcast content over the Internet in real time, with premium-quality video and audio. StreamFactory is a plug-and-play solution in a space-saving 1-RU (rack-unit) package. Simply connect a VTR, video camera, or other A/V source and convert to popular streaming formats in real time. With future options, you can also convert compressed digital video files directly to Internet-streaming formats. The aim is to provide a powerful, reliable, and cost-effective way to stream high-quality live content to the Web or to convert archived content for on-demand viewing.

StreamFactory accepts all popular video and audio input formats, including your choice of serial digital (SDI), YUV, Y/C, DV (IEEE-1394), and composite video. StreamFactory also imports media files via its integral 10/100 Mb Ethernet port. For crystal-clear audio, StreamFactory features balanced stereo analog audio via twin XLR connectors, plus digital AES/EBU audio and embedded SDI audio; a good feature for all Internet broadcast facilities. In addition, integrated support for Real Networks SureStream or Microsoft Windows Media allow users to stream multiple data rates concurrently.

EIGHT STEPS TO OPTIMUM LIVE STREAMING MEDIA

You've created your content and are ready to Webcast it. But just as a chain is only as strong as its weakest link, that's also true of the "chain" of connections between you and your audience. This section outlines suggestions for improving the A/V quality of Webcasts, using eight principal links in the streaming-delivery chain. By optimizing each of these, end-users will see the best possible picture, limited only by their bandwidth. What's the best encoding format to stream to end-users? What encoding settings can maximize your output quality? And what other parameters can impact the end quality of content? For these and other answers, read on. (Although we focus on live Webcasting, most of the discussion also applies to streaming-on-demand applications.)

QUALITY METRICS

What are the measures of Webcast quality? As we all know, "Beauty is in the eye [or ear] of the beholder"; the perceived quality of your streamed content is a strong function of how viewers judge it. One viewer may enjoy a small video display with unnoticeable video artifacts whereas another may want to view the same material at full screen at the expense of some observable artifacts.

AUDIO QUALITY. Because "half of what one sees is what one hears," audio quality is paramount. Fortunately, the quality measure of audio is less complex than that of video. In general, if the audio has few if any artifacts (noise, garbling, poor frequency response) then most users are satisfied. Of course, there is no end to improving audio quality in fine detail. If, however, a broadband Internet delivery channel can support an audio portion of ~16 Kbps (stereo encoder, compression rate), end users are satisfied for most current applications. Usually, a streaming A/V encoder gives preference to audio over video in terms of bandwidth utilization.

As broadband rates improve, audio artifacts should become history. Audio codecs are very mature, and excellent versions are widely available.

VIDEO QUALITY. End-viewer video quality is influenced by these five measures:

- *Image (spatial) quality (IQ).* Faithfulness to the original, sharpness of detail, color retention, lack of artifacts in a video frame.
- *Temporal quality (TQ).* The more frames per second, the better motion may be rendered. Streaming frame rates range from greater than 1 fps (frame per second) to 30 fps.
- *Encoded image size.* Viewers generally prefer large images. Sometimes a larger displayed image comes at the cost of more image artifacts. The encoded image size (640 × 480 HxV) is always less than the original source size (~720 by 480, for NTSC), so the entire image is resampled before the encoding operation.
- *Encode bit rate.* The higher the available bit rate, the better IQ and TQ may be. The streamed delivery rate should be greater than or equal to the encode rate.

- *Packet loss and jitter.* As packets traverse the Internet they may encounter congestion. Packet loss and jitter degrades the end A/V display quality.

One way to visualize four of these quality measures is shown in Figure 13.4. Image quality, frame rate, image size, and encode rate are related in the cubelike total quality space. For any given video encode rate (say 80 Kbps), there is a trade-off between image size, image quality (IQ), and frame rate (TQ). The video encoder decides how to spend its precious bits across the different parameters. The more bits available, the better the IQ, or TQ, size, or all parameters.

But which is more important? How should the bits be spent? Consider a thin slice taken from the total quality space. The square in the lower right of Figure 13.4 shows such a slice. Let's assume that it is a slice from encode bit rate R1 = 80 Kbps. Assume, too, that the media encoder input is a series of frames measuring 640 × 480 pixels and the temporal rate is 2 fps (original source is 30 fps, so 1 per 15 frames is chosen). So, about 40 Kb are used per frame to encode the material. If we

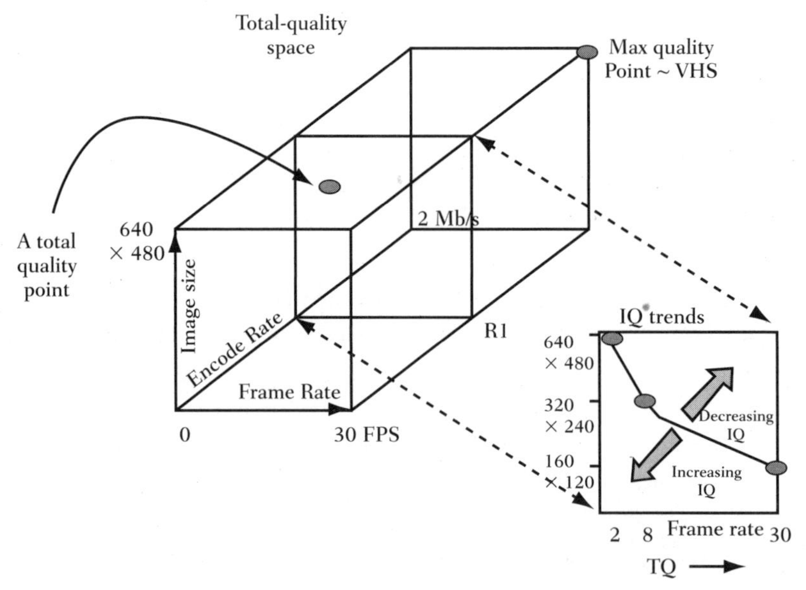

FIGURE 13.4 What is steaming quality?

change the encoding rules so that the entire image size is first decimated to 320 × 240 and a frame rate of 8 fps is used, then we have 10 Kb to spend per frame, but each frame is one quarter of the size of the first case.

The IQ per frame remains about the same for both cases. So, it's possible to draw a constant IQ contour curve in the square. As shown, the IQ can be increased or decreased by choosing different parameters. More encode bits per second moves the slice backward in the cube. Also, as the decimated image size continues to shrink (160 × 120), the image quality will suffer because of the heavy downsizing before encoding starts.

By adjusting encoding parameters, users may place the total quality point in the cube at their discretion. It's easy to waste precious encode bits; judicious choice of parameters puts the available bits to better use. Figure 13.4 is an ideal view of quality; real-world encoders do not allow for perfectly setting IQ or TQ. Also, because of the nature of encoding algorithms, coded material does not have a fixed number of bits per frame. (Incidentally, the cube point in Figure 13.4 labeled "max quality point" is about equivalent to VHS quality with a ~2 Mbps encoding rate.)

The Quality Delivery Chain for Live Streaming of Web Media

Figure 13.5 is the basis of further discussion. There are eight links in the delivery chain from original live content to delivered and displayed content.

Links 1 through 3 are discussed in relation to Pinnacle Systems' StreamGenie and StreamFactory products for live streaming (www.pinnaclesys.com), but these principles apply to other products of a similar nature. Let's now examine each link and outline what it takes to optimize their individual performance. Ideally, if all links are optimized, the client viewable A/V quality is at peak.

Link 1: Input-Capture Devices. Capture devices and production tips:

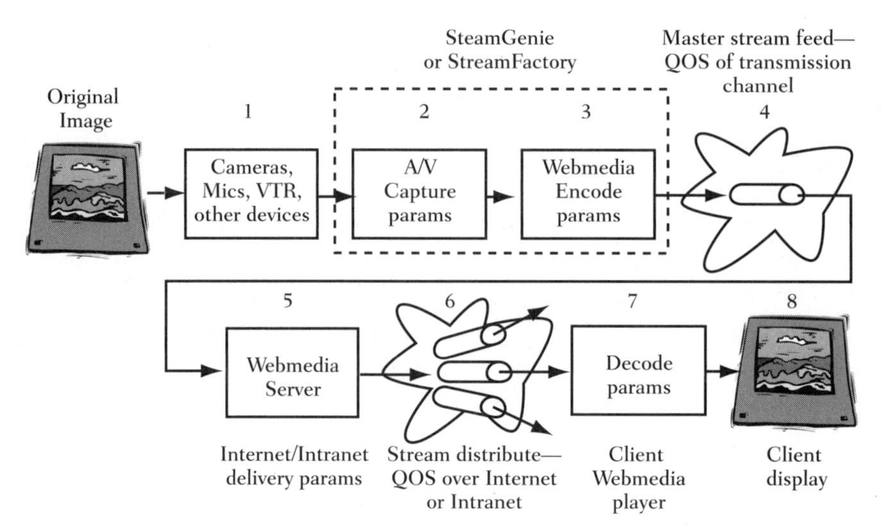

FIGURE 13.5 The quality delivery chain.

- If you captured your original material on a small DV-format (or similar camera) use image stabilization when shooting.
- Use a tripod to reduce shake and unintentional motion.
- An S-video output provides better quality than a composite output.
- Use as good a format as possible (DV VCR, a DVD) for playback of external material.
- Use a good microphones—wired or wireless.
- Optional use of external audio processing improves quality (noise gates, limiter, gain compressor).
- Use balanced audio links when possible to reduce noise, hum.
- Use tight shots when shooting people and keep motion to a minimum.
- Straight cuts are better than fancy effects.
- Graphics should total one quarter of the content.
- Keep your program brief; under 10 minutes.

The worst enemy of A/V codecs is noise. Video compression algorithms remove correlated spatial and temporal detail, and noise wastes bits. Take care to reduce the amount of noise of any kind. What helps? Good-quality cameras and sources, plenty of lighting, reducing the number of active mics, and/or using automatic noise gating. An event environment with low levels of background sounds translates to audible garbled noise at the end client. Also, unintentional image jitter reduces IQ and TQ. Aimless source movement wastes valuable encode bits.

LINKS 2 AND 3: A/V CAPTURE PARAMETERS AND WEB MEDIA ENCODING. Here we examine some present guidelines and best practices for the operational settings of capture cards and web media codecs. (Pinnacle's StreamGenie and StreamFactory are typical devices that perform the operations of Links 2 and 3. This discussion assumes that the Web media content is in either Microsoft or Real Networks format, but the same basic reasoning applies to MPEG or other compression formats.) Figure 13.6 shows how the various capture and encoding parameters affect the final encoded content.

When adjusting parameters for Link 2, select the image size depending on the total quality desired. The image size is also adjustable in Link 3 for most codecs. If possible, the value set for Links 2 and 3 should be the same. If the sizes differ, then

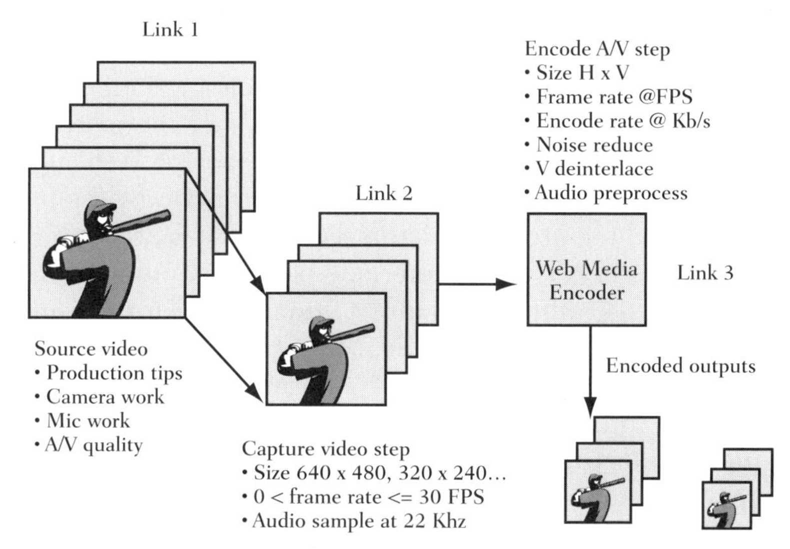

FIGURE 13.6 The video capture and encoding process.

the encoder will resize (most encoders are software-based and CPU power is wasted) the image before encoding. So why have two size adjustments? The Microsoft Web media codec can encode an identical source to multiple simultaneous streams, each at different bit rates. This is called *Intelligent Streaming*. Real Networks' version of this concept is called *SureStream*. For a given encode format, each encoded stream may (depending on the codec chosen) have a different image size and frame rate. Usually the stream with the largest encode rate has the biggest size.

Make the frame rate in Link 2 as large as desired, keeping in mind the total image quality tradeoffs discussed. Link 3 may, however, have a frame rate adjustment too (codec dependent). The Link 3 frame rate adjustment should be equal to the Link 2 value, if possible. The SureStream and Intelligent Streaming modes decimate the frame rates as needed when generating multiple simultaneous streams.

Some capture cards support *deinterlacing* in hardware, which is discussed next. Also, some Link 2 steps include noise reduction and audio-processing functions.

Link 3 usually is the software-based media encoder portion of the chain. Both the Microsoft and Real codecs offer slightly different encode-parameter choices, and as with most quality issues, you must decide on the settings, because there is no perfect tradeoff. Additionally, both codecs drop frames (lowering the TQ) to achieve better IQ when source motion is high. Usually, the smaller the allocated encode bit rate the more variation in delivered frame rate should be expected. Web media codecs usually offer a variety of preprocess functions, the most important of which are video deinterlace and noise reduction.

Both PAL and NTSC are interlace-based scanning formats; PC displays (which the typical target end user has for a display) use *progressive scanning*. The ideal input to a codec is a progressive format, because progressive formats encode to better quality than interlace formats, especially if the source has motion. So, selecting the deinterlace option is important. What is the penalty? CPU cycles are consumed.

Noise reduction allows for more encode bits to be allocated to the signal and not the noise. Noise reduction is best when performed in hardware. Software-based noise-reduction wastes CPU power that is better spent in the actual encoding of the web media.

So, which codec is better, Real's or Microsoft's for Link 3? This is a "religious" question with no clear answer. It is not just a matter of A/V content quality: Selection also depends on media server platform choice (Microsoft's media server only runs under NT or Windows 2000) and end-client platform type. Many subtle factors make the choice a difficult one.

Link 4: The Master Stream. Once a stream has been encoded, it must be sent to the Web media server for general distribution. Both Microsoft and Real provide Web-media servers. On the surface this sounds like a simple proposition, but the problem lies in the quality of service (QOS) of the transmission in Link 4. If enough packets are dropped on Link 4, then all end clients receive poor audio and video quality. Therefore, link QOS must be excellent, which implies that a general Internet connection may not be sufficient because IP packet loss is not uncommon. In fact, most professional event Webcasters use the following types of Link 4:

- ISDN at 128 Kbps (excellent QOS, good pricing, fair data rate)
- Dual ISDN links, inverse mux mode to achieve 256 Kbps
- T1 or fractional-T1 link (excellent QOS, but pricey)
- Satellite link
- Corporate LAN access
- Internet access over DSL or cable modem (least desired of all choices)

Each link has its own advantages and disadvantages. These choices are listed in descending order of their approximate suitability for live Webcasts. (Of course, the link must have access to your Web-media server of choice. All major Web-media server providers (Yahoo! Digital Island, Globix, etc.) offer T1 and/or ISDN access for this link.)

Link 5: The Web-Media Server. Web-media servers are mature software packages that run on NT/Windows 2000,

Solaris, Linux, and other platforms. If the server is under control of an IT department (intranet), then you can configure it to meet your needs. If the server is in the Internet, then you will need a contract to use it. Provisioning Internet servers can be a complex and time-consuming task. (Pinnacle Systems offers instant Internet Web-Media Server provisioning using our www.castconnect.com portal.)

So what parameters affect streaming quality? For one, server capacity. It should be able to serve up to the expected number of clients (say, 1,000 max) and not drop any logged-on clients under any loading situation. The server should have an adequate streaming connection (DS3, OC3) to the Internet to avoid packet loss under heavy loading. The actual hardware platform should be robust and have a guarantee of service time. You must discuss your specific needs with a service provider.

LINK 6: INTERNET DISTRIBUTION. The Web-media server distributes streams to clients throughout the Internet or an intranet. Two issues to deal with here are quality of services (QOS) and access speed. The QOS of the Internet is an ever-changing and debatable factor. But as DSL and cable modem access spreads to the masses, the dream of broadband will become reality. Many market studies predict that more than 15 percent of the U.S. homes will have installed greater than 384 Kbps access by 2003. This will push streaming media into the everyday lives of tens of millions of people. (Of course, the higher the bandwidth the better the final viewed quality.)

To address quality problems, if there is unrecoverable packet loss (due to congestion, usually), client players try to conceal the loss by repeating previous frames and using other strategies. Prolonged congestion can cause a player to hang or disconnect from the stream. Future player releases will be more forgiving under adverse conditions.

LINK 7: CLIENT PLAYER. The client player decodes the received stream. Both Microsoft (Windows Media Player) and Real Networks (Real Player) are common players that do not have many adjustable parameters. The following should, however, be noted:

- Screen size is selectable, but the bigger the screen the more artifacts will be apparent.
- Some players artificially limit the stream-decoding bit rate. If you suspect this, look for user controls that raise the minimum stream-decoding rate.
- All players have a "statistics" screen that monitors the decoded frame rate and bit rate.

You may find these measures useful when experimenting with various encoding parameters. There is another not-so-obvious issue regarding end-to-end delay. For reasons of packet-loss recovery and media-server optimization, there are stream buffer delays along the chain. In fact delay can add up to over 25 seconds from end to end. For this reason, traditional video conferencing is not practical with current technology. It is not inevitable that the delay be so large, but reducing it to less than one second is not an easy prospect; for the near term, Webcast applications must account for this delay.

LINK 8: CLIENT DISPLAY. How faithfully does the display render the decoded video? It depends on the age of the monitor and how clean the RGB chain is. One of the most common problems with monitors is the inaccuracy of displayed colors. In practice, a Webcaster has little control over the end-user monitor.

SUMMARY

When all is said and done, what's important to remember is that the world of streaming media is set to explode. The new media will influence many aspects of traditional video content creation. The basics covered in this chapter should give you a jump-start into this new world.

Webcasting your content is not a trivial matter. There are "gotchas" aplenty, and hopefully this chapter will help you avoid them. Forewarned is forearmed: By applying the simple caveats and suggestions described, your next Webcast should be first class.

FOOTNOTES

1. Although Windows Media Technology offers a compliant MPEG4 video encoder, its media multiplexing file structure is not compliant to MPEG4 standards. The multiplexing and streaming format is called *Advanced Streaming Format* (ASF).
2. QuickTime 4 supports a variety of A/V codecs, but the Sorenson Video codec is widely used.

CHAPTER FOURTEEN

THE IMPORTANCE OF WEB-SITE DESIGN

NICOLA GODWIN

Web-site design is changing to meet the evolving demands of the Internet's ever-increasing legion of users, as well as to take account of such elements as 3D and video, which are increasingly commonplace.

The Web continues to grow steadily as a vehicle for media-rich content, and the technology that supports it is at least keeping pace. But despite intense competition for the attention of Web users, it's amazing how many really ugly-looking sites exist—sites where little or no thought has been given to presentation, where the design was obviously a purely technical exercise in construction rather than a creative endeavor of any kind. As proven in so many other areas (magazines, packaging, advertising), design is a crucial key to attracting readers, buyers, consumers. For the purposes of this chapter I've spoken to several Web designers who make their living designing the type of "sticky" Web sites that not only stand out in the crowd, but also attract users time and time again (no mean feat given the sheer number of sites out there). I've also considered interactive elements to see what impact they are having on the realm of design.

ART AND SCIENCE COMBINED

Creating effective Web sites is part art and part science. It draws on some of the fundamentals of graphic design while also requiring an increasingly broad grasp of underlying technologies. Both aspects are integral to the creation of a good Web site.

"Strong Web design is a combination of several things," asserts Steve Wilkie, director of design at DesignerCity, a new-media facility based in London. The ideal, he maintains, is to create a site that looks good but is nevertheless easy to access and simple to use. "With the tools available today," Wilkie continues, "there's no end to the levels of complexity you can add to a site; it's very tempting, but there's no point unless the visitor can understand and make use of it." In his opinion, a well-designed site makes use of design elements to guide the user through his or her interaction, so that, although design is aesthetic at one level, it remains entirely functional at another.

Deb George is director of DW Design, a Web-design facility also based in London. She agrees with Wilkie on the key issue to consider when designing effective Web sites:

"Personally I think the most important thing is to keep it simple and intuitive," George says. "Remember that even though it might take some fairly advanced technology to make it work, at the end of the day you're dealing with people."

The amount of time spent on most Web sites is woefully short, which is why George places such emphasis on her next point. "Make sure people know they are at the right site straight away," she says. "Make the logo or brand really obvious and then repeat it throughout the site." This latter tip is especially good advice when you consider that users may find their way to your Web site by virtue of a search engine and land on page 10. How will they know it's the site for them if proper identification is only present on your home page?

"The other really important thing is to make sure users can get to where they want to quickly and easily," George continues. "It's essential that the navigation is clear because the average person's attention span wanes pretty rapidly, and if they feel like they are getting nowhere fast there are usually plenty of other sites offering similar information—people won't hesitate too long before heading off to find them."

Like all other businesses, cyberspace is a "dog-eat-dog" world. As Wilkie puts it: "The Internet is still at a stage where the graphical style is secondary to the key purpose of delivering content. But increasingly, the way a site looks is playing an important role in attracting people's attention and making them feel that this is the site for them, that the style of the site somehow suits them."

So aesthetics is firmly in the service of functionality. Combine this with a dash of flair to attract people's attention in the first place and—presto!—an effective Web experience. But what do the experts have to say about bad design?

"Bad Web design can include so many things," Wilkie says. "It's very easy when constructing a Web site to lose sight of the most important element—the user. There are so many ways to build a site—will it be static, dynamic, all Flash, no Flash? Will it cater for all browsers or just one? Will it be for high bandwidth only? Without someone carefully focusing on the users of a site, their needs, and what the typical connection speed is that they have, the site design will undoubtedly fail. The most important question a site designer can ask is: Why am I adding this element? Is this the best way to build it or am I adding it simply because I want to?"

Deb George lists her top-three design no-no's: "Too much text, large files that don't download quickly, and bad navigation. Also be wary of having too many pointless effects just for show. Remember, even though it's nice to have flashy buttons and things popping up here and there, you can overdo it, and on the whole it doesn't really serve a purpose. The novelty value can wear off really quickly." This is a good point, given that Web site creation is dominated by a handful of popular packages—when you get the latest upgrade with some cool new feature, remember that thousands of other designers have it, too.

THE TECHNOLOGY

So what of the technology? Are all the sophisticated, feature-packed programs currently available a waste of time, given that the good Web designer's job is to keep things simple? Not exactly, bearing in mind that technological transparency tends to accompany deepening feature-sets. Hopefully, by the time you are reading this you have some familiarity with the key technologies available to Web designers, and of the trend to incorporate Web capabilities or compatibility in existing 2D and 3D packages. But beyond individual packages, technology plays a broader role in shaping our experience of the Web. New browsing devices, for example, expand the audience not only in number but in type; e-commerce technology makes commercial transactions possible via the Internet while also bringing a new dimension to Web site design. Suffice to say the technological knowledge required of today's Web site designer is on the increase (there is, for example, HTML, JavaScript, CSS, style sheets, SSI, ASP, PHP, Meta tags, databases, XML, SWF, QuickTime, Flash, etc.), and this leads to important issues about workflow management, which can further impede the cause of good design.

A sound knowledge of the technology, however, can also enhance rather than hinder the Web site construction process. Take the key issue of download ease and speed; there are plenty of tips and tricks for generating leaner HTML, layout "cheats" that improve performance, methods for faster browser rendering, and image considerations that impact on Web-server performance. Even prior to this process, there's a range of techniques for streamlining your site at the graphic design stage. Graphics can be optimized for faster download, for example, and productivity can be boosted by automating functions.

Additionally, certain tools are specifically designed to aid workflow. Macromedia's Fireworks and Dreamweaver come bundled together precisely because most Web designers switch frequently between graphics and layout packages. Fireworks HTML can be used in Dreamweaver, file updates can be tracked across programs, and behaviors can be swapped.

As for compatibility built into non-Web–specific packages, the latest upgrade of Photoshop is a good example of this. Photoshop is a proven design tool for the worlds of print and imaging, but increasingly it's being used to enhance the design of Web sites, bringing with it an assemblage of creative and technical standards and conventions.

Photoshop version 6.0—along with ImageReady version 3.0—includes a range of integrated Web tools that have been specifically designed to automate and streamline the Web-design process. It can be used as a layout tool, as well as for building navigation bars, buttons, and other aspects of the interface, but it's worth learning how best to transport Photoshop elements into your construction program and to consider the pros and cons of, for example, working with image maps as opposed to separate graphics.

Other technical issues that today's Web designer might like to consider include Meta tags (making sure the right people get to see your site) and embedded scripting. The latter can seem to be a mystifying jumble of acronyms, but the likes of server side include (SSI) and application service provider (ASP) can, in fact, perform useful design tasks and significantly lighten your workload.

Finally—and perhaps most important—planning is a crucial part of the Web design process, from proposal development to budgeting and scheduling. Although not a technical issue itself, the efficient Web designer will be aware of technical considerations from the planning stage so that she can pitch and budget effectively and make her job easier during production.

INTERACTIVE ELEMENTS

One of the most important factors pushing forward the development of the Web is audience expectation. Once a novelty, interactive elements are now commonplace.

"I think interactivity between the site and the user has been one of the biggest evolutions in the past few years," states Deb George. "From simple chat rooms through to games, more and more sites are including some element of interactivity beyond the old style 'e-mail us' kind of feedback."

Ian Holding is sports director of TEAMtalk, a Web site that caters to sports enthusiasts and is distinguished by its high volume of content and interactivity. "It's a content-oriented site," says Holding. "We have a huge journalistic resource and occupy the position we do in the market because of the sheer amount of information we provide."

This volume of information, combined with the site's rapid turnaround and the incorporation of the interactive elements that the TEAMtalk audience has come to expect (the site receives in excess of 2,000 e-mails a day), has posed a particular challenge to the site's design team.

"Our interactive features include a push e-mail service, daily stats and score-lines, audio features, and live radio," Holding says, "and the site's layout has to cater to all these elements by ensuring that the various layers of information and services available are not hidden. Design has to be very simple to support this kind of resource without overwhelming the user."

Clearly, Web-site visitors want more for their money but they also want good value and don't want to be bamboozled with pointless effects.

"The Internet is now a part of everyday life; the novelty has worn off," Wilkie says. "When people go online these days they do so for a reason, and if content isn't easy to get at, then the site can't really expect any loyalty and that all-important return visit."

What's essential is not using new technologies just for the sake of it, but using them because they add something to the visitor's experience of the site. Steve adds: "Devices such as Flash, 3D, or streaming all contribute something to a Web site provided they are not being used as a token bit of technology. Lots of sites have a Flash intro, for example, but often this is simply for the sake of it and because it has become something of a norm, even when there's no good reason for it."

So, are the likes of Web 3D and streamed video just design affectations with no real purpose, or can they make a genuine contribution to the Web-site experience? Nigel Hunt is the creative director of Glowfrog Studios, a digital visual communications agency. He says of Web 3D: "It brings familiarity to the physical world while also allowing people to have fun and learn and experiment with objects and environments."

Bengt Starke, president and CEO of Cycore, manufacturers of the Cult3D software for Web 3D applications, says: "We have done studies that show visitors spend two-and-a-half times longer on a site with interactive 3D elements than they do on one without. Basically the sites offer more opportunities for exploration and thus visitors stay longer."

"We are seeing a natural progression on the Web from flat 2D graphics to 3D and finally to compressed streamed video," states Nigel Hunt, regarding the impact of interactive elements on design. "For designers it will push their roles into the realm of the broadcast industry, with Web directors potentially becoming world famous—one day there'll be a Speilberg of the Web."

Then—as now—sites that are attractively designed, easy to view, and not confusing will most likely prove the most appealing and successful. Not coincidentally, the same is true for every other form of digital content as well.

CHAPTER FIFTEEN

DATACASTING

RICK DUCEY,
SPECTRAREP, INC.

In this chapter, we'll consider the relatively *nontraditional* components of the emerging marketplace for terrestrial digital television. In a sense, this is all about including the market beyond the television receiver (whether NTSC or DTV) because that is where the industry can develop entirely new revenue streams. The major foci of interest addressed here cluster around emerging Internet opportunities for digital television and are grouped into three areas: Receiver/client devices, applications and platforms other than traditional broadcast television, and distribution infrastructure.

DTV AND INTERNET BANDWIDTH

Because the infrastructure for internetworking of any scale is based on or at least supports TCP/IP protocols, Internet Protocol (IP) packets can be encapsulated into the MPEG-2 transport stream of ATSC DTV, thereby allowing DTV to offer Internet bandwidth to the market.

There are 7,288 Internet Service Providers (www.ispworld.com) distributed around North America (as of March 2001). DTV broadcasting applied to the Internet is a particularly elegant solution to providing maximum bandwidth if broadcasters partner with Internet Service Providers (ISPs) to identify locally cached (i.e., highly redundant client requests for initiation) unicast and multicast streams for rebroadcast over DTV spectrum rather than clogging the scarcer, wired TCP/IP infrastructure. DTV delivery of high-demand and high-volume IP packets to end-users, such as in the case of streaming video services or large files, can solve a huge last-mile bandwidth infrastructure problem by using its inherent multicast and broadcast nature.

Broadcast infrastructure can fully integrate into the Internet infrastructure and offer "ready-to-go" bandwidth of up to 19.39 Mbps, which is particularly suited to UDP/IP multicast applications, and to "push" applications where a unidirectional stream can efficiently replace a large number of IP unicast streams.

This is significant because streaming and multicasting are fast-growing applications on the Internet. As these applications become more common in the marketplace, the strain on internetwork infrastructure of supporting bitstreams that are both highly bandwidth-intensive and highly redundant is beginning to wear users', not to mention ISPs', patience a little thin.

Bandwidth management solutions must be developed and phased in soon. For broadcasters, this involves understanding and quantifying mixed revenue models of providing high definition (HDTV), standard definition (SDTV) multicast, interactive (ITV), and datacast services. Broadcasters must balance business opportunities with public interest obligations to derive the maximum benefit from their 19.39 Mbps packetized data stream.

BANDWIDTH TO THE END USER

Deploying higher bandwidth pipes to end-users is a major strategic activity of the telephone, wireless, satellite, and cable industries. Although this will add to infrastructure capacity, the joint constraints of capital requirements, build-out rate, and eventual take-up rate forecast over the next 5 to 10 years place significant bandwidth in well under half of U.S. homes. Although pricing is coming down, accessing new, high-bandwidth providers could be an issue for many potential consumers and other end users who face economic barriers to entry into the high-bandwidth marketplace.

Arguably, the more significant short-term opportunity for broadcasters (2001–2005) might well be in the enterprise bandwidth market and not in the consumer market. The broadband wireless market is projected to grow to $17 billion by 2005, according to IGI Consulting. Streaming media alone will account for a $6 billion by 2005, J.P. Morgan estimates. However, these forecasts are dependent on end users having the bandwidth.

It is relevant to note that small- to medium-sized enterprises account for 85 percent of U.S. firms, 40 percent of the employment, and 33 percent of the economic output, yet only 6 percent have broadband connections. This is a limiting factor! And even among the "broadband-enabled" firms, there is a hidden trap. Suppose a medium-sized enterprise of 100 employees has a T-1 line for Internet services—this provides 1.544 Mbps and definitely qualifies as broadband. When the CEO's speech is streamed at 500 Kbps to provide a high-quality feed, exactly 3 percent (i.e., 3 clients × 500 Kbps consumes 1.5 Mbps) of the user base can watch this video before exhausting the available bandwidth. Delivery of streaming video packets via DTV solves this last-mile problem for the price of a cheap TV antenna and a DTV data tuner card.

The broadcast television industry can readily participate in this rapidly emerging bandwidth marketplace. Indeed, broadcasters are favored with several inherent competitive advantages, including currently deployed network, wireless distribution, ubiquity in the local market, cost-effectiveness in scale, and the ability to support IP multicasting (which is, after all, a receiver-based protocol, something broadcasters know about).

THE INTERNET IS MORE THAN WORDS AND GRAPHICS

The Internet is fast moving from its text-based origins and graphic enhancements and into streaming audio and video services—full-fledged multimedia. Except for bandwidth and data compression issues, users no longer care if something is stored on their hard drive or in a drive array of some remotely located server farm to which they are connected via the Internet.

Bandwidth and data compression, however, are the key and likely lasting barriers to realizing the objective of universal multimedia delivery. Internet-based applications typically presume a PC client, although this is changing with the introduction of non-PC clients or Internet devices that include personal digital assistants, cellular and PCS phones, and an increasing array of consumer electronics devices. If we concede that the Internet and PC clients form one of the endpoints of the emerging multimedia marketplace, perhaps broadcasting and TV-set clients can be seen as the other endpoint.

To establish a simple frame of reference, the traditional model for television involves the one-way delivery of linear audio and video program content via RF terrestrial spectrum over broadcast (i.e., one-to-many) open network mode to a client population of TV receivers.

The multimedia market in the broadest sense embraces delivery of analog and digital multimedia content via arbitrary packaged and networked means (e.g., CD-ROMs, DVDs, and various telecommunication pathways) to arbitrary devices (TVs, VCRs, set-top boxes, PCs, PDAs, etc.). Of course Internet (TCP/IP) is the runaway winner for distributing multimedia content to PC-based one-to-one and one-to-many connections.

AND TV IS MORE THAN PICTURES

As set-top box and Web-TV fortified TV receivers plunge television audiences into the Internet world, broadcasters have come face-to-face with the Un-TV market, a.k.a.—convergence. Broadcasters now recognize that the systems they build and operate are actually part of a larger information and telecommunication market. Their ability to distribute content beyond traditional services has far more utility and economic value than they may have anticipated. Good news to broadcasters worried about paying the DTV transition bills and desperate for new revenue streams!

In short, anything distributed via the Internet can be distributed via broadcast television—using MPEG-2 packets in the ATSC DTV transport layer.

"INTERNET SPACE" AS KEY PIVOT POINT FOR BROADCASTERS

The "Internet space" is a strategic pivot point to open nontraditional markets to broadcasters. For broadcasters to capitalize on this opportunity they must reconsider what their population of "client" devices is, think of their signal in terms of data bandwidth and not television channels, and reconsider their basic business model of traditional television broadcasting—a single linear program with embedded advertising.

UNDERSTANDING THE GENERIC NEEDS

For broadcasters, the first trick is to understand the *generic needs* to which they can apply their businesses. This is sort of like the classic challenge proposed by Theodore Levitt's *marketing myopia* test presented in his *Harvard Business Review* article back in 1960: "Are we in the railroad business or are we in the transportation business?" Railroad cars may come and go (pun intended?) but *transportation* will always be around. Companies that picked "railroad" as the answer have long since dwindled away.

REDEFINING THE BUSINESS MODEL

Once the audience's generic needs are defined, broadcasters must redefine their basic business model in a couple of ways.

First, broadcasters are in the spectrum bandwidth business. One use of this resource is traditional television, but there are very highly valued spectrum bandwidth markets that are growing far faster than the traditional television market and, indeed, which exceed its total value.

Second, broadcasters must redefine their retail distribution network from dedicated, single-purpose devices known as *television receivers* to anything capable of detecting, receiving, decoding, and processing encoded signals distributed via their spectrum emissions.

The best long-term bet for broadcasters looking to not only hold their own but dramatically expand their market is to start thinking of *client devices* as the targets of their broadcasts and not just television receivers. They need to think about the Un-TV market.

CLASSES OF CLIENT DEVICES

What kinds of client devices are useful for broadcasters to consider in their strategic market planning? There are several classes of devices worth considering.

- *TV Clients.* TV client devices include two familiar classes; standard television receivers, and MTS devices—multipurpose but still relatively unintelligent devices like Multichannel Television Sound (MTS) television receivers and VCRs. This type of client device extends to set-top boxes for dedicated applications like closed-captioning, video description service, secondary audio program (SAP), professional (PRO) channel services, and teletext services.
- *PC Clients.* Another type of client devices, which is decidedly not traditional for broadcasters, is PC-based devices like TV-card enabled PCs, WebTV boxes, and car PCs.
- *Consumer Electronics.* The final type of client device is the consumer electronics (CE) microprocessor-empowered devices, which range from pagers and cellular or PCS phones to personal digital assistants. These are known as non-PC devices. Because of the ability of non-PC, microprocessor-powered devices to feature TCP/IP connectivity, these devices are becoming known as *Internet appliances.*
- TV client devices offer a few more options and features over and above the basic television viewing experience, but

they are hard-put to serve as the basis for the development of any brand-new, "killer products" with fundamentally new features.

- Non-PC devices are viewed as the next logical platform to expand into, as are variations of the first two device classes. The PC industry is constrained and relatively mature; companies are looking for new horizons to explore (television broadcasters might learn from this example) as PC sales growth declines significantly.

Using the PC industry as a model, broadcasters should look to expand beyond the familiar TV client devices and into the PC and non-PC device classes. These devices have nowhere to go but grow, and the rapidly growing deployment of Internet appliances can only be seen as good news in this context.

To support high bandwidth downloads (i.e., unidirectional data flows via IP tunneling, MPEG2 data transport, etc.), technical standards for connectivity and interoperability must be developed and used at the client and service provider sides.

ENHANCED TELEVISION

Enhanced television services are those that primarily rely on TV client devices. They fall into two main categories: added-value content and data delivery.

The first category of enhanced TV provides additional functionality and content to augment the traditional viewing experience. This might include data streams encoding Electronic Program Guides; program information such as title, running time, main characters, plot summary; or even program-related games and quizzes.

In advertising, it might include additional information such as local distributors, maps, and other locally relevant details. This type of service links directly to the main service broadcast content in a time frame that is either real-time or intended for consumption within a narrow window of a main channel programming events. This process is sometimes referred to as *TV hyperlinking*, or as one company refers to it, *channel hyperlinking*, given its conceptual similarity to Hypertext Transport Markup Language (HTML) protocols.

The second category of enhanced television service uses basically the same technology, but is not limited to time-sensitive links between the data service and on-air content. This process is very familiar to Europeans (among others), who have made use of teletext over the years.

BEYOND TV

When we explore the kinds of services supported by the "Un-TV client devices" (i.e., PCs and consumer electronics Internet appliances or "Un-PCs"), there are two more considerations that television broadcasters may find interesting.

First, by adding a relatively inexpensive (U.S. $100 or so) TV tuner card into a PC, that computer can in effect become a television set. This really adds no technical capabilities, aside from perhaps the ability to perform video captures, but because TV-card enabled PCs might extend the viewing experience into "workplace viewing," they create a potential new demographic that the broadcast industry finds quite attractive. Broadcast audience measurements might then consider the expanded viewing universe of not just "sets-in-use" but "screens-in-use"—something for the Nielsen ratings company to fuss over.

Second, and more significant, is the idea that television broadcasters can deliver more than traditional television content (or even enhanced content) to PCs and consumer electronic devices. DTV broadcasters have enormous data-carrying capacity and it does not have to be limited to audio or video data.

As the price of microprocessor-enhanced devices becomes more attractive, DTV broadcasters will have a growing "audience" of PC and consumer electronic devices with DTV tuner chips built into them. These chips will offer scaled services, perhaps not HDTV-capable, but certainly capable of detecting, decoding, and displaying data packets representing different service classes.

Services could be developed for DTV/Internet applications targeting a range of these client devices. Video destined for an HDTV client device might be 1080I, but a scaled-down version

(480P) might be displayed on a desktop PC; there may also be a multimedia (text and animated graphics) version for an Internet appliance such as a cellular phone, PDA, or even a smart pager.

Distributed services could deliver part of the content via the wired Web and part of it via DTV datacasting to enhance customer service and ease Internet congestion.

For example, a key moment in the history of Internet streaming video was when Broadcast.com featured a major promotional event in January 1999, the Webcasting of the annual Victoria's Secret fashion show. Traffic contention exceeded the capacity of the infrastructure to deliver. Broadcast.com could support about a half million simultaneous users—several times this many tried to initiate streams and were rejected. As a promotional event, this certainly scored a success; as a demonstration of Internet streaming media scalability, it was a failure (albeit one damned by success).

Similar "flash" traffic bottlenecks happen with any major event, particularly those promoted on the air (e.g., the Super Bowl, President Clinton's video testimony) and even occur with nonstreaming media (e.g., the release of the Independent Counsel's report in the Starr Investigation).

A highly scalable and infrastructure-conserving alternative would have been to deliver these applications over DTV bandwidth, wherein users do not contend with each other for streams.

And, it gets even better than this: DTV stations are deployed to serve local markets, and can be interconnected in a variety of ways to achieve a custom demographic area larger than a single market. By comparing what local and regional Internet Service Providers decide to cache on their servers, DTV broadcasters have access to real-time data on what local users want. Setting up ISP/DTV cooperatives or partnerships could provide win–win opportunities for ISPs, DTV operators, and of course—users.

DISTRIBUTION INFRASTRUCTURE

The final topic to be explored in this overview is the notion of how DTV and Internet can work together in terms of distribution infrastructure. To begin with, DTV broadcast facilities boast numerous inherent advantages in the infrastructure marketplace for serving client devices running on diverse platforms.

These advantages include broadcasting's in-place deployment of DTV transmission facilities; wireless signal propagation with scheduled and ad hoc regional, national, and international networking; low incremental costs to enter and operate in the bandwidth marketplace; ubiquitous and wireless distribution in a local market; and the ability to interoperate with a variety of back-channel technologies for full-duplex operations.

A relatively simple model for exploring how DTV can add value to the Internet infrastructure is to consider the current lack of scalability for unicast and even multicast streaming applications. The Internet infrastructure is not currently equipped to meet high-volume demand in narrow windows.

The *migration* of Internet content to DTV broadcast media could occur at two levels: content or application. Content-level migration refers to digital audiovideo streaming, and would provide additional information addressed in the program. Application-level migration may be distributed—some of the elements may be delivered via high-bandwidth unidirectional gateways, while other elements are served via bidirectional Internet links.*

DTV–ISP EXAMPLE

In practice, a potential application might be a case where a DTV station partners with a local ISP during a major event heavily cross-promoted between television and Internet. For example, the half-time show at the Super Bowl, the streaming video coverage of the Victoria's Secret Fashion Show, or even popular and large text or multimedia files that are not streamed (e.g., Clinton testimony and Starr report).

From the ISP's perspective, they could well be damned by the success of content that is *too* popular. ISPs are in the band-

*Philip's Jan van der Meer has a very insightful paper posted on the Internet (www.w3.org/audiovideo9610_workshop/paper27/paper27.html) that considers "Opportunities for Migration of Internet and DTV Applications."

width and connectivity business—and both bandwidth and ports are scarce resources.

ISPs manage bandwidth by locally caching popular downloads so that they do not have to continually reach across the Internet to download the same resources again and again. They go get it once, and store the files on their own server farms for local redistribution. Exactly which files are locally cached varies by market. Even using this bandwidth conserving practice, caching covers only the peer-to-peer backbone network segments and not the ISP-to-end-user segment. Here, the ISP must still worry about how many ports are available to satisfy peak demand.

A worst-case scenario for an ISP occurs when it cannot meet user demand for bandwidth (cannot push packets out fast enough) or provide enough open ports for connectivity. DTV bandwidth offers a welcome and locally relevant solution to both of these ISP challenges. DTV bandwidth is neither switched nor port-based; it is a true one-to-many delivery system that offers infinitely scalable architecture for delivery of IP packets. Music to the ears of ISPs.

By distributing content and application elements between the DTV signal and interactive Internet connections, bandwidth infrastructure is optimized. Telephone dial-up or other lower bandwidth connections can support back channel communications. Client-server data flows are typically asymmetric. Client-side requests typically represent mouse clicks to select URLs or perhaps to send e-mail (which are typically small files, unless they come from a college student and include video file attachments wishing Mom a happy birthday—and please send more money!). The larger bandwidth requirements typically reside with the DTV signal originator.

Of course, there are some technical requirements to be met for these applications to work. Internet/DTV client devices, suggests der Meer, must be user friendly, platform independent (in terms of API implementation), and cost-effective. But once met, Internet/DTV devices can support wide-ranging applications leveraging both the power of Internet client–server interactivity and DTV high bandwidth, albeit unidirectional, delivery.

In the Super Bowl or Victoria's Secret examples, some content could be distributed via the "traditional" Internet but when file sizes or caching parameters reach predefined limits, the 5,000 ISPs around the country could *gate* part of their data flows to DTV station partners for local data broadcast to properly equipped client devices.

SUMMARY

Broadcasters have significant opportunities available to them in the datacasting market. Capitalizing on these opportunities requires new and more sophisticated business models than broadcasters have employed to great success in the past. The transition to digital television is expensive, risky, and complicated, but, exploring entry into new markets such as datacasting may well help make this an extremely worthwhile venture for digital television broadcasters.

CHAPTER SIXTEEN

THE VIDEO "PRINTING PRESS"

LARRY JAFFEE

DVDs are a rich opportunity for digital content creators. In journalism there's an old adage that the only free press belongs to those who own a printing press. And for video content creators today the same could be said in terms of the production, storage, and distribution advances provided by DVD technology.

In addition to its relatively young track record as the fastest growing consumer electronics product in history, according to the Consumer Electronics Association (CEA), the DVD format's applications lie far beyond DVD-Video players showing Hollywood movies in the living room. (A misperception is that DVD stands for *digital video disc*; it can also mean *digital versatile disc*, thus indicating that the same optical medium can be read in a computer's DVD-ROM drive as well as by a DVD-Video player.)

For video content creators, such versatility opens new realms of possibilities for production work. The latest strides in digital camcorders enable users to record images and store them with a built-in DVD-R burner. Apple's newest iMacs include a DV pro digital video editing suite and iDVD authoring software. When combined with a DVD-R duplication unit that can crank out hundreds of fully printed discs in several hours for an investment of several thousand dollars, the independent video content creator can virtually become a movie studio, postproduction house, authoring facility, and replication plant all rolled into one. Assuming, of course, that the content creator possesses the talent to make products that sell.

There are other low-cost do-it-yourself options when it comes to authoring software, which are largely cut-and-paste icon-driven programs. Hence, a content creator needn't be a computer programmer to create a basic DVD disc. A company called Sonic Solutions has had much success bundling its DVDit authoring software package with personal computers. As a stand-alone product, the software costs $500.

The flexibility and interactivity that DVD presents is a far cry from the linear, static nature of VHS, which admittedly has become ubiquitous over the past 25 years and is fairly easy to duplicate. But when compared to the superior picture quality and interactivity of DVD there's little reason not to immediately take advantage of this readily available new technology. Now that the hype of Internet distribution of video has died down some (did anyone think of coming up with a business plan?), a low-cost piece of physical media, such as DVD, is looking much better than the intangible, ethereal domain known as cyberspace.

Although DVD was launched in March 1997 by Hollywood (spearheaded initially by Warner Home Video in conjunction with a consortium of the leading consumer electronics manufacturers), the driving force obviously was theatrical movies, which could be presented in ways previously unavailable to the average consumer. First and foremost is picture quality, which is excellent.

From a producer's perspective, DVD provides an opportunity for an interactive experience accessible via a remote control or a computer mouse. Furthermore, the same disc can hold multiple language tracks—a boon for international-minded productions, as well as multichannel, surround sound.

Personally, I think the turning point that ensured DVD's future was about two years ago when the New York-based consumer electronics and home entertainment chain, The Wiz, took the discs from behind the counter and into their own dedicated bins. This merchandising decision served as a regional microcosm of what was to come nationally.

DVD'S SALES TRAJECTORY

Since the inception of the format in 1997, U.S. software shipments of DVD have reached more than 380 million units, according to figures released in April 2001 by the DVD Entertainment Group (DEG). Based on retail and manufacturer data, hardware players sold to U.S. consumers have reached more than 16 million units.

Whereas many industries are suffering from a downturn in the economy, DVD hardware and software sales continue to grow at an accelerated pace. In the first quarter of 2001, an additional 2.4 million DVD-Video players were shipped to retail, bringing the total units shipped since the format's launch to nearly 16.5 million, according to the CEA. If growth continues at this pace, 17 million DVD-Video players are anticipated to ship to retailers in 2001. Add to that another 18 million computers equipped with DVD-ROM drives, and this is a consumer format quickly approaching critical mass with no sign of its exponential growth slowing down.

A DVD-ROM drive can read DVD-Video content, but a DVD-Video player generally can't play most DVD-ROM material, which means that video content creators most likely would want to use DVD-Video as their principal multimedia canvas. Additional DVD-ROM material could be placed on the same disc, adding, for example, a directory that could be updated periodically via a Web link.

DVD's trajectory is also quite a feat considering other entertainment optical disc formats that either failed outright in the U.S. (e.g., CD-i, Video CD, and MovieCD), or never made it beyond a niche, enthusiast base (i.e., LaserDisc). DVD's relative quick success can be attributed to the fact that it won support from not only the Hollywood studios, but also the consumer electronics manufacturers and the computer industry.

According to figures compiled by Ernst & Young on behalf of DEG, nearly 70 million DVD movies and music video titles shipped in the first three months of 2001, more than double the number shipped in the same quarter last year.

But even with numbers like these, U.S. household penetration for DVD-Video players is at a mere 10 percent. Needless to say, there are still huge growth opportunities for the format, as evidenced by the DVD-awareness rating of 98 percent—half of those surveyed are planning to purchase a DVD player in the next 12 months, according to research from Warner Home Video.

At this relatively early stage of the format's evolution, it's safe to assume that DVD is well beyond the "early adopter" stage. In other words, your customers should already be able to play back DVD, and you'll look like you're on the cutting edge, if they don't have something to view your production. It might just very well be the impetus for them to go out and buy a player, which is as low as $99 these days. At that price, you might even want to buy a player for your customer—it'll certainly be money well spent if they hire you to produce a disc for them.

Of course, like all consumer electronics, you pretty much get what you pay for. An up-scale consumer DVD player for $500 features *progressive scanning* to provide an enhanced digital picture at 60 frames per second and 720 lines of resolution—a superior picture when compared to the typical DVD player's interlaced scanning of 600 lines of resolution. That's not to say you can't get a fairly decent player for $200.

DVD'S "BIT BUCKET"

So now that we've established that DVDs are worthy of the interest of video content creators—and the sooner the better—what's the best way to do your first DVD project? Of course, the answer, like everything else in this business, depends on the extent to which you or your organization wants to devote in-house resources to such an endeavor, or tap the expertise of a consultant who specializes in such work.

Think back six years ago, when corporations raced to get onto the Web. As in the go-go, dot-com days, there's certainly no shortage of firms looking to provide end-to-end DVD solutions. One option is to use the services of a replication plant that also offers in-house authoring, compression, and encoding. This will likely cut down the time to get a project completed, as opposed to allowing different entities to handle different phases. With the profit margin dropping out of the pure replication task, be forewarned that some replicators might try to make it up through ancillary services, which include printing and packaging. A DVD-5 disc costs about 75 cents per unit to replicate and package; in contrast, a clunky VHS tape costs more than twice that to duplicate and package.

DVD is often described as a "bit bucket." The DVD specification contains several varieties of a disc that essentially increases the size of the bucket. For example, the plain-vanilla *DVD-5* refers to a single-sided, single-layer disc that contains 4.7 GB (gigabytes) that can hold 9 hours of VHS-quality video or 2 hours and 15 minutes of high-quality video.

A *DVD-9* provides 8.5 GB on a two-layer single side or over 4 hours of high-quality video. The other (non-data) side of the disc for DVD-5 and DVD-9 can be used for art. The double-sided disc can come in the form of *DVD-10* (9.4 GB, one layer per side), or *DVD-18* (17 GB or two layers on each side), which can actually hold over 8 hours of high-quality video or 33 hours of VHS-quality video.

Keep in mind that all of these are referring to the various forms of a DVD-Video disc. A DVD-ROM boosts significantly the capacity of what a disc can hold. And for comparison sake, it's also important to note how much more a DVD-ROM can hold when compared to a CD-ROM, which holds 650 MB (megabytes)—seven times less than what a single-layer DVD holds. The National Geographic Collection takes up 31 CD-ROM discs, but only four DVD-ROM discs.

LINKING TO THE WEB

DVD's interactivity is best realized in a computer drive, and when linked to the Internet. Essentially, such connectivity, generically known as *WebDVD*, offers a broadband environment without the high-speed Internet hookup. Interactual

Technologies, Inc.'s PC-friendly software enables large-scale content that's too unwieldy for today's Internet speeds to be "remote controlled" via the Internet. This software has the ability to display both DVD-Video and Web pages in the same browserlike player environment, thereby extending DVD movies with computer graphics, animation, and Web links. Interactual's PC Friendly software has been used for more than 300 movies, on 30 million DVD discs, and accessed by some 2 million online users.

Besides Interactual, other companies offering WebDVD solutions include Microsoft, which taps its existing Windows Media Player and DirectShow technology; and Spinware, which adds an additional digital layer to a DVD-Video disc that allows users to access encrypted video and audio content in real time through an Internet browser. Content providers benefit from using SpinWare by establishing a two-way communication with content consumers, which leads to additional revenue after the initial point of sale.

SpinWare's newest technology, iControl, is a robust Web-based media-playback technology that allows content creators to control DVD movies and other high-bandwidth disc-based content over the Internet. The software also enables control and playback of a DVD movie from within PowerPoint slides.

AUTHORING TO REPLICATION

Meanwhile, Apple has practically cast its iMac line as synonymous with home movie-making. According to Apple, iDVD makes authoring easy and quick. With iDVD in play, it takes only a couple of hours, rather than the typical 8 to 12 hours, to prepare a single hour of video. The company is bundling iDVD with fully DVD-equipped G4 Macs, including both recordable software and drives (manufactured by Pioneer), priced at roughly \$3,500 as of March 2001. Additionally, Apple sells a standalone retail DVD-authoring package for less than \$1,000.

For the non-Mac inclined, Compaq also has bundled a combination DVD-R/CD-RW drive in its Presario line of consumer desktop PCs, including its MyMovieSTUDIO 7000 model (MSRP $2,399), driven by a Pentium IV processor and a package of consumer DVD moviemaking software. MyMovieSTUDIO transfers video from digital camcorders via an IEEE 1394 connection, and includes StudioDV editing software, as well as Sonic Solutions' DVDit! authoring software, which is available in both PC and Mac versions. Other low-cost authoring software alternatives are available from Spruce Technologies (www.spruce-tech.com), and the lesser known INTEC America, which in February 2000 offered a desktop solution called DVDAuthorQuick for $2,500.

Obviously, producing a DVD can be as complicated as you want to make it. For example, you can skip the menu phase, thereby removing some of DVD's inherent interactivity. But no matter how involved (everything from a Hollywood blockbuster to a corporate video), the project must be mapped out (*branching* in DVD vernacular) so that navigation makes sense and its interactivity doesn't lead to a dead end.

As in every other stage of producing video, making a DVD has many hidden traps, with repercussions that reach back through the production chain. Invariably, it takes longer and costs more than originally thought. Proper planning will help the process move more quickly and more smoothly, and will, in the long run, save money and result in a superior product.

Any error during the many phases of authoring, mastering, or encoding and decoding through the digitization process from editing to the final master on a digital linear tape (DLT) will cause glitches. Things can also easily go wrong when multiple parties are responsible for different aspects of the project. Hopefully errors will be caught before replication, thus avoiding a costly mistake. But the best way to avoid such pitfalls is quality control (QC) early and often—the longer you wait to check, the costlier the fix will be.

RECORDABLE DRIVES AND MORE AFFORDABLE MEDIA

To get an idea of how quickly the DVD-recordable landscape has changed, in 1998 an external DVD-R drive would have cost a cool $17,000. As of May 2001, Pioneer's DVR-103 drive sells for less than $1,000, and prices are expected to drop even more dramatically over the next few years.

Disc prices are also set to drop. Whereas blank DVD authoring discs sold for upwards of $30 each in the fall of 2000, Apple sells packs of consumer DVD-R media for $49.95 at its online store (www.apple.com).

DVD recordable media comes in three varieties: Pioneer's DVD-R (write-once) and DVD-RW (rewritable), Panasonic's DVD-RAM (rewritable), and a Philips/Hewlett-Packard–led consortium featuring what they call DVD+RW, which was still available on the market as of mid-2001. Keep in mind that these competing recordable technologies are not compatible (i.e., playable in each other's drives).

There are even some DVD authoring software packages that allow you to burn your production onto a CD-R disc, certainly a much less expensive endeavor.

Meanwhile, the world's leading CD-R manufacturers are poised to enter the DVD-recordable market when the demand becomes evident, and blank DVD-recordable media prices are expected to drop in the same manner as CD-recordable media did. (True, Napster spawned a free downloadable music frenzy; it remains to be seen if Hollywood content owners will allow the same kind of free-for-all mentality that occurred with music.)

THE DIGITAL TELEVISION CONNECTION

DVD is also sharing a symbiotic relationship with digital television sets. Indeed, CEA president and CEO Gary Shapiro reported in May 2001 that by the end of the year approximately $5 billion will be spent on digital television equipment. "Digital television is being driven by DVD. The primary use of digital television [now] is playing DVD," he said.

Shapiro noted that the technical standards, as well as marketplace reality, for DTV and HDTV are still far from being settled. Specifically, both DVD and HDTV must be in sync with the two-thirds of U.S. households that receive their TV signals through local cable system operators.

This suggests that, whereas DVD offers a significantly improved video picture over VHS playing on an analog set, the quality bar is only going to improve, which is good news for the future. (It should be noted, however, that at the time of this book's last edition, it was expected that an HD-DVD standard would be hatched within a year; that still hasn't occurred.)

DVDs weigh very little, when compared to VHS and other tape formats, and so they're cheaper to distribute. They offer interactivity, which can be exploited as much or as little as the program content demands or the client for the program desires. There is no time-consuming tape rewind on a DVD, so repeated playback is a cinch, a valuable feature that makes the discs well suited for training videos. DVD may well be the first ubiquitous standard for consumers and businesses alike. For content creators, the format is a rich opportunity waiting to be exploited.

CHAPTER SEVENTEEN

THE DTV TRANSITION

MICHAEL GROTTICELLI

Television broadcasting is being transformed as never before. Much is still uncertain, especially the nature of the opportunities for content creators, but of all the markets for digital video content creation, broadcast television remains one of the largest. Prior to the proliferation of cable television networks during the past two decades, broadcast TV was by far the largest market for video and nontheatrical filmmaking. Today, however, the targeted programming alternatives provided by an ever-increasing number of cable outlets are putting increased competitive pressure on broadcast networks and call-letter television stations. As in every other area of communications, digital technologies are transforming the way in which content is created and distributed. Digital cable and direct-broadcast satellite (DBS) services provide viewers with more choices than ever before—and in so doing represent an ever-growing market for digital video content creation. But television broadcasting is also being transformed by digital: In the United States, the Federal Communications Commission (FCC) has mandated that digital television broadcasting eventually replace the analog system in use since the 1950s. Furthermore, the new digital television (DTV) system will include features that enable broadcasters to better compete with their cable rivals.

Understanding these features and the nature of what will probably be a long transition to DTV is important for digital content creators interested in addressing this potentially explosive new market.

A BRIEF HISTORY

In 1996, there was a great deal of optimism surrounding the adoption of a new terrestrial DTV system, as developed by the Advanced Television Systems Committee (ATSC). After several years of industry fighting about whose technology worked best (principally broadcast interests versus computer interests), the ATSC eventually settled on a hybrid system developed by a group of manufacturers and research organizations now known as the Grand Alliance.

This ATSC system was approved on April 3, 1996 by the FCC for use by approximately 1,600 licensed, full-power U.S. broadcasters who wished to employ the new digital system over the air. Their optimism was felt most by the National Association of Broadcasters (NAB), which spent several years and millions of dollars lobbying the U.S. Congress to have the American government simply give digital spectrum (airwaves) to commercial analog-broadcast stations that are currently on the air. Many felt that the biggest bargaining chip the NAB had was that broadcasters promised to use the digital spectrum for a new, high-resolution service known as high definition television (HDTV). (Spectrum, by the way, is a finite resource owned in the United States—at least in theory—by the American people. One of the reasons for the existence of the FCC is to regulate the licensing of spectrum to radio and TV stations and other users so that its orderly use is ensured.)

This was attractive to Congress because they wanted the U.S. to lead the world technologically, and they were made to feel confident that HDTV was what TV viewers all over the world really wanted. Congress also desperately wanted to transition broadcasters off their existing analog spectrum so that it could be auctioned off to cellular phone companies and others to help balance the Federal budget. The NAB was happy because the new digital spectrum that broadcasters would get in return for their analog spectrum would not be going to cellular companies, which represented (and still do) competition in this digital era.

To make sure broadcasters kept to their word, the FCC instituted a schedule that would mandate when the various Designated Market Areas (DMAs) would have to be on the air with a digital signal, or risk losing their new license to do so. Realizing that the larger markets in the U.S.A. typically had the most money to make the transition (each station will need to spend approximately $2 million to $5 million just for the necessary transmission equipment), the FCC's schedule started with them first and worked its way down to TV stations in smaller markets.

The mandated schedule states that commercial TV stations in the top ten markets, affiliated with the top four U.S. broadcast networks (ABC, CBS, Fox, and NBC), were to have applied for their DTV construction permits (CP) as of May 1, 1998 and be on the air with a digital signal—but not an HDTV signal—by May 1, 1999.

Commercial stations affiliated with the top four networks in the 11th to 30th markets were to have applied for a CP by April 1, 1998 and be on-air by November 1, 1999.

Finally, all remaining commercial TV stations must have applied for their CP by May 1, 1999 and have until May 1, 2002 to get on the air with a digital signal using the ATSC specifications for video and audio. Noncommercial TV stations, such as member stations of the Public Broadcast Service (PBS), were given extra time to transition, because of the fact that these taxpayer-funded organizations typically have the tightest budgets. Those stations had until May 1 of 2000 to apply for a CP and are not scheduled to shut off their analog NTSC service until May 1, 2003.

The FCC finalized the basic DTV transition process on April 3, 1997, by releasing its Table of Allotments, which provides a digital channel allotment to each existing NTSC station and anyone who had a construction permit for a station not yet built. After months of exhaustive research and experimentation, the FCC put forth a Replication Principle behind the assignments, which means that each DTV channel assignment is intended, as much as physically possible, to replicate a station's existing NTSC service area. Each station's DTV and NTSC licenses are paired, so that the two run for the same term; a station cannot sell one without selling the other.

Once the facilities are constructed, simulcasts of digital programming are supposed to begin. The FCC also adopted a phased-in simulcast requirement for DTV stations so that viewers do not continue to depend solely on NTSC facilities. The Commission said that broadcasters are free to accelerate the schedule, which is as follows: By May 1, 2003, 50 percent of on-air programming must be digital; by May 1, 2004, 75 percent should be digital; and by May 1, 2005, all of a station's programming should be digital.

After this transition has been completed (whatever that year might be), the U.S. government will effectively shut off NTSC service and those companies that have won the analog spectrum in auction will then begin their own operations.

Feeding off the excitement, the Harris Broadcast company (in Mason, Ohio), a company mainly known for its high-power transmitters, staged a high-profile demonstration of the technology for the broadcast of a baseball game in Arlington, Texas on September 16, 1997. This demonstration was successful, although several stations in other parts of the country, which had hoped to receive the signal, were not able to.

WHAT HAPPENED

At the NAB's annual convention in 1997, there was a great debate about HDTV formats, after CBS and NBC announced that they were going to distribute HDTV programs in the 1080-line interlaced format, whereas ABC and Fox said they would use 720 lines scanned progressively. This splintering caused much confusion among independent broadcasters. The reasons given were that CBS and NBC felt that the highest quality signal was most important, whereas ABC and Fox said they wanted to save some room within their allotted 6 MHz signal to distribute other things—such as electronic data (newspapers, coupons, recipes, etc.) and other content.

Both ABC and Fox also argued that a "720P" picture looked as good or better than one that's "1080I" to most consumers, and that the progressive format was better suited to the eventual convergence of the Internet and television.

Just before the November 1, 1998 start date for DTV (on October 29th), over 40 stations considered themselves "early adopters" and broadcast a live, HDTV (1080I) transmission of the John Glenn NASA space shuttle launch (the most televised NASA event ever), from Cape Canaveral, Florida. It's interesting to note that among these 42 stations, several were PBS stations that were not required to broadcast digitally.

The transmission was greeted with applause, but it was also not without its technical difficulties, which were attributed to engineer inexperience with the technology and "prototype" encoding equipment.

By most accounts, the "voluntary" November start date for DTV was a concession by the U.S. broadcast industry to silence critics who cried "corporate welfare" when Congress voted to allow stations to operate on the digital spectrum for free.

In devising its DTV plan for the U.S., the FCC had hoped to initially have at least three digital stations on line in each of the top-10 markets, to help stimulate DTV consumer set sales. The Commission was successful in meeting that goal in all but three markets: New York, Chicago, and Detroit. (Several stations in those markets are still experiencing difficulties in finding adequate space to install their digital antennas.)

HDTV was seen as a victory for broadcasters, even though there were very few digital HDTV sets to receive the signals in consumers' homes. Many of these early adopter stations, however, said at the time that they were getting on line early for competitive reasons, stating that they felt HDTV gave them an edge with consumers over other stations in their market as well as over cable TV and DBS services, which were beginning to sign up an ever-increasing number of subscribers. They also felt that the more experience they had with this new technology, the better.

THEN WHAT?

The next step was for broadcasters to wait as consumers bought the new generation of DTVs and HDTVs. The Consumer Electronics Association (CEA), which represents all of the major TV set manufacturers, displayed a number of these new sets at its annual convention in January 1999, but not many people bought into the idea. These new sets were big (45–65-inch), rear-projection models that included a DTV decoder and cost anywhere from $8,000 to $15,000. In addition, the few sets that were on the market were not working as well as broadcasters had hoped, often requiring a broadcast technician to personally go out to a consumer's home to adjust the off-air antenna and DTV tuner correctly. Very few people bought the new sets, except for R&D labs and various broadcasters themselves to test their off-air signals.

Soon broadcasters were blaming consumer electronics (CE) companies for not offering an affordable alternative to the current NTSC TV sets, which were just as big but cost less than half the price. The CEA, in response, chided broadcasters for not putting any programming on the air to watch on the new DTVs. This was true initially in that ABC, CBS, and NBC only had 5 to 10 hours of scheduled programming on their digital channels per week, and not every affiliate that could distribute HDTV was doing so.

Executives at ABC, Fox, and several independent station groups quickly realized that the digital signal that the FCC's rules required them to put on the air could be as simple as just a 480-line digital test pattern, if they so chose. These executives went before Congress to answer questions about what broadcasters actually planned to do with the spectrum the government had given them at no cost. When broadcasters said they were considering multicasting—that is, splitting up their 6 MHz signal into several lower-resolution channels—Congress was not pleased. Dreams of beautiful widescreen HDTV signals were being watered down by the reality that a TV station could instead show popular cable networks.

Today, the situation has changed somewhat on both sides. More programming is now available from both the networks and local stations, and increasingly less expensive DTV models and PC-reception cards have emerged at retail. The cost of digital HDTV displays has also come down to the sub-$3,000 range. There's even a widescreen digital set from Zenith that retails for under $1,000, although it can only receive standard-definition digital broadcasts (480P). A separate set-top receiver is required for these sets to receive native HD programs.

Therein lies the basic problem. The CEA often reports increased numbers of digital television sets sold (the latest figure from the CEA states that over 600,000 sets have been sold in the United States, as of Spring 2001) but only 27,000 of those DTVs have over-the-air tuners inside them. With virtually no one buying receivers, the cable TV and satellite industries continue to be the place where most Americans get their digital television programming.

Indeed, the fact that only about 20 percent of Americans today watch TV with a roof antenna has the broadcast industry lobbying hard for "digital must-carry" to be mandated for cable and satellite providers. Under such a law cable and DBS companies would be required to transmit DTV; currently cable providers, by law, must carry NTSC broadcasts—rather ironic, considering that cable is the way most audiences receive over-the-air broadcasting. The CEA still stands firm in its belief that if more HDTV programming was available, more sets would be sold.

TODAY

Today, most broadcasters are at least considering some type of multicast situation, whereby they might present HDTV-quality movies during their prime-time schedules (8–11 p.m.) and DTV quality (480 lines) during the "dayparts." This is because there is no clear business model for single-channel HDTV operation in a world where cable TV and DBS are providing subscribers with hundreds of digital channels and improved multichannel audio.

Because of current economic conditions (and the fact that network ratings numbers have continued to decline), many broadcast companies are considering some type of "central-casting model," whereby one stations feeds programming to two or more stations within the same ownership group. Companies such as The Ackerley Group, USA Networks, and NBC have all created some type of centralized "hub-and-spoke" or distributed playout model among their remote stations to reduce the cost of the transition to digital. For stations used to the traditional methods of single-channel NTSC broadcasting, it's a whole new world, where video and audio are just another data type to be accessed from a central server.

As per the FCC, all commercial stations—whether networked or not—are required to broadcast at least one free television program service "of a quality at least equal to the quality of a standard definition (NTSC) signal." If broadcasters decide to multicast their signal, it looks like Congress will vote to levy some type of fee to use the spectrum, as they currently do with other digital services.

In fact, the Telecommunications Act of 1996 requires the FCC to assess a 5 percent fee on any gross revenue gained from ancillary and supplementary uses of the digital spectrum. Some see this requirement as a way to discourage stations from multicasting, but the reality is that stations need additional revenue streams to stay competitive. If a cable provider can offer multiple channels, why not a broadcaster, the argument goes.

MODULATION DEBATE

As if the DTV transition wasn't problematical enough, there's an additional challenge that's further hampering progress. The DTV transmission technology chosen in the United States doesn't seem to work especially well. These technical issues have caused the FCC's DTV system to be hotly debated. For example, the modulation portion of the ATSC standard has come under a cloud of controversy. Modulation refers to the process (or result) of changing information (audio, video, data, etc.) into information-carrying signals suitable for transmission and/or recording.*

*Mark Schubin *HDTV Glossary*, Union Square Press, page 45, 1988, 1990

The modulation scheme chosen in the United States is 8-*VSB*, but now that several stations are on the air with a digital signal, they are finding that 8-VSB is not able to adequately replicate their existing NTSC service. This has caused tremendous concern among the smaller independent broadcasters who have not yet invested in their DTV plants.

To these broadcasters, the European Digital Video Broadcast (DVB) system, with its *COFDM* modulation specification, appears to hold more promise, both for its ability to reach a wider coverage area (albeit at higher power levels) with shorter, multiple antennas, and its success using a broad range of portable devices that are coming onto the market. These devices could provide new revenue streams for stations that might want to transit data and other RF services in the future.

Thus far, based on a number of field tests done in the United States and in other parts of the world, 8-VSB does not seem to be able to transmit signals to portable devices in a reliable fashion. To save face and potentially lucrative patent royalties, engineers at companies that make 8-VSB decoder chips—such as Motorola, NextWave, Oren, and Zenith—are working feverishly to keep the ATSC standard intact.

The ATSC established a special Task Force in 2000 to look into mobile reception and NTSC replication with 8-VSB modulation and the possibility of including COFDM into their DTV standard. The NAB, in conjunction with MSTV, performed a number of side-by-side modulation tests and released a report late in 2000 that found the performance of both systems to be lacking in specific urban areas where multipath problems exist. *Multipath* refers to TV signals bouncing off buildings and causing *ghosting* on TV screens. The FCC achieved similar results in its own test in and around Washington D.C. at the end of 2000.

Due in large part to the debate surrounding these unresolved technical issues—which many people feel delayed the digital transition by at least a year—many broadcasters are now calling for the FCC's NTSC shut-off date of 2006 to be extended. There appear to be legitimate claims to support this point of view, especially because there are very few digital receivers in the marketplace. As this book goes to press, the new FCC Chairman,

Michael Powell, appears to recognize the situation and seems willing to extend the deadline. The FCC has released a number of statements hinting that an extension is possible, but they have also urged broadcasters to get on the air with DTV or risk a revenue tax for the continued use of the analog spectrum.

The NAB seems to have considered the possibility of extending the deadline when it lobbied to have selected exemptions written into the Balanced Budget Act of 1997, which relies heavily on the proceeds from the analog spectrum to fill its coffers. Indeed, specific rules pertaining to a station that might need an extension before vacating its analog channel are:

- One or more stations affiliated with one of the top four networks is not operating digitally, because the Commission has granted the station(s) an extension of time to complete construction;
- Digital-to-analog converters are not generally available on the market; or
- 15 percent or more of the homes in the market cannot receive a DTV signal either off air or via cable, or do not have either a digital set or an analog-to-digital converter attached to an NTSC TV set capable of receiving DTV signals in their local market.

Looking around the broadcast industry, one can find many examples of these cases to warrant an extension of the 2006 NTSC shut-off date. As a matter of fact, numerous stations that have not even started construction on their new digital facilities are counting on the FCC to extend the date to somewhere around 2015.

The other aspect to the transition delay is that most U.S. stations are currently enjoying healthy profits from advertisers on their NTSC channels and there's no compelling reason for them to give that up anytime soon. Although the number of broadcast-TV viewers has declined over recent years, operational costs are stable (unlike the huge investment that DTV requires, which thus far shows no signs of a return) or are being reduced through station consolidation or switching.

On a more positive note, broadcasters are increasingly aware of the value of their allotted spectrum and are interested in innovative ways in which to leverage it. Several companies—such as iBlast Networks, in Los Angeles—have worked with station groups and persuaded them to "lend" differing amounts of their spectrum during selected parts of the day. Having signed up several of these stations groups (to 10-year contracts), such third-party companies then pool all of the spectrum together and sell it to advertisers. The station then splits the revenue with the company.

All of the networks have recommended that their owned and operated (O&O) affiliates not join with these new so-called *bandwidth brokers*, but other independent groups have said that this is their only hope for an immediate return on their DTV investment. In fact, some are hoping to fund their transition to DTV with money gained from selling parts of their 6 MHz channel.

Broadcasting over the Internet, or *Webcasting*, is another area where TV stations stand to gain a foothold in the Digital Age. Yet, this too has been difficult for most broadcasters, who are typically experienced in distributing just video and its accompanying audio over the air. Stations now find they have to become expert in file-transfer technologies and compression formats for the World Wide Web. Many stations have hired specialized companies to stream their newscasts and other video programs onto the Internet. Other larger stations have set up dedicated divisions within their companies to host the station Web site.

What every broadcaster has found out about the Internet is that content must be compelling, full-motion (30 frame per second) video, not just text and pictures, if they want to keep their audience's attention. Slow Internet connections in consumers' homes have hampered this, but the gradual increase in Internet bandwidth to those homes may eventually make this a workable alternative for TV stations to distribute content.

This being said, nearly 200 stations in approximately 64 DMAs are currently on the air with a digital signal (as of June 2001), according to the NAB, and that number is steadily increasing. About half of those on line are noncommercial stations.

LOOKING AT THE COMPETITION

While the broadcast industry seems to drag its feet with the DTV transition, multiple system operators (MSOs), which represent the major cable companies in the United States, are not even considering HDTV until there is a sustainable market for that type of programming. Their approach thus far has been to test HDTV is select markets, such as New York and Los Angeles, with one or two channels. These include Home Box Office and Showtime Networks, which are both distributing two channels of upconverted feature films. These are also being presented in the 16:9 widescreen format, with Dolby Digital 5.1 channel surround-sound audio, when available.

In some markets, such as New York, Time Warner's cable system is the only place to reliably see high-quality movies with Dolby Digital. CBS and Fox are on the air in New York with a digital signal, but given the high multipath areas in that city, reception with an antenna is problematic at best.

There has been some dispute between broadcasters and the cable industry about how these HDTV movies are displayed. In most cases, cable operators have said they will not pass through full, 1080-line signals because they eat up too much bandwidth that might be used for other, lower quality channels. In any event, cable operators have found few subscribers for their HDTV channels (which require a special set-top box to receive them), so they have not been aggressively advertising the service.

There had also been some problems getting a standard interface for "DTV-ready" sets approved among CE manufacturers, but this was resolved in May 2001 when the CEA announced that its members were unanimously supporting the IEEE 1394 (sometimes called *FireWire*) connection standard. As this book goes to press, however, a TV that can receive a DTV signal through the cable in the home without a set-top box is still not available to consumers. There are also program copyright issues that have yet to be solved.

The term *broadband* is being talked about as never before within the cable industry, as evidenced at the industry's annual convention in June 2000. Cable TV operators express optimism about the future of video-on-demand (VOD) movies, interactive games, digital music channels, and Internet access. It is their belief that U.S. consumers want more such channels, so they've been busy upgrading their infrastructures to accommodate two-way access into subscribers' homes. Such access is broadband in nature in that it involves increased bandwidth for more video, audio, and data information. Fiber optics, which offers greater bandwidth than traditional coaxial cable, is often the solution. Installing fiber is a costly undertaking that can take years to complete, however; most of the major MSOs have planned to be completely digital by the end of 2001.

There's also been a lot of talk about providing long-distance telephone service through the cable line, but this has not materialized and there is skepticism that cable companies can handle all of the required billing and other services necessary. By far the biggest growth market for cable is in the Internet-access sector, where high-speed modems and a monthly subscription (separate from video channels) is becoming the norm in many consumers' homes. These cable modems have successfully replaced earlier telephone modems that operate at 56 Kbps maximum and often get bogged down when network traffic is high.

DIRECT BROADCAST SATELLITE (DBS)

Of all the competitors to the broadcast industry, DBS providers (DirecTV and EchoStar, known as the DISH Network) have been the fastest growing, with more than 15 million subscribers, as of February 2001. According to Sky Trends Research (New York), DirecTV now has approximately 9.8 million customers, whereas Echostar claims slightly more than 5 million. It is generally acknowledged that consumers are becoming increasingly tired of their reliance on cable TV and are making the switch to DBS in record numbers. In addition, the DBS industry is in the best position to be flexible, in terms of digital technology, and can add and subtract channels quickly, as required.

A big hurdle for the DBS industry has been its inability, by law, to retransmit local TV broadcast signals into consumers' homes. This meant that consumers often had to subscribe to a basic cable package just to get the news from their local stations. With the passing of a new Satellite Act in 2000, DBS now has the authority to retransmit a local station's signal, as long as a retransmission consent agreement has been reached with the broadcast station or station group.

As with the infamous Time Warner/Disney disagreement on May 1, 2000 (in which ABC was pulled from Time Warner's New York cable TV systems because of a contract dispute, causing both companies much embarrassment and lost customers), several markets have seen their local stations cut off their DBS provider's system until a carriage agreement with the specific station groups could be worked out. It's interesting to note, however, that the DBS industry gained many customers during and soon after the Time Warner/Disney dispute, but little such negative consumer reaction has resulted from similar DBS disputes with local broadcast station groups.

That being said, retransmission of local broadcaster signals is now available from DirecTV in 27 DMAs, and from EchoStar's DISH Network in 121 DMAs, according to both companies. This has helped them experience accelerated growth in core urban and suburban markets.

In addition, both DirecTV and EchoStar are currently distributing HDTV channels from HBO and Showtime, but they, too, have found few customers. Consumers who want the service must buy an HD receiving dish separate from the one for their existing NTSC service, and they must pay a higher subscription fee. There's talk of offering movies in HDTV on a pay-per-view basis, but that hasn't happened yet because there are so few customers.

TELCOS

Although they've had their problems deploying service, no one should underestimate telephone companies (telcos) and their ability to distribute video, audio, and data along with reliable phone service. One need only look at the fact that AT&T now owns one of the major cable operators (the former TeleCommunications, Inc.) to see that telcos are into cable.

Also, the telcos' digital subscriber line (DSL) high-speed Internet access service has become very popular with U.S. consumers. The technology initially took off in places where cable modems were not available but has since gone head-to-head with cable modems and has done quite well. Will the telcos become greater competition for TV broadcasters? Time will tell.

THE DIGITAL CONSUMER WANTS CONTENT

As the selling of new DTV and HDTV sets continues, it has become apparent that the better pictures and sound that DTV promises may not be enough to get Americans to pay considerably more for a set. As with such current digital services as DBS and digital cable, what consumers seem to want most from DTV is more channels, shopping services, pay-per-view opportunities, and Internet access.

TV broadcasters are in the process of transitioning to digital, but it's expensive, and many small-market stations can't afford the cost of converting their infrastructures. What will they do?

The answer lies in the various interactive services that have been introduced over the last two years. Companies such as Wink Communications and OpenTV are bringing the value-added concept of more profits to television, which previously had been missing from the DTV model. By offering these services in their DMA, broadcasters can provide their viewers with more kinds of content (game-playing, home shopping, Internet access) while also transmitting multiple standard-definition (480 lines of resolution) channels to provide more choices for the viewer.

The DTV revolution is about real estate—screen real estate, that is. Although currently in its infancy, once stations learn how to utilize their spectrum more efficiently—packing data and interactive services into the same bandwidth—enhanced television could be the key to their future success.

That's the inherent beauty of DTV. Not only can digital broadcasts include different kinds of video content, they can also feature multiple video programs simultaneously in a single transmission. If an NTSC broadcaster is transmitting a documentary, for example, that show must be simulcast on the same broadcaster's DTV transmission, according to the rules of digital transmission laid down by the FCC. Yet, that broadcaster can also carry a movie, a kid's show, and weather updates as well. Again, more content.

During a cooking show, the actual recipe being prepared can be downloaded to a consumer's TV. The educational activities of PBS can be that much more enhanced via data broadcasting as well. The opportunities are vast, and the investment can be lessened if stations partner with program producers and technology providers.

In addition, personal video recording (PVR) devices from such companies as TiVo and ReplayTV have come on the scene, enabling consumers to watch only what they want (PVRs can, for example, find and record every available show on fishing, or cooking, or movies about World War II). PVRs may make VCRs obsolete or change consumers' viewing habits. They may make it possible for broadcasters to sell more screen real estate and ancillary merchandise. PVRs are still complex to use, but they may become part of future set-top boxes in mass deployment. Clearly, the age of DTV is redefining television, and what we have known as "broadcast TV" is in for major redefinition. Where the areas of opportunity lie for content creators is still to be determined, but demand for new content of one kind or another seem assured.

THE INTERNET

The value of the Internet in the DTV puzzle can't be overstated. The Fox Network, for example, has developed an animated, interactive link to the Web sites of its owned and operated stations that will include ways for its stations to generate income via online merchandising. Every other major U.S. network has similar activities already working. Watch a show and buy a sweater just like the one the star is wearing. See a documentary and click on an icon to purchase a VHS copy. Like the product advertised? Click, enter your credit card number, and it's in your mailbox later that week.

There's no doubt that people are using the Internet. A University of Texas study has found that over $300 billion was spent online in 1999. Once the convergence of TV and the Internet becomes transparent to the end user—as many experts predict—TV stations (or cable companies or DBS providers) stand to reap rich rewards.

There's a huge potential for DTV and the Internet, but creative ways of stimulating people's desires to use them for monetary gain have yet to be fully developed. (The success of those online companies that have found a way to generate interest among the clutter, is not easily copied.)

The terrestrial television industry is wrestling with these issue at the same time that it's faced with investing in converting to DTV. And DTV isn't simply a revised version of what broadcast television has always been—it's a delivery "pipeline" carrying multiple forms of content, some of which are interactive or otherwise new. Control of that distribution pipeline into the home is the most important part of broadcasting, cablecasting, DBS, and every other form of content distribution in the new millennium. Perhaps HDTV will ultimately be a premium VOD service for viewers with state-of-the-art home A/V theatres. (Consider: Advertisers have thus far been unwilling to pay more for their commercials to be shown in HD because there are so few viewers.) Whatever happens, digital content creators must keep an eye on the DTV transition and be mindful of the fact that its outcome may challenge them to supply programming of a quality and quantity radically different from what has gone before.

C H A P T E R
E I G H T E E N

FORGET HDTV; GET HDTV!

MARK SCHUBIN

A funny thing happened on the way to the live broadcast of the 100th Anniversary concert of the Philadelphia Orchestra in the autumn of 2000. A different funny thing happened on the way to the taped broadcast of the Metropolitan Opera's *Semiramide* a decade earlier. The difference between those funny things suggests that HDTV production equipment has come a long way.

Semiramide was shot in high-definition television (HDTV). That prevented the show's director from having as many cameras as he would have liked to have used and also restricted their positions. The restriction on the number of cameras was based, in part, on the extremely limited HDTV production switcher. Even the few cameras used stretched the capability of the switcher so much that one camera had to replace the "black" signal. The director couldn't fade to black, but that wasn't a big problem, because the show would later be edited. Another problem was the cost of the HDTV equipment. Each camera cost more than $300,000. Some of the lenses used cost even more, and they had to be flown in from Japan.

The *Semiramide* HDTV cameras used multicore camera cables with a maximum-length restriction. That, in part, governed where the cameras could be placed in the auditorium.

In 1990, when standard-definition cameras all used solid-state imaging chips, HDTV cameras used tubes. They were much less sensitive than their standard-definition counterparts and exhibited the many flaws that camera tubes were known for: stickiness, lag, comet-tailing, blooming, etc.

The open-reel digital HDTV videotape recorders cost even more than the cameras or lenses, and they were huge and heavy. A forklift was needed to get an extra machine into the production truck, a vehicle so small that foam had to be placed on the ceiling above the director so that he would not hurt himself if he got too enthusiastic.

The digital HDTV tape stock cost as much as $1,500 an hour, and it needed to be *burnished* before use: Each reel had to be run through the machine slowly without recording, so that the spinning heads could scrape the rough surface of the tape. Then the tape had to be rewound and the heads cleaned thoroughly before normal recording could take place.

Every HDTV connection or patch in that 1990 HDTV truck involved three separate signals. A misconnection wouldn't necessarily be obvious during recording, but it could be deadly in editing. When it came time for the editing, the necessary personnel had to fly to Japan. As was the case at the time, the HDTV signals had 1,035 picture-carrying scanning lines, as specified in a standard of the Society of Motion Picture and Television Engineers (SMPTE) that had been *opposed* as an American national standard. The signals also had 60.00 images per second instead of standard-definition video's 59.94. That last characteristic meant that different time codes had to be used for the HDTV tapes and the downconverted VHS viewing tapes. Unusual connections and settings were required to ensure that separate audio recordings could be synchronized to the video.

The end result, as seen by U.S. TV viewers, it was generally agreed, was not as good as it might have been had the show been shot with the Met's usual standard-definition facilities. So, why shoot it in HDTV? The show was a co-production with Nippon Hoso Kyokai (NHK), the Japan Broadcasting Corporation. NHK was promoting their HDTV transmission service and wanted the opera in HDTV.

Ten years later, many of the same people who'd worked on that Metropolitan Opera HDTV production of *Semiramide* wound up working on the Philadelphia Orchestra's 100th Anniversary concert. Again, NHK sponsored the show—this time as a live broadcast — but they did *not* request HDTV; an ordinary National Television System Committee (NTSC) feed, of the sort used in the United States since 1953, would be fine. The local PBS station, WHYY, also carried the show live and transmitted it by satellite to PBS—again, as ordinary NTSC. The show had the same director as did *Semiramide*, but this time he wasn't restricted by the technology. Not only could he safely jump up and down in the control room without hitting his head, but he could also have as many cameras as he wanted, put them wherever he wanted, and have the sorts of lenses he wanted. He could fade to black at will, and there was plenty of room left over in the switcher.

There was no need to increase stage lighting for insensitive cameras, and the very thought of having to deal with lag, comet-tailing, and other tube-related artifacts was essentially a joke. All of the cameras, of course, used high-sensitivity solid-state imaging chips.

Naturally, the signals were recorded onto relatively inexpensive videocassettes inserted into small, lightweight recorders, not gargantuan open-reel machines demanding pre-burnishing of expensive tapes as a condition of possible success. There was no difference in frame rate between the VHS recordings and the more professional formats, so there was only one time code. Video patching required only one wire per path, but most signals simply passed through a routing switcher. There were no opposed standards in use. In short, the concert was a standard, multicamera, live event. The ordinary, NTSC pictures looked gorgeous— better than the look of any other show from the same venue that anyone could recall. But the show was shot in HDTV.

Why? The name on the side of the All-Mobile Video production truck said it all. Originally, when it was the second of three HDTV mobile facilities, the vehicle had been called HD-2. When All-Mobile Video got it, they renamed it "HD, Too." In other words, it's a full-fledged, high-end production truck, fully equipped to provide standard-definition tapes and feeds. If one wants HDTV, however, it's "HD, Too."

What's going on? HDTV has finally come of age. All of the cameras and lenses in the truck are HDTV. The production switcher is HDTV. There are plenty of HDTV recorders. But, thanks to built-in downconverters in all of the cameras, recorders, and switcher, there are also plenty of *standard-definition* recorders and feeds. The truck is special in not treating HDTV as being special. It's where we've already come in color.

Suppose you're shooting video for a director who wants everything in black and white. Do you shoot with a black-and-white camera and record on a black-and-white recorder? Of course you don't!

There are only a few black-and-white video cameras available for purchased these days, and none of them are of the quality necessary for video production. There may no longer be *any* black-and-white recorders available for purchase.

That wasn't always the case, of course. From the mid-1950s through the mid-1960s, it was much harder to find color video equipment than black-and-white. And the use of that color equipment usually resulted in worse black-and-white images than would have been obtained with black-and-white shooting equipment.

Times have changed. Video in 2001 is not the same as it was in 1956. There's little question that color equipment would be used to shoot essentially *anything* today, including black-and-white video. In 1956, color equipment needed the adjective *color*; today, *black-and-white* is the unusual adjective.

Someday, the same will be true of HDTV. Someday, *all* video production will use HDTV equipment in the same way that all video production today uses color equipment.

Today is not yet that day, but something extraordinary happened at the National Association of Broadcasters (NAB) convention in April 2001. Sony offered the HDW-2000 HDCAM HDTV videotape recorder for a list price of $40,000. At the same time, they were selling the DVW-500 Digital Betacam standard-definition videotape recorder for $40,600. Yes, the HDTV model was (and is) actually priced *lower* than the standard-definition model.

That was not the case in all of Sony's product categories. Consider camcorders, for example. Back in 1990, when *Semiramide* was being shot, there were no HDTV camcorders of any kind, but the closest equivalent to a Digital Betacam camcorder would have been a separate camera and recorder costing around $700,000 combined. At NAB 2001, an HDW-750 digital HDTV camcorder had a list price of $65,000, less than a tenth as much as the 1990 combination. That's not quite as low as a roughly equivalent Digital Betacam version, the DVW-790WS at $60,100, but it's awfully close. (Panasonic has even less expensive HDTV camcorders, but, for the purposes of this comparison, it's probably best to stick to products from the same manufacturer.)

Before the introduction of their new multiformat-switcher product line at NAB 2001, a Sony HDTV switcher with three mix/effects banks (M/E) and two channels of digital video effects (DVE) cost $1.3 million. Now, a 3-M/E 2-DVE Sony multiformat switcher that can be set to handle either HDTV or standard-definition signals lists for just $500,000, not far from the $475,000 of Sony's pre-NAB-2001 4-M/E 2-DVE standard-definition digital switcher.

HDTV prices are not yet entirely magical. Sony's new Xpri nonlinear editing system can cost $150,000 for an HDTV version versus $90,000 for a standard-definition version, because of the additional memory required for HDTV. Sony's highest-end 20-inch standard-definition evaluation monitor, the BVM-20E1U, is just about $10,000; although the 20-inch BVM-D20F1U HDTV monitor is only $12,900, it's actually not quite as sharp as the standard-definition version (another amazing fact about HDTV in 2001). The nice-and-sharp BVM-D24E1WU HDTV monitor is $23,000, but it looks better and does more. At the Philadelphia Orchestra show, for example, one such monitor in the control room was set to show HDTV detail; another was set to show true NTSC (not merely standard-definition video, but color-subcarrier–affected NTSC), so that there would be no question of what was being transmitted to homes.

What was transmitted to those homes looked beautiful, thanks both to the oversampling inherent when an HDTV camera is used for standard-definition video and to improved technology and technique used to convert HDTV signals to standard-definition. Even after HDTV cameras started using sensitive imaging chips and produced beautiful HDTV pictures, their downconverted imagery sometimes looked bland when viewed on an NTSC TV. Today, appropriately downconverted HDTV looks better than the best standard-definition–shot material—even when viewed on an ordinary TV set from an ordinary NTSC broadcast.

Contrary to some belief, there is no U.S. government requirement for any U.S. broadcaster to carry any HDTV at any time, now or in the future. All commercial television broadcasters are supposed to be transmitting *digitally* by May 1, 2002, but that digital transmission need not be HDTV. In fact, WJLA-DT, the digital ABC affiliate in Washington, D.C., is equipped to transmit only standard-definition programming and is in accord with all laws, rules, and regulations in doing so. Noncommercial broadcasters have until May 1, 2003 to transmit digitally, but they don't have to carry HDTV either. Cable operators are under no obligation to transmit HDTV (but if they carry the digital signals of a broadcaster that is transmitting HDTV, they're supposed to deliver that signal as HDTV). Satellite-service providers also have no HDTV obligations.

In theory, 2006 was to have been the date when NTSC analog television stations were to have been shut down, leaving only digital broadcasts. There may no longer be a single person who believes in the 2006 deadline, but, even if it were, none of the digital broadcasts would have to be in HDTV.

If that's the case, why should anyone even consider HDTV now? First, it offers improved picture quality, visible even on ordinary TV sets receiving ordinary (non-HDTV) signals. Second, it helps make programming "evergreen." A program may not have any HDTV distribution today, but will that always be the case? HDTV also looks better when converted to a foreign image format than does standard-definition TV, and it's good for extracting photographic still images.

Third, HDTV is unquestionably the wave of the future. As the pricing and availability at NAB 2001 shows, it is well on its way towards become tomorrow's "color." Fujinon showed the longest zoom-range lens at NAB 2001, an 87:1 following on the heels of the 86:1 that Canon introduced last year. Both are HDTV lenses (usable on standard-definition cameras). It's no longer the case that HDTV means using a more restricted range of equipment.

Still, HDTV is not yet for everyone. Buying a single HDW-2000 to replace an aging DVW-500 is probably not such a good idea. The new deck may be $600 cheaper and include a built-in downconverter, but it won't play existing Digital Betacam tapes. The HDW-M2000 will (as well as Betacam, Betacam SP, Betacam SX, and MPEG IMX), but it costs $60,000, not $40,000.

It seems almost as foolish, however, *not* to consider HDTV when making a major equipment purchase. Is it wise to buy a routing switcher today that can't handle HDTV? Might it be a smart investment to get an HDTV camcorder and a multiformat player? Once upon a time, not so very long ago, the only reason to buy HDTV equipment was to distribute HDTV programming. Now it's arguably the highest-quality way to make even standard-definition programming. Sony's multiformat decks and JVC's and Panasonic's backwards-compatible recorders, as well as a broad range of converters from many manufacturers, allow HDTV and standard-definition equipment to coexist. So they live happily ever after.

COMPROMISES

Today's HDTV cameras, like high-end standard-definition cameras, use $^2/_3$-inch imaging chips (which do not have a single $^2/_3$-inch dimension). That's good, because it allows standard-definition lenses to be used on HDTV cameras in a pinch. That may be bad, however, because the optical path appears to have lost some of its quality.

Today's HDTV recorders usually use $^1/_2$-inch or even $^1/_4$-inch tape in convenient cassettes instead of 1-inch tape on open reels. That's good. But the old recorders captured uncompressed signals with 1,920 picture elements (pixels) of brightness detail across the picture and 960 of color detail; most of the newer recorders capture (in 1080I mode) compressed signals with either 1,440 and 480 (HDCAM) or 1,280 and 640 (D-9 HD and DVCPRO HD). That's probably not as good.

Two versions of HDTV cassette recorders, HD D-5 and D-6 (the latter using 3/4-inch tape), capture the full 1,920 and 960. That's good. But neither is available in a camcorder. That's bad.

The compressed D-5 HD format is more expensive than DVCPRO HD, D-9 HD, or HDCAM—enough to discourage a videographer from buying HDTV equipment. That's bad. But HD D-5 is downright inexpensive compared to D-6. That's... YIKES!

THAT WAS THEN; THIS IS NOW: 1991–2001

60.00 images/second—59.94 images/second

Different time code—same time code

Tube cameras—chip cameras

Poor sensitivity—good sensitivity

Lag, stickiness—clean images

One-inch image format—2/3-inch image format

Limited-length multicore cable—fiberoptic cable

Limited lenses—big range of lenses

Huge, heavy open-reel recorder—small cassette recorder

Burnish expensive tapes—pop in inexpensive cassettes

No camcorders—range of camcorders

$1M+ camera, recorder, lens—under-$100k package

Tiny production switchers—normal production switchers

Three-line connections—one-line connection

Limited downconversion—built-in downconversion

Poor downconversion—excellent downconversion

CHAPTER NINETEEN

THE PROMISE OF DIGITAL INTERACTIVE TELEVISION

JERRY C. WHITAKER, EDITOR

The television industry is entering a new era in service to the consumer. Built around two-way interactive technologies, the rollout of the digital television (DTV) infrastructure opens up a new frontier in communication. Two worlds that were barely connected—television and the Internet—are now on the verge of combining into an entirely new service: namely, interactive television. Thanks to the ongoing transition of television from analog to digital, it is now possible to combine video, audio, and data within the same signal.

This combination leads to powerful new applications, with limitless possibilities of great commercial potential. For example, computers can be turned into traditional TV receivers and digital set-top boxes can host applications such as interactive TV, e-commerce, and customized programming.

INTERACTIVE TV

Two terms are commonly used to describe the emerging advanced television environment: *interactive television* and *enhanced television*. For the purposes of this chapter, we will use the term interactive television (ITV) and define it as anything that lets a consumer engage in action with the system using a remote control or keyboard to access new and advanced services. General categories of use include the following:

- The ability to select movies for viewing at home
- E-mail and on-line chat
- News story selection and archive
- Stock market data, including personal investment portfolio performance in real-time
- Enhanced sports scores and statistics on a selective basis
- On-line real-time purchase of everything from groceries to software without leaving home

There is no shortage of reasons why ITV is quickly gaining momentum—and will continue to do so as new technologies take hold. The backdrop for ITV growth comes from both the market strength of the Internet and the technology foundation that supports it. With the rapid adoption of digital video technology in the cable, satellite, and terrestrial broadcasting industries, the stage is now set for the creation of an ITV segment that meets the test of sound economic principles and introduces to a mass consumer market a whole new range of possibilities.

For example, services are becoming available that offer interactive features for game shows, sports and other programs, interactive advertising, e-mail, and Internet access as a package deal. Rather than concentrating just on Web services, the goal is to deliver a better television experience. An important component of such services is the hard-drive–based video recorder. It is practical, with video compression, to store as much as 30 hours of television programs locally at the consumer's receiver.

As the price/performance ratio of hard drives continues to move forward, even more storage will be practical. This capability opens the doors to numerous innovations and features never before possible.

Other services of the DTV era offer customized channel guides that make it easy to search for shows that subscribers want to watch, or to select future TV shows to record. Such services also make it possible for consumers to create their own customized channels filled with their favorite TV shows or custom channels that contain movies by favorite actors or directors.

Soon the entire video distribution system will handle digital media only. These digital media offer content providers and consumers much higher picture quality. Unlike analog, this quality is ensured no matter how many copies are made. The digital content can be distributed in an MPEG-2 compressed format over high-bandwidth digital satellite, cable, terrestrial broadcast, Internet, or by fixed media. In the home it can be played via set-top boxes, other consumer electronic devices, or PCs, all connected by an IEEE 1394 high speed home network bus.

The potential benefits of the digital environment go beyond picture quality. The data packets associated with digital content give the content provider access to more functions and business opportunities. For example, video can be tagged with index pointers that allow the viewer to jump to a particular scene or to links that access supplementary information on a local disk or anywhere on the Internet. *Conditional access* methods allow the content provider to control viewing according to their business paradigms; pay-per-view can be enhanced by controlling who views and how much viewing they are entitled to; what price they pay, depending on loyalty and other personal criteria; when and/or where they are; and so on.

Digital media of this type requires an investment in new infrastructure at the level of the content provider, distributor, and consumer. This infrastructure will help content providers and distributors take advantage of the ongoing business opportunities created by the new digital media.

COPY PROTECTION

In digital technology, every copy of a program is a perfect copy. The digital era introduces new challenges to content producers and content providers. Their intellectual property must be protected so that their investments are not compromised. Also, in order for consumer electronics devices to be successful in the marketplace, they must be relatively low cost and easy to use. And yet, most consumers are not pirates and any scheme protecting the content must also take into account the perceived rights of the consumers and the ease with which they obtain, store, and continue to use the media they purchase.

OVERVIEW OF INTERNET PROTOCOLS

The Internet architecture uses three core transport protocols:

- *Internet Protocol (IP).* A network protocol that defines, among other things, the structure of an IP *datagram*. A datagram has a payload and a header. The header contains a source and a destination address that has global significance. These addresses allow datagrams to be individually routed throughout the entire network on a "best-effort" basis. There is actually no guarantee at the IP layer that a datagram is correctly transmitted or that a sequence of datagrams will arrive at their destination in the same order that they were sent. Above the IP protocol, the User Datagram Protocol and Transport control Protocol serve to provide end-to-end functionality between a pair of host systems.
- *User Datagram Protocol (UDP).* A service above the IP layer that provides port multiplexing and a checksum that may or may not be used. The UDP does not require the addressee to acknowledge whether or not data transmitted actually arrived properly. As such, it is not a reliable protocol, but it is still to be preferred in the context of real-time streaming applications, such as video and audio, when the ability to time packet transmission against an external clock source is of much greater importance than avoiding possible data losses.

- *Transport Control Protocol (TCP).* A service that adds more functionality than UDP at the price of a longer header. Along with IP, TCP guarantees a reliable, serialized, and fully duplexed channel. A stream of bytes generated by the sender will be passed across the Internet by TCP so that it is presented o the addressee as the same sequence of bytes and in the same order as that generated by the sender. Unlike UDP, TCP requires a return path to send acknowledgment messages.
- TCP and IP are commonly paired for networking purposes, referred to as TCP/IP.

DATA BROADCASTING

Data broadcasting (*datacasting*) offers the potential for entirely new revenue sources. Using modern encoders, even a high-definition (HD) broadcast requires only 16 to 18 Mbits/s of digital bandwidth, out of the 19.38 Mbits/s available in a 6 MHz broadcast band under the ATSC DTV standard. This leaves a significant amount of bandwidth that can be used for arbitrary data. Figure 19.1 illustrates one possible scenario.

There are basically two ways in which broadcasters can use this available bandwidth to generate additional revenue:

- Use the excess bandwidth to broadcast data that enhances the appeal of their TV programming and/or TV advertising in an attempt to attract more advertising dollars.
- Lease the excess bandwidth to other enterprises that want to distribute data to large numbers of users in the broadcaster's viewing area.

In practice, both of these approaches will likely be used to varying degrees by different broadcasters.

For datacasting applications targeted to consumers, a key requirement for success is that large numbers of consumers have a DTV receiver that can receive and use the broadcast data. Utilizing broadcast data requires not only standards for

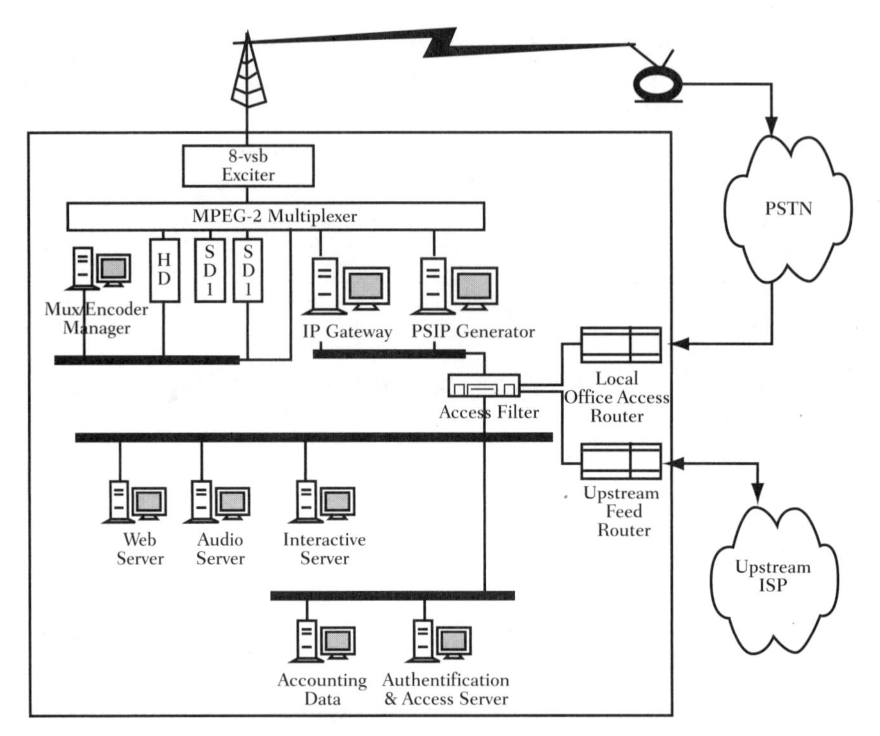

FIGURE 19.1 A broadcasting and datacasting terrestrial station.

encoding the data in the broadcast stream, but also standards for applications in the receiver to operate on the data. This could be in the form of specifications for one or more standard applications in the receiver, such as a standard HTML engine. Alternatively, it could be in the form of specifications for a standard execution environment in the receiver, so that a variety of applications could be downloaded from the broadcast stream and executed.

Datacast applications targeted to consumers can be further classified by the degree of coupling to the normal TV programming:

- *Tightly coupled data* is intended to enhance TV programming in real time. The viewer tunes to the TV program and receives the data enhancement along with it. In many cases the timing of the display of the broadcast data is closely synchronized with the video frames being shown.

- *Loosely coupled data* is related to the program, but are not closely synchronized with it in time. For example, an educational program might send along in the broadcast some supplementary reading materials or self-test quizzes. These might not even be viewed at the same time as the TV program. They may be saved in the DTV receiver and perused later.
- *Noncoupled data* is typically contained in separate "data-only" virtual channels. This may be data intended for real-time viewing, such as a 24-hour news headline or stock ticker service, or it may be data intended for use completely outside the DTV context.

COMPUTER APPLICATIONS

One of the characteristics that set the ATSC digital television effort apart from the NTSC (conventional television) efforts of the past is the inclusion of a broad range of industries—not just broadcasters and receiver manufacturers, but all industries that have an interest in imaging systems. The most visible of these allied industries is the computer business. With computers integrated into television receivers, consumers will have a host of new options and services at their fingertips, hence facilitating the public interest, convenience, and necessity.

In an industry that has seen successive waves of hype and disappointment, it is not surprising that such visions of the video future are treated skeptically, as least by broadcasters who see these predictions by computer companies as an attempt to claim a portion of their turf.

The reality that has emerged, however, is one of cooperation among a host of industries ranging from broadcasters to cable companies to computer hardware and software manufacturers.

HDTV AND COMPUTER GRAPHICS

The core of HDTV production is the creation of high-quality images. As HDTV emerged in the early 1980s, a quite separate and initially unrelated explosion in electronic imaging was also under way in the form of high-resolution computer graphics. This development was propelled by broad requirements within a great variety of business and industrial applications, including:

- Computer-aided design (CAD)
- Computer-aided manufacturing (CAM)
- Printing and publishing
- Creative design (such as textiles and decorative arts)
- Scientific research
- Medical diagnosis

A natural convergence soon began between the real-time imagery of HDTV and the non–real-time high-resolution graphic systems. A wide range of high-resolution graphic display systems is commonly available today. This range addresses quite different needs for resolution within a broad spectrum of industries. The ATSC DTV system enjoys a good fit within an expanding hierarchy of computer graphics. This hierarchy has the range of resolutions that it does because of the varied needs of countless disparate applications. HDTV further offers an important wide-screen display organization that is eminently suited to certain critical demands. The 16:9 display, for example, can efficiently encompass two side-by-side 8 × 11-inch pages, important in many print applications. The horizontal form factor also lends itself to many industrial design displays that favor a horizontally oriented rectangle, such as automobile and aircraft portrayal.

This convergence will become increasingly important in the future. The use of computer graphics within the broadcast television industry has seen enormous growth during the past decade. Apart from this trend, however, there is also the potential offered by computer graphics techniques and HDTV imagery for the creation of special effects within the motion picture production industry, already demonstrated convincingly in countless major releases.

ATSC DATACASTING

Although the primary focus of the ATSC DTV system is the conveyance of entertainment programming, datacasting is a practical and viable additional feature of the standard. The concept of datacasting is not new; it has been tried with varying degrees of success for years using the NTSC system in the United States, and PAL and SECAM elsewhere. The tremendous data throughput capabilities of DTV, however, permit a new level of possibilities for broadcasters and cable operators.

In general, the industry has defined two major categories of datacasting:

- *Enhanced television.* Data content related to and synchronized with the video program content. For example, a viewer watching a home improvement program might be able to push a button on the remote control to find more information about the product being used or where to buy it.
- *Data broadcast.* Data services not related to the program content. An example would be current traffic conditions, stock market activity, or even subscription services that utilize ATSC conditional access capabilities.

Effective use of datacasting could have far reaching effects on advertising and commercial broadcaster business models. A new generation of intelligent ATSC receivers with built-in Internet browsers and reverse communications channels will be able to seamlessly integrate Internet services with broadcast television.

DIGITAL HOME NETWORK

It is expected that in the near future, data for audio, video, telephony, printing, and control functions are all likely to be transported through the home over a digital network. This network will allow the connection of devices such as computers, digital TVs, digital VCRs, digital telephones, printers, stereo systems, and remotely controlled appliances. To enable this scope of interoperability in home network devices, standards for physical layers, network information, and control protocols must be generally agreed upon and accepted. Although it would be preferable to have a single stack of technology layers, no one selection is likely to satisfy all cost, bandwidth, and mobility requirements for in-home devices.

From a broadcasting perspective, the ability to provide unrestricted entertainment services to consumer devices in the home is a key point of interest. Also, ancillary data services directed to various devices offer significant marketplace promise. Control and protocol standards to enable the delivery of selected programming from cable set-top boxes to DTV sets using IEEE 1394 have been approved by the Consumer Electronics Association (CEA) and the Society of Cable and Telecommunications Engineers (SCTE).

PREDICTIVE RENDERING

Traditionally, television-based content producers have a great deal of control over how their product appears to the customer. The conventional television paradigm would be problematic if there were no assurances that a content producer could predict how their content would be rendered on every receiver-display combination in use by consumers. The requirement for *predictive rendering*, then, is essentially a contract between the content developer and the receiver manufacturer that guarantees the following parameters:

- Content will be displayed at a specific time
- Content will be displayed in a specific sequence
- Content will look a certain way

The presentation engine supports these requirements through a well-defined model of operation, media synchronization, pixel-level positioning, and the fact that it is a *conformance specification*. The model of operation formally defines the relationship between broadcast applications, native applications, television programs, and on-screen display resources.

Pixel-level positioning allows a content author to specify where elements are rendered on a display. It also allows content authors to specify elements in relation to each other or relative to the dimensions of the screen.

The presentation engine architecture consists of five principal components:

- *Markup language*, which specifies the content of the document
- *Style language*, which specifies how the content is presented to the user
- *Event model*, which specifies the relationship of events with elements in the document
- *Application programming interfaces*, which provide a means for external programs to manipulate the document
- *Media types*, which are simply those media formats that require support in a complaint receiver

DTV PRODUCT CLASSIFICATION

In the analog days, the interconnection of a television set with a peripheral device was a relatively minor task. In the digital era, however, understanding what devices will work together is not a minor consideration. For this reason, the Consumer Electronics Association announced new definitions and labels for DTV products. The definitions were expected to be incorporated in to manufacturers' television marketing materials as DTV receivers entered retail channels.

The CEA Video Division Board resolved that analog-only televisions (televisions and monitors with a scanning frequency of 15.75 kHz) should not be marketed or designated to consumers as having any particular DTV capabilities or attributes. In a second related resolution, the Board agreed that the new definitions for monitors and tuners should be used by all manufacturers and retailers to replace general, nonindustry terminology like "DTV-ready" or "HDTV-ready." They also defined minimums for HDTV displays as those with active top-to-bottom scan lines of 720 progressive or 1,080 interlaced, or higher. Manufacturers were also to disclose the number of active scan lines for a high-definition image within a 16:9 aspect ratio "letter boxed" image area if the unit has a 4:3 HDTV display.

The CEA digital television definitions as follows:

- *High-definition television (HDTV).* A complete product/system with the following minimum performance attributes: (1) receives ATSC terrestrial digital transmissions and decodes all basic ATSC video formats (commonly referred to as Table 3 formats); (2) has active vertical scanning lines of 720P, 1080I, or higher; (3) is capable of displaying a 16:9 image; and (4) receives and reproduces, and/or outputs Dolby Digital audio.
- *High-definition television monitor.* A monitor or display with the following minimum performance attributes: (1) has active vertical scanning lines of 720P, 1080I, or higher; and (2) is capable of displaying a 16:9 image. In specifications found on product literature and in owner's manuals, manufacturers were required to disclose the number of vertical scanning lines in the 16:9 viewable area, which must be 540P, 1080I or higher to meet the definition of HDTV.
- *High-definition television tuner.* An RF receiver with the following minimum performance attributes: (1) receives ATSC terrestrial digital transmissions and decodes all ATSC Table 3 video formats; (2) outputs the ATSC Table 3 720P and 108P/I formats in the form of HD with minimum active vertical scanning lines of 720P, 1080I, or higher, and (3) receives and reproduces, and/or outputs Dolby Digital audio. Additionally, this tuner may output HD formats converted to other formats. The lower resolution ATSC Table 3 formats can be output at lower resolution levels. Alternatively, the output can be a digital bitstream with the full resolution of the broadcast signal.
- *Enhanced definition television (EDTV).* A complete product/system with the following minimum performance attributes: (1) receives ATSC terrestrial digital transmissions and decodes all ATSC Table 3 video formats; (2) has active verti-

cal scanning lines of 480P or higher; and (3) receives and reproduces, and/or output Dolby Digital audio. The aspect ratio is not specified.

- *Enhanced definition television monitor.* A monitor or display that has active vertical scanning lines of 480P or higher. No aspect ratio is specified.
- *Enhanced definition television tuner.* An RF receiver with the following minimum performance attributes: (1) receives ATSC terrestrial digital transmissions and decodes all ATSC Table 3 video formats; (2) outputs the ATSC Table 3 720P and 1080I/P and 480P formats with minimum active vertical scanning lines of 480P; and (3) receives and reproduces, and/or outputs Dolby Digital audio. Alternatively, the output can be a digital bitstream output capable of transporting 480P, except the ATSC Table 3 480I format, which can be output at 480I.
- *Standard definition television (SDTV).* A complete product/system with the following performance attributes: (1) receives ATSC terrestrial digital transmissions and decodes all ATSC Table 3 video formats, and produces a usable picture; (2) has active vertical scanning lines less than that of EDTV; and (3) receives and reproduces usable audio, No aspect ratio is specified.
- *Standard definition television tuner.* An RF receiver with the following minimum performance attributes: (1) receives ATSC terrestrial digital transmissions and decodes all ATSC Table 3 video formats; (2) outputs all ATSC Table 3 formats in the form of NTSC output; and (3) receives and reproduces, and/or outputs Dolby Digital audio.

These industry standard definitions were intended to eliminate the confusion over the product features and capabilities of television sets and monitors intended for DTV applications. The agreement promised to spur the sale of DTV-compliant sets by injecting a certain amount of logic into the marketing efforts of TV set manufacturers. The consumer electronics industry had come under fire early in their DTV product rollouts because of the use of confusing and—in many cases meaningless—marketing terms. For example, terms such as "DTV-ready" mean different things to different people.

CABLE/DTV RECEIVER LABELING

On the heels of the CEA DTV product classification agreement, the FCC adopted a Report and Order requiring standardized labeling of DTV receivers that are marketed for connection to cable television systems. The order specified that such receivers offered for sale after July 1, 2001, must be permanently marked with a label on the outside of the product that reads: "Meets FCC Labeling Standard Digital Cable Ready (DCR) *x*," where $x = 1, 2$, or 3.

The Commission prohibited marketing of receiving devices claimed to be fully compatible with digital cable services unless they have a functionality of one of the three categories defined. The new rules cover any consumer television receiving device with digital signal processing capability that is intended to be used with cable systems. The rules permit marketing devices with less capability, provided full compatibility is not claimed.

DCR 1 refers to a consumer electronics television receiving device capable of receiving analog basic, digital basic, and digital premium cable television programming by direct connection to a cable system providing digital programming. This device does not have an IEEE 1394 connector or other digital interface. A security card or point of deployment module provided by the cable operator is required to view encrypted programming.

DCR 2, a superset of DCR1, adds a 1394 digital interface connector. It is clear from the FCC order that other digital interfaces also could be present. The FCC noted that connection of a DCR 2 receiver to a digital set-top box (also presumable a level 2 device) may support advanced and interactive digital services and programming delivered by the cable system to the set-top box.

The distinction asserted for DCR 3 is the addition of the capability to receive advanced and interactive digital services. A device with this label is not required to have a 1394 connector.

Because such services are not defined, the meaning of this distinction is unclear. The FCC did state that additional industry work was required to design specifications for the DCR 3 category of receivers, and that it would keep the record open in this proceeding, reserving the option to incorporate these specifications into the rules at a later date.

The FCC required the consumer electronics and cable industries to report their progress on developing technical standards in two other areas: direct connection of digital TV receivers to digital cable television systems and the provision of tuning and program schedule information to support on-screen program guides for consumers. The FCC said these two issues had been substantially, but not completely, resolved in an earlier agreement. Reporting requirements were consolidated into a single reporting timetable than began on October 31, 2000, and every six months thereafter until October 2002.

When these rules were announced by the Commission, it was unclear how much the Report and Order would aid in the delivery of DTV over a cable system to a DTV receiver, other than preventing some grossly misleading marketing practices. The actual connection of the cable systems' coaxial cable to the DTV set is not covered. Presumably the products will use the relevant CEA or Society of Cable Television Engineers standards for the coax interface, which were being harmonized as this book went to press. The FCC also stopped short of selecting the standard for transport of signals over the IEEE-1394 physical interface (technically only requiring a connector, not any signaling through this connector). It is unlikely that a manufacturer will put a 1394 connector on a product without some implementation of a protocol using one of the 1394-based protocol standards. Unfortunately, TV receivers with the DCR 2 label are not assured to work with set-top boxes with the same label because of multiple 1394-based protocols.

CHAPTER TWENTY

INTERACTIVE ENHANCED TELEVISION—A HISTORICAL AND CRITICAL PERSPECTIVE

Tracy Swedlow
Interactive TV Today,
American Film Institute–Intel Enhanced Television Workshop

What is still a broadcast, passive, linear, entertainment viewing experience for millions of people around the world, television is now becoming an on-demand, participatory, nonlinear, infotainment, advertising-targeted, broadband, two-way communications platform. When fully realized on a mass scale which will require that digital terrestrial signal technology become standardized around the world, an occurrence

This chapter is excerpted from a white paper originally commissioned by the American Film Institute–Intel Enhanced Television Workshop for distribution to participants of its 2000 activities and on its Web site. The purpose of the paper is to provide technical and historical background for the television and creative community who are not principally concerned with technical issues.

still steeped in tremendous controversy—our current experience of television will drastically transform. For the first time, possibly, TV can become something a viewer can control and use for information and communications. At some point in the future, viewers at home may have their own mini-interactive television (ITV) production studio in their living rooms. If that happens, television will not remain a passive delivery vehicle for programming delivered solely by major networks.

In these early stages of a fully integrated ITV environment, a viewer is able to read more about the topics presented during a show at the time he scheduled the show to play back or broadcast; download and store related media files or special interactive documentation for later viewing or perusal; purchase goods associated with a program; share, in context and in real-time, his knowledge or interpretations about the broadcast through various communications applications; use banking, betting, or video-on-demand applications; and finally, participate in competitive or cooperative group activities in association with video content. This is just a beginning.

Those producing ITV shows and applications, however, will soon discover that not one, but hundreds, thousands, or even millions of viewer interest groups will form around the context of shows—each with a different perspective, agenda, and style of communications. Ultimately, this will encourage and eventually require television producers to create shows that consider the shared group communications dynamic experience (possibly for many related groups independently at once) and not the individual or the mass audience solely as a viewer unit. Community or public television, in other words will, potentially—at last—emerge when ITV technologies make video and data content a platform for discussion and participation.

On the other hand, cable and satellite media providers and operators, also called Multiple System Operators (MSOs), are hoping the public will want and pay for basic ITV services such as walled-garden information (interactive content contained or "walled-off" from direct access to Internet users) and Internet portals, video-on-demand, or banking services only. These are near-term, conservative applications; will future content, technology, and business developers envision something more powerful and creative for this new medium? One can hope.

WHAT IS INTERACTIVE TELEVISION?

Interactive TV is, essentially, video programming that incorporates some form of "give-and-take" action—be it with data on video, graphics on video, video within video, or video programming retrieval and recording to digital hard disk drives for further use. To the viewer, "enhancements" appear as graphic and sometimes purely informational overlay elements on the screen, (Some technologies, such as Hyper Video, actually incorporate data enhancements in the video MPEG stream.) These overlays are opaque or transparent, or semitransparent. Common overlay elements are icons, banners, labels, menus, interface structures, and open text fields in which the viewer can insert his email address, forms to fill out to buy a product, or commands to retrieve and manage video streams and graphics on a relevant Web page. Interactive or accessible information data, of course, is the most important new addition to the television landscape.

If the producer has done his job adequately, these enhancements will be part of the television program. In some cases, the viewer may want to access information that is irrelevant to the current programming such as news, stocks, scores, weather, and so on. To understand what these graphic elements look like, visualize the way semitransparent banners with statistics printed on them appear on TV now during basketball games, car racing events, or golf tournaments. During a recent broadcast of the Milwaukee 250 car race, for instance, semitransparent graphics boxes appeared from time to time featuring information about which drivers were in the lead, their backgrounds, racing factoids, speed statistics, commentator information, and other informational tidbits. These producers were clearly aware that audiences hunger for more supportive data in their entertainment programming and copped an interactive television approach. Good for them.

Another long-standing example of ITV-like video programming is the data boxes that appear in the corner of the TV screen during music videos on MTV or when a gameplayer sets up a Nintendo, PlayStation, or Sega console experience. Here, the player navigates graphic or text elements with a keyboard or joystick to select the difficulty level of the game or learn about its rules. Another good example of ITV-like programming is the analog TV Guide Channel (a scrolling TV listings service) still carried on many cable and DBS systems. Although not interactive, this guide demonstrates a type of interface that interactive TV technologies will exploit. Here, the video screen is reduced to one corner of the screen while the viewer browses a data schedule. Rather rudimentary in its creativity, an interactive digital program guide now exists, although many complain about its rigidity. Many MSOs have deployed these electronic programming guides (EPGs)* in association with TV Guide or Gemstar (these companies have now merged).

Concerning the navigation of such ITV platforms as the EPG, viewers use buttons on the remote control to tab from place to place, type commands or words with a wireless keyboard, or use a PC-like roaming mouse resident on the keyboard. In the United Kingdom and in Europe, remote controls (called handsets there) and wireless keyboards aggressively exploit primary-colored buttons (red, green, yellow, blue) called Fast Keys. These simple buttons provide a consistent navigational infrastructure—something U.S. manufacturers have yet to exploit. In the future, we may also see voice commands via one's remote control or cell phone, speaker-driven commands for sound-sensitive TVs, touch screens on consumer televisions, lab devices for more complex interaction, and more.

CONTENT NOW AND IN THE FUTURE

Certain types of ITV programming are beginning to thrive in a commercial setting even in these early stages EPGs: synchronized

*This is one of the most hotly contested areas of ITV development. Due to the fact that EPGs are easily becoming the portal application to the new television experience, many companies are trying to develop their own versions of the EPG outside of Gemstar's purview. At this time, Gemstar controls many of the patents surrounding this technology and vigorously pursues litigation. In the future, producers and even individual viewers may be able to design and build their own. If this occurs, adoption of ITV will explode. One such company which has developed this technology is iSurfTV.

TV applications and integrated interactive TV programming such as interactive news, sports, 3D games and game shows, home shopping, court programs, weather channels, educational documentaries, and advertising. In the future we may see new "content platforms" in distance learning, live town-hall meetings, real-time voting statistics during political conventions, interactive situation comedies (remember *Fahrenheit 451*?), financial programs, and documentaries. The possibilities are endless.

As already mentioned, the most well-viewed and used example of ITV technology is the EPG. EPGs are widely available to millions of subscribers on digital cable and DBS systems, especially in the United States. These EPGs appear interactively when the viewer calls it by pushing a button (e.g., Menu or Guide) on the remote control, or by some other method. Once displayed, the EPG allows the viewer to easily navigate or search for programming by time, theme, channel, and so on. Summaries of shows are often included. Those companies leading the development and deployment of EPGs include Gemstar, TV Guide, GIST, ReplayTV, and TiVo. These last two have aggressively reinvented the concept of the EPG by coupling an online data service broadcast through the phone line to a set-top box with a recordable digital video hard disk drive. These units, called digital video recorders (DVRs), have a profound influence on the way the average person becomes familiar with and knows about the availability of such video-based interactive services.

Looking ahead, MSOs and set-top box makers are planning for that time when they can support EPGs and other types of interactive entertainment programming alongside digital voice streams. A realization of the Video Phone—promised to technophiles for decades, and now called "Video Telephony"—may prove to be a real killer app. Producers and developers must consider how to integrate video telephony into their programming as soon as it is feasible.

THE LATEST TECHNOLOGIES

A few important underlying ITV technologies have appeared that enable producers to "author once, broadcast everywhere," the professed goal of industry developers. These are based on the HTML-based 1.0 specification developed by members of the Advanced Television Enhancement Forum (ATVEF—pronounced "atveff" by some), the Digital Video Broadcasting—Multimedia Home Products protocol (DVB-MHP), the JavaTV application programming interface from Sun Microsystems, and XML for controlling metadata supported by the TV Anytime Forum. These technologies are continually mired in controversy because not one of them has become a worldwide standard. Companies in Europe and American firms such as Liberate, Sun Microsystems, OpenTV, Excite@Home, ICTV, PowerTV, Microsoft, and interoperability organizations like CableLaboratories (CableLabs), promote discussions and workgroups to invent the answers, but no firm conclusions have been made. Meanwhile, such technologies are finding acceptance on a regional basis. ATVEF has found favor in North America. In Europe, companies have embraced DVB-MHP. XML is just now gaining real popularity within the Internet community, which may carry over into ITV. Unless a standard can be agreed upon, however, ITV technologies and, therefore, content programming will be expensive to produce.

PAYING FOR ITSELF

Production budgets on ITV projects are escalating. Today, a producer might spend perhaps $70,000 to $3 million to get it right. Before long, these budgets will increase as technology improves and audiences demand more functionality and enhanced shows. Production companies are developing new business models that reflect complex revenue sharing arrangements between producer, set-top box vendor, software provider, MSO, shopping vendor, Internet Service Provider (ISP), and billing vendor, in order to get commitments on a project. Some revenues may come from the viewers, of course, through tiered subscriptions. Additional revenues will come from interactive and/or targeted advertising. T-commerce (interactive commerce on television), on the other hand, will be the greatest revenue generator, or so the theory goes. Content producers are discovering that, in order to pay for new ITV programming, they will have to build a store, too. Before embarking on an ITV production, therefore, it is important to develop a business plan that considers T-commerce. Although some producers might shudder at this prospect, viewers may enjoy the accessibility of products relevant to programming. On the other hand, this close relationship between consumerism and content can be abused. Our hope is that we will build a medium that outshines its initial promise as a "revenue generator" and becomes an influential medium of personal and cultural expression. But to understand where ITV may go, let's first discuss its origins.

EVOLUTION

Years before television was invented, people spoke as if film, radio, and the telephone would some day converge. Sound familiar? In the early part of this century, terms like "Radiovision" and "Telephone Eye" were used to express a future device that might provide an integration of services. Although the electronic transmission of pictures ("television") is what they got, the idea that these technologies could be combined into one device became a deeply embedded dream back then. Today, our PC-TVs, NetTVs, or even video and teleconferencing services echo that long held vision. Real ITV is something much more complicated, though: inventing it has taken great investment, innovation, trial, error, and further exploration. One of the earliest innovations to have an influence on the generation that would invent ITV was a show called *Winky Dink*.

THE TELEVISION AS TOY

Some people may remember *Winky Dink*—a program first broadcast in October of 1953 in black and white, on the CBS network. Created and hosted by future *Joker's Wild* game show host, Jack Barry, *Winky Dink* featured the adventures of a cartoon character named *Winky Dink* and his dog Woofer. The simply drawn character was a small boy (voiced by Mae Questal of "Betty Boop" fame) with ragged hair who appeared on a TV set next to Barry. Winky Dink talked intermittently with Barry and the kids in the studio audience. During the program, Winky Dink went on dangerous cartoon adventures and got into a lot of trouble. To save him from his perils, Barry came up with a unique gimmick: the Winky Dink Kit. Inside the kit, there were sheets of transparent plastic and several crayons. When prompted on the show, kids would place the plastic on the TV screen and draw a bridge or rope across a cavern or river, as examples, from which Winky Dink could escape. At the end of the show, kids would also be able to connect the dots at the bottom of the screen to find a secret word. Cheap by today's standards, the kits cost $.50 a piece if sent in the mail or $2.95 at toy stores. The show and the kits were a big hit. Into the early 70s, Barry kept things going until popularity waned. *Winky Dink* is still remembered today by the people building emerging ITV industry. At conferences, often, someone will remember their days spent in front of the set watching the show. (A company called Hollywood Ventures is selling new kits called "Winky Dink and You" for $19.95. In the kit, you get a 30-minute video with three cartoons, magic screen, a wipe-away "woobie," and five magic crayons. You can buy them via their Web page at http://www.bennysmart.com.)

EARLY INNOVATIONS

In the early 1970s, it became obvious to television engineers on both sides of the Atlantic that the TV was a powerful portal delivery vehicle for data communications. Playing around with the analog signal, these engineers soon developed a way to send data through the Vertical Blanking Interval (VBI) which could then be displayed on a screen. In the United States, work to develop closed captioning for the deaf also led to innovation.

The VBI is represented by the black stripe at the top and bottom of a TV picture. Physically, it constitutes 21 lines of a total 525 lines transmitted to the set per second using the analog National Television Standards Committee (NTSC) TV signal. In the U.S., use of Line 21 began when the U.S. National Bureau of Standards funded early experiments in cooperation with the ABC network to send out exact timing information over the TV signal. Fortunately, this experiment failed to provide needed results so ABC suggested text captions for the deaf instead. This and other experiments throughout the 1970s, in cooperation with the Public Broadcasting Service (PBS) affiliate WGBH (on programs like the *The French Chef*, *The Captioned ABC News*, and ABC's *The Mod Squad*), led to more developments. In cooperation with PBS, ABC developed early in-room decoder devices to interpret and display data sent within the VBI. It wasn't until public television station WETA broadcast and encoded data reliably on line 21 of the VBI that closed captioning became possible on a mass scale. Only in 1997 did the FCC mandate that closed-captioning become available within all video broadcasts "regardless of distribution technology.

In England, engineers from the British Broadcasting Corporation (BBC) developed teletext, a news and information broadcast display technology for TV. The BBC launched a teletext service in 1974 called CEEFAX. These text broadcasts continue today (although there are many services available from other broadcasters) as pages of information on the TV that the viewer can access by punching in a three-number code on a remote control and use of the "Fast Keys." The "home page" (for want of a better term) of CEEFAX is 100. To receive these pages, the viewer's television set must have the proper decoder built in. Broadcast content includes news headlines, sports scores, racing results, gardening tips, travel reports, film and theater reviews, weather, and much more. Some new companies in the United States have found a way to take advantage of this available technology. Bloomberg TV terminals, for example, have sent out news headlines and stock prices through the VBI for some time, and cable networks broadcast TV schedule information for EPGs. Early developers of ITV platforms such as Intel, WebTV, Wink Communications, and WorldGate (discussed later), explored new types of broadcasting over the VBI in the mid and late 90s and continue today. ATVEF, mentioned earlier, was set up to exploit the uses of this technology. When the digital signal broadcast over the air (terrestrial) and through other types of networks becomes more widely adopted as the mass broadcast standard, data sending analog signals through the VBI will become an anachronistic practice—in theory. The transition to digital, as it is commonly called, may take a lot longer to achieve than the original legislators in Washington D.C. intended. Analog platforms will persist for a long time because new digital equipment is still expensive, and local broadcast stations are facing astronomical costs and challenges during this upgrading period. Furthermore, the controversy surrounding the adaptation of one type of broadcast modulation scheme over another (e.g., 8VSB vs. COFDM) continues to rage. For now, broadcast data streams must travel over the VBI to digital set-top boxes or other data receivers. The origins of these very first two-way set-top boxes can be found in the history of the first commercially deployed ITV network, QUBE, deployed in Columbus, Ohio.

The First Two-Way Set-Top Box

In the mid-70s Warner cable executives in Columbus, Ohio (my hometown) decided they needed something to attract and keep people subscribed to their newfangled cable TV network. At the time, Warner offered 36 channels—a large number then—but did not really have much programming to provide. Subscribers, ultimately, did not want to pay for an empty service and began to cancel their subscriptions. In 1977, executives quickly recruited a production team whose mandate was to develop 8 hours of original programming for the service now being called QUBE. Producers knew it was possible to build a two-way system at the time and thought it would be a great gimmick to attract people's interest. With help from Pioneer, the only cable plant with early two-way technologies, the QUBE team put together a set-top box featuring buttons viewers could push during shows. While watching, viewers could vote, select movies to watch, play along with game shows, and more. QUBE was an instant hit. I can attest to the fact that all the kids in town wanted QUBE—although most did not even have cable TV, yet. Word of mouth traveled fast. Early programming featured home shopping, children's shows, a movie channel, and music videos—formats that all morphed into QVC, Nickelodeon, The Movie Channel, and MTV networks.

Eventually, QUBE came to an end because of economic complications. Looking back, QUBE was a significant achievement: It strongly demonstrated that viewers wanted and would pay for interactive ITV programming.

Meanwhile, post-1983, the basic one-way, analog set-top box decoder began to appear in everyone's home. Today, some 90 million homes have several boxes and/or are upgrading to advanced analog or digital cable services because of aggressive marketing efforts by cable providers. Plain old cable TV, though, introduced an important interactive technology to the masses: the remote control unit.

Remote Control

A small, handheld device that offers a variety of buttons and functionalities, the remote control unit (affectionately called "the clicker," "the pusher," the "zapper," or "the changer") permits the user to easily manipulate the TV set and its accessibility to content. The television industry's complete adoption of this handheld technology, compression technologies, and the introduction of new content networks has created the phenomenon known as channel surfing. Channel surfing, of course, is a metaphor that describes the casual, almost mindless activity of skipping between channels by punching buttons. An early form of interactivity with the television, the remote control effectively and quickly teaches the viewer to be selective and interactive. Today, firms like Interlink Electronics offer remotes with pen input pads, wireless keyboards, and integrated mouse pointing devices. Veil Interactive has technology that can send data through the light of the TV set to a viewer's remote and to a Palm Pilot. Another company, Pace Micro, a set-top box developer in the United Kingdom, has developed a remote control handheld unit that doubles as a smart home device. Called the Shopping Mate, this "remote" has a large screen and a clickable interface featuring all kinds of software. Soon, these devices will offer voice input control of the TV and, of course, telephony. In advance of that day, some companies are also exploring the use of wireless cell services to transmit text to a TV show (as mentioned before). Producers will want to take advantage of all these opportunities.

REVOLUTION

In the early 1980s, just when cable programming became serious competition for the film industry, Japanese representatives from NHK introduced a High-Definition Television (HDTV) technology to Hollywood called NHK Hi-vision, a type of digital signal technology. NHK Hi-vision was revolutionary because it was able to transmit better pictures and sound inside a wider screen—something the film industry had always been keen to provide. Unfortunately, the High-Definition Television (HDTV) signal, as it came to be called, required several times the bandwidth (20 MHz) than an NTSC analog signal (6 MHz). In addition to spectrum usage and compression problems, there were many incompatibilities with the present analog system. For the next 16 years, until 1996, the standardization debate raged while Japan put a non-NTSC analog version of HDTV in place. More than 23 well-funded international proposals from corporations and educational institutions were submitted to the FCC. These proposals tried to answer:

- Would an HDTV transmission be analog, a mixture of analog and digital, or purely digital?
- How would the signal transmit: over terrestrial broadcast, satellite, or cable?
- What part of the broadcast spectrum would HDTV occupy?
- What video compression scheme would be used to fit a 20-MHz signal into a 6-MHz one?

Finally, in 1990, after much haggling, the FCC decided an HDTV purely digital signal would be simultaneously broadcast until the current analog signal was phased out. To receive this signal, people would be required to buy either a digital TV set, a digital set-top box , or digital TV tuner card, and all broadcast stations would have to upgrade their facility to digital. Eventually, four proposals seemed serious, but no one emerged as the winner, so a suggestion was made to form a "Grand Alliance" between AT&T, General Instrument, MIT, Philips,

Sarnoff, Thomson, and Zenith. After much discussion, in 1996, the FCC adopted the scheme backed by the government-appointed Advanced Television Systems Committee (ATSC). This digital standard was based on an MPEG-2 compression scheme proposed by the Grand Alliance. It was also in that year that the Telecommunications Act of 1996 was passed. The Telecommunications Act of 1996 mandated the adoption of the digital signal by 2006. In 1997 the FCC began to allocate pure digital spectrum, not analog or a blend, to broadcasters. The FCC also instituted a graduated schedule with 2006 as the cut-off date for all broadcasters. (The slow conversion of TV stations to DTV and recent FCC delays in spectrum auctions are indications that the 2006 deadline will likely be abandoned.) Many countries, however, have adopted the Digital Video Broadcast (DVB) or the Association of Radio Industries and Businesses (ARIB) schemes. Their deadlines vary somewhat from that of the U.S.

Important broadcast digital signal technologies include:

- Advanced Television Systems Committee (ATSC)—North America, Taiwan
- Digital Video Broadcasting (DVB)—Many countries
- Association of Radio Industries and Businesses (ARIB)—Japan

As the industry progresses towards the deadline, the FCC is taking back analog spectrum from broadcasters and reallocating and auctioning it to other companies for new data services, such as mobile communications and datacasting. Some new companies have emerged to buy up this analog spectrum and are rapidly building data broadcasting networks. Although not discussed much in the press, these new data networks will have tremendous growth opportunities. Such data services, for example, are likely to serve wireless handheld devices, set-top appliances, and add to digital signal transmissions, especially HDTV. Over the coming years, we will see many developments in this area that will directly compete or collaborate with cable and satellite broadcasters.

In the meanwhile, the big networks, some cable broadcasters, PBS and its local affiliates, and some local stations are broadcasting HDTV programming in the United States, even though there are few digital TV sets or receivers to display them. Prices of HDTV sets are simply too high at the moment for the average person, and there is not much HDTV programming available. In fact, some TV manufacturers such as Sony and Konka (Korean) delayed their DTV set production until the programming issues are worked out. Some programmers have also cut back their HDTV broadcasts until the same. In general, digital terrestrial broadcasting is experiencing very slow growth. In the United Kingdom and elsewhere in Europe, ONdigital began broadcasting November 18, 1998. Rebranded to ITV Digital in July 2001, this network is currently experiencing a lot of turnover and subscriber churn.

NEW MEDIA AND THE INTERNET CHANGE EVERYTHING

During a long hashing out period (1970s–1990s), many other important developments occurred to bring the new media and, eventually, the ITV revolution forward:

- The cable industry established analog networks around the country that encouraged the emergence of general and specialized interest cable channels.
- The PC revolution gave the television and film producing community powerful suites of software to digitally edit and manage work, especially those from Adobe, Macromedia, and Avid. The CD-ROM industry moved powerfully forward to become the stalking horse for the development of interactive content and applications.*
- A mix of analog and digital consumer electronics devices appeared, such as music CDs, VCRs, camcorders, laser disks, and digital video disks (DVDs). More information on this topic is available at http://www.itvt.com.

*This was the first time the author of a digital product used the same machine to make the final product the consumer bought to play it.

Nothing was as influential as the arrival of the Internet, however.

For the next several years, usage of the Internet grew and a new industry was born. Although many people quickly migrated to jobs at Internet companies, because "that's where the money was going," a few rebellious projects, hidden from view, began to explore the possibilities of ITV and large companies began to invest hundreds of millions of dollars to find out if interactive technologies could also be applied to the medium of television.

CONVOLUTION

One small start-up company called Telemorphix (I was a member of the team) formed around 1992–1993 and tried their luck on a 1-hour live show they developed called *21st Century Vaudeville*. A show which played once a week on a leased cable channel from Viacom (a cable franchise now owned by AT&T) in San Francisco, *21CV*, as it was called in short, became a cult hit in town. *21CV* ran for 6 months there, for a year in Boston in partnership with local station WMFP, and traveled in the "Electronic Carnival" with the Lalapolooza tour. Viewers who saw the show could call in via an 800 number that flashed on the screen. Once in the virtual green room, the viewer could request to become another cartoon character, from which there were hundreds to choose—the Punk, the Fork, the StickMan, etc. Viewers could also fax in a drawing of their own, which would then be animated by the producers. Eventually, it would be the viewer's turn to appear on a TV intercut with Jack. In between music videos that were submitted by viewers, Jack and a character appeared one at a time. When the viewer talked into their telephone mouthpieces their voice came out of the TV set, which also triggered the mouth of the character to move (the mouth was preanimated by producers offline). Each character interacted with Jack in an ad hoc way. Some viewers would take on the personification of the character, or were simply themselves. The effect was marvelous. In some ways, this was truly interactive TV in the purest sense (e.g., personalities interacting with other personalities).

Unfortunately, the company suffered a financial breakdown and everything had to be sold. Additional experiments like this and other start-ups experimenting with ITV technologies such as the beleaguered NetChannel (eventually bought by AOL to serve their AOLTV platform), TV Answer, and Interactive Television Network (ITN) lost a lot of money and never got off the ground. Although many industry pundits term these experiments "failures," those in the industry know that this work paved the way for future developments and insights.

LARGE ERRORS

Perhaps because of the potential that corporations saw in Internet interactivity and commerce and new developments in video compression, some cable and telecommunications firms began to test very large, very expensive, multiservice, high-speed ITV systems to see if ITV was possible. (At that time, of course, everybody called it "interactive television"—a term still in use today.) Trials were sponsored by one company or a coalition of companies—sometimes from different industry sectors. In general, competitiveness and a belief that ITV would be a lucrative industry drove these companies to explore new methodologies and business models. Services tested in the field included movies-on-demand (now called video-on-demand or VOD), "walled-garden" services featuring news and personal information portals, interactive gaming, home shopping, commerce applications, and interactive educational programming. Tremendous errors in judgment were made, however. Available technologies at that time were too expensive to support such an undertaking. According to Daniel Levy, now vice president of client services at RespondTV, an ITV infrastructure services provider working with many of the biggest names in the business, these lessons "still reverberate today."

Levy was in charge of the service style guide for the ill-fated Time Warner Full Service Network (FSN) project, which launched on December 14, 1994 in Orlando, Florida. Over 4,000 homes, made up of mostly family units, had access to this service via a fiber network. Services available included VOD, shopping, games, an EPG, and postal services. Not a free ride by any means, subscribers were billed on a pay-per-use basis via their credit cards. All told, the project cost many millions of dollars—possibly up to $100 million, says Levy. Most blamed this excess on the fact that the technology simply cost more than the deployment could support. "It was far too expensive, but we knew that going in," he said. Contrary to public assumptions about the project, however: "We knew FSN would eventually become deployable much later on. It wasn't a wasted effort: we learned a lot." Levy also points out that a few things gleaned from the experience were invaluable. They were:

- The service itself must be available free to the customer
- Different tiered pricing models do not work
- VOD is a very popular application of the technology
- People really want simple interactive options

FSN eventually closed its doors in 1997. Throughout the mid-1990s, over 21 trials were launched following Source Media's 1993 inauguration of the Interactive Channel in Denton, Texas.

Table 20-1 shows a small sampling of large trials that took place around the United States in the mid-1990s.

A CONFUSING TIME

While big corporate cable trials came and went, the Internet continued to grow. More and more people were connecting online with modems. Telecommunications companies could barely keep up with the desire to surf. In the mid-1990s to about 1997, TV-related browser plug-ins and video applications (a good one was CUSeeMe, still available today) were released to the market. Terms like "PCTV," "InternetTV," "NetTV," and others became the buzzwords in every trade and consumer computer publication. All claiming to be "the" video solution that would enable a user's computer to display digitally compressed

TABLE 20-1 ITV Trials

COMPANY	NAME	LOCATION	TECHNOLOGY	SERVICES
Bell Atlantic AT&T	FutureVision TelITV	Dover Toms River, NJ	Phillips set-tops nCUBE servers Switch Digital Video	Near VOD Pay-per-view Shopping
Bell Atlantic AT&T	Stargazer	Fairfax, VA	Stellar One set-tops nCUBE servers AT&T ADSL	VOD Internet
Time Warner	Full Service Network (FSN)	Orlando, FL	Fiber to curb	VOD, games, shopping, postal
TCI Microsoft	MS Network	Redmond, WA	General Instrument Hewlett Packard NEC	VOD, games
Cox Cable	no name	Omaha, NE	Zenith set-tops Hybrid Fiber Coaxial	VOD, NVOD transaction
Southwest Bell	Little Richard	Richardson, TX	Fiber to curb	VOD, games, 60 channels

Data provided by: S.Churchill

MPEG data streams downloaded from the Internet, what was actually seen were jerky video images that did not synchronize to the audio. Cable video infrastructure developers like Scientific-Atlanta (S-A) and General Instrument (GI), meanwhile, continued to develop advanced set-top box systems without much notice by the media. Also at this time, new TV shows emerged to cover news about computers, technology gadgets, and "cyberlife." ITV was nowhere to be found, however.

At the same time, telecommunications companies returned to their labs and marketing offices to figure out what new broadband technologies would continue to feed the desire for a faster Internet. Across the fence, the cable industry—seeing new business being built through ISP subscriptions, consulting, online advertising, ad banners, and media investments—started to invest and build hybrid fiber coax (HFC) and two-way optical fiber networks. These cable operators were determined not only to bring their cable legacy systems up to speed, (a task still in process today), but were motivated to offer more pay-per-view programming packages and explore new interactive TV technology platform strategies to maintain and expand their subscriber base. Both cable operators and the telecommunications industry, in an atmosphere of intense rivalry, touted their respective network methodologies as the best offering (this fact continues to fuel the Open Access vs. Forced Access controversy). Following the Telecommunications Act of 1996, a few cable industry players built their own high-speed data networks (@Home, MediaOne, RoadRunner) to compete in the telecommunications space. Telecommunications companies have only recently begun to experiment with residential gateways—set-tops that use broadband IP technologies to bring entertainment programming and many other services into the home.

NEW COMPETITION

As the cable and telecommunications industries invested hundreds of millions of dollars in more flexible and high speed networks, the Digital Broadcast Satellite (DBS) industry (e.g., Sky Broadcasting, DirecTV, and EchoStar) emerged as a strong competing provider of entertainment programming and interactive digital services. DBS firms, when they first came out in the early 1990s, offered consumers more video and audio channels than ever before. Using transponders that could compress digital video efficiently and the fact that consumers could buy an 18-inch dish and set-top receiver in retail stores, the industry grew (to the chagrin of the cable industry). Marketing their service as "digital," DBS services attracted thousands and then millions of new subscribers in a few years. Reasons for the switch seemed to be that customers were dissatisfied with cable service and control over content, were inconveniently located in apartment buildings not outfitted for cable or in rural areas, or they wanted clearer images and audio. (DBS set-top boxes are, for the most part, analog receivers that convert the digital signal to display programming; new digital receivers are also available.) Today, the DBS industry has over 14 million paying subscribers.

DBS subscribers may be, in fact, the first en masse consumers to get access to the first ITV applications. Both EchoStar and DirecTV, for example, have made EPGs available on their set-top receivers. They have also been aggressively building alliances through strategic technology and marketing partnerships with companies such as OpenTV, Microsoft's WebTV, and TiVo. America Online has invested $1.5 billion in General Motors' Hughes Electronics, which owns DirecTV, and struck partnerships with ITV middleware providers such as Liberate and TiVo for their AOLTV platform. Sky Broadcasting has aggressively launched Open Interactive (also called Open or Open…) to rave reviews and growing profits. We may even see Rupert Murdoch of the Sky empire buy DirecTV soon to enlarge it. Other competing services to Open such as Energis, will deploy in 2001. The strategy behind all this is to attract further dissatisfied cable and telecommunications customers and, perhaps, get a jump on that industry for T-commerce opportunities. Industry rivalries such as these keep the pressure on for the development of more ITV backend architectures.

SOLUTIONS

From the very late 1990s to the present, the ITV industry has finally started to take shape. Why? Bandwidth is finally achieving the speeds necessary to process digital video, computer technologies are becoming device-centric, innovations in software have paved the way, lessons have been learned, and early forms of content and applications are proving the concept. In this environment, not only are we seeing more established companies embracing ITV as a long-term strategy, newer ITV companies are emerging—a few have gone public (e.g., ACTV, Liberate, OpenTV, TiVo, WorldGate, and Wink). Large mergers are also taking place interindustry and intraindustry (e.g., Liberate–MoreCom for $521 million, and OpenTV–Spyglass for $2.1 billion). To capture the evolution of this phenomenal growth, general media and trade publications are making room for coverage of the topic (e.g., my email newsletter *InteractiveIV Today [itvt]* focuses exclusively on news, trends and analysis of

ITV). In the end, however, it's the actual solutions—the stand-alone retail products, set-top boxes, middleware platforms, data broadcasting services, and other types of ITV applications being deployed commercially—that will determine whether the industry is taking root and will thrive in the years to come. The strategic combination of these and their associated business models will establish what and how content producers can realistically begin to produce and broadcast.

STAND-ALONE RETAIL PRODUCTS

Early in the history of ITV (1995), Intel released a stand-alone product to the shelves called Intercast that featured software bundled with commercially offered TV tuner cards. Intercast gave the public its first example of interactive data and television in a unified environment viewable on a PC. Content developers included CNN, CNBC, Lifetime, QVC, M2, Lifetime, and The Weather Channel. NBC, an early partner, broadcast a special Intercast version of the 1996 Summer Olympics—a project for which the network had high hopes. Unfortunately, Intercast found few buyers as TV tuner cards were still such a new concept, difficult to install, and there was not enough content offered to capture consumers' interest. In 1999, Intel withdrew its support for the product in order to refocus their efforts on developments within ATVEF. After that, Intel became very involved in the evolution of that commercial organization and invested aggressively in projects in association with the Public Broadcasting Service (PBS), although even these strategies have greatly shifted as Intel repositions its investments around information appliances in general rather than ITV specifically.

WebTV established itself as an important leader in the industry when it premiered a stand-alone set-top box with an information and Internet service in October, 1996. Grabbing the media and consumer's attention with low prices and "couch-based" access, WebTV (purchased by Microsoft in 1997 for $400 million) now has over a one million subscriber base. Although this number and rate of growth has been disappointing to industry observers, WebTV has created a bit of a cult following with users, obtained space on several MSO set-tops in

Europe and Japan as well as a recent deal with EchoStar for their DISHplayer box; it is being mainstreamed into Microsoft TV, a new total ITV platform for MSOs. Ultimately, WebTV has maintained its leadership position with respect to VBI analog advanced broadcasting. Additional services offered include picture-in-picture (PIP), a "walled-garden" service, JavaScript support, banking and bill payment services, surveillance software, an EPG, and Web page building tools. WebTV's set-top products include Classic, Plus, and WebTV for Windows 98—the last two feature two-way VBI broadcasting. Producers of ITV content over a WebTV Plus or WebTV for Windows 98 can use third-party authoring tools, such as Dreamweaver, and the services of various ITV data broadcasters, such as Wink Communications, Mixed Signals, and RespondTV, to provide subscribers with interactive enhancements. In the second half of 2000, a slew of new ITV, ATVEF-compliant, VBI-based programming was available over the WebTV platform. WebTV also made a deal with the CBS network to "enhance" a range of different programs.

In 1999 ReplayTV and TiVo introduced retail stand-alone products that offered viewers advanced VCR-like digital recording capabilities via hard disk drives. These DVR units also feature software that enhances the EPG model (including software that enables viewers to skip commercials) as well as data broadcasting content services. At once proclaimed as important developments by the press and analysts, broadcasters continue to invest millions of dollars in these companies because the boxes clearly give the viewer a powerful tool of control; broadcasters want to have some influence on the evolution of the software that will control it; these companies want to own a piece of that action. Strangely, these units have not met with great success. Reasons for this lack of acceptance may be because the units are complicated to set up, too expensive (each has their own business model, but neither gives the box away), and there hasn't been enough advertising or brand development, although this fact has recently changed as both companies have launched aggressive advertising campaigns, rebate programs, and other special promotions around the world. Certain deals made by ReplayTV and TiVo may bring their technology to a wider public. TiVo will be featured on a later version of America Online's AOLTV stand-alone box. (TiVo has also launched in the United Kingdom) ReplayTV is now packaged as Panasonic's "Showstopper," debuting the first industry 60-hour drive in October, 2000. (It has been purchased by SONICblue.) Microsoft TV's Ultimate TV platform has its own system. These companies are also beginning to evolve their software and data broadcasting services to create more powerful ways to select programming interactively and even remotely (e.g., MyReplayTV, which controls from the Web). Without skipping a beat, the DBS industry has again taken notice.

The DBS industry has always deployed stand alone retail products. This fact has enabled them to sell or give away digital set-top receivers and interactive services mostly in Europe where DBS service is strong in advance of the cable and telecommunications industries there. These companies (too many to mention) clearly understand the potential of ITV to deliver a unique service, targeted advertising, and T-commerce. Each provider is trying to determine which box and set of services are strategic winners. In the United Kingdom, BSkyB has deployed Open, an ITV service that features middleware from OpenTV. DirecTV also offers its own ITV boxes featuring Wink Communications' enhanced service and features that include DVR units.

Another company, which has yet to deploy, but might have another bright idea, is Telecruz. This chip manufacturing company hopes its all-in-one ITV-on-a-chip series for embedding in televisions or even set-top boxes may prove the ultimate stand alone product. Announced late in 2000, Telecruz received an additional $35 million round of funding on top of a previous $34 million to drive their strategy. Partnering with TV manufacturers such as Zenith, Aiwa, Samsung, and Videocon in India, Telecruz hopes consumers will want to avoid the multiple set-top box dilemma. But what of digital STBs?

DIGITAL SET-TOP BOX DEPLOYMENTS

Although the set-top box has long been associated with the cable industry, the FCC mandated, in the Telecommunications Act of 1996, that set-top manufacturers must separate TV tuner controls from security modules in those boxes sold at retail stores after July 1, 2000. For the most part, there aren't any available to date. (The exceptions are, of course the stand alone products named earlier.) Although the cable industry continues to be slow to comply, Cable Laboratories recently announced that the OpenCable initiative is now supporting the development of an open specification for an interoperable middleware platform supported by Sun's JavaTV API, Liberate, and Microsoft TV. The cable industry has until 2005 to offer their own retail interoperable digital set-tops. In the meanwhile, those cable companies in the United States (and there are too many to mention) that smartly invested in new digital and two-way cable networks in the 1990s have already deployed noncommercial, or are beginning to deploy trial or commercial, digital set-top box networks. These deployments require the renting or free distribution of the boxes. Two box series that seem to be the most popular with cable providers are S-A's Explorer 2000 and GI's DCT 1000, 2000, and 5000+ series. Both companies are now coming out with more advanced boxes, but these are still in the early stages. Set-top boxes from companies such as Pace Micro, Mitsubishi, Sony, Thomson, uniView, Grundig, and others contain video and audio microprocessors, memory, conditional access technology, a cable modem, middleware to control or to enhance their capabilities, and host of other technologies. Unfortunately, hardware inside the boxes will soon become obsolete—although flash downloading of software makes upgrades possible. Some operators may elect to sidestep the box and offer ITV services over a digital video server located at the cable headend, or the central office in the case of telecommunications companies. Here, all software, games, and other resource-intensive applications can be streamed to the viewer. One company, ICTV, which received an $87 million investment from Motorola, ACTV, OpenTV, and others, offers such a system. It will be interesting to see if this company is able to deploy quickly to those operators already invested heavily in digital set-top boxes. Most cable operators will continue to experiment with a variety of middleware platforms now available.

Middleware Everywhere

The software inside a set-top box is becoming more robust and, therefore, takes up a larger footprint inside the box. Besides managing video display or other basic television functions, middleware serves applications such as the EPG, access to the Internet, email, interactive graphic walled-garden environments, video-on-demand, and a variety of services such as multicamera digital video switching and others. Popular middleware platforms from Liberate, OpenTV, Spyglass, Microsoft TV, MediaHighway from Canal +, and PowerTV have become the popular way to provide these services. Each company has a strategy to offer: Liberate's modular platform, based on open Internet and international broadcast standards, features their TV Navigator browser, which communicates with the TVConnect server at the headend. To proliferate their platform, Liberate established the PopTV and Variety Pack programs to educational and package content developers, respectively. Liberate also made strategic mergers with MoreCom (a more complex IP-over-cable middleware platform) and SourceMedia (middleware for thin set-top boxes; i.e., those with a small hardware footprint, such as DCT-1200 and 2000s) to extend capabilities. Liberate has had much success selling to cable operators who expect and prefer choice (Cable operators have the privilege of buying cable franchises from local communities without facing much competition because of the fact that installing cables underneath the street, stringing cable in the air, and running a network is expensive. This allows them to deploy on their own schedule and be more cautious in their investments—a fact upon which the modular platform of Liberate is able to capitalize.)

OpenTV, on the other hand, has had much success installing their proprietary middleware platform on DBS networks in Europe, Latin America, and now, with EchoStar, in the United States. DBS providers seem to want a range of ready-to-go applications that they can deploy. OpenTV's merger with Spyglass (which develops small-footprint IP-based middleware) was formed to give that company the ability to offer access to the Internet and to appliances such as Telecruz's set-tops and televisions. Telecruz in fact, upgraded to Spyglass' Device Mosaic 4.1.0.

Microsoft, on the other hand, has developed a complete end-to-end ITV backend architecture platform called Microsoft TV (MSTV), which incorporates Windows CE (a small device operating system) and WebTV software. To ensure deployment with a big-name cable operator, Microsoft paid AT&T $5 billion and has invested in networks around the world. In September of 2000, however, Microsoft announced that it was unable to deliver MSTV to AT&T or to UPC in full, causing both companies to sign with Liberate. We have yet to see whether many cable operators will truly buy in to this Windows-based resource-intensive system despite the fact that many integration companies are working hard with that platform.

MediaHighway from Canal + and JavaTV from Sun Microsystems are having varying luck with set-top box deployments. MediaHighway has been released mostly in Europe, but has one U.S. site in Florida. The JavaTV API, however, has been adopted by the Digital Video Broadcasting–Multimedia Home Platform group in Europe and now provides the application development environment for OpenCable.

PowerTV, from the company of that same name, is available primarily on S-A's Explorer 2000 box, which has been deployed in a variety of locations.

The newest middleware provider to date is WorldGate. A year in development, WorldGate's CableWare 2000 hopes to provide the combined package deal of middleware and data broadcasting through integration with their data broadcasting service, Channel Hyperlinking and "walled-garden" capabilities.

DATA "ENHANCED BROADCASTING"

One of the earliest ITV data broadcasting services out the door, Wink Communications was the first to use the term "enhanced broadcasting" as a way of differentiating itself from the bad hype surrounding the term "interactive TV." Today, Wink's services include downloadable software to a set-top, a proprietary and ATVEF-compliant data enhancement broadcasting service, and a special backend tracking and billing environment called the Wink Response Network. Wink has made many partnerships with big branded media companies like MSNBC, the Discovery Channel, the Weather Channel, E! Entertainment Television, DirecTV, and others and received multimillion dollar investments from Microsoft and Paul Allen's Vulcan Ventures. The general philosophy at Wink is to present a limited, yet interactive choice to the viewer. When the "i" icon is presented on the screen, for example, (as on WebTV) the viewer can click on it to bring up the interactive enhancement from an advertiser (for example, requests for more information) or content providers. A competitor in this space, WorldGate delivers a URL trigger within network broadcasts, which can be clicked using the remote. Called Channel Hyperlinking, WorldGate's system is proprietary, although they have announced support for ATVEF.

Two other companies focused on ITV broadcasting infrastructure technologies and services are RespondTV and Mixed Signals. RespondTV has had much success signing advertisers and content providers to their services such as Bloomberg, HGTV, MSNBC, Domino's, Purina, and others. Mixed Signals has a special arrangement with Sony's Columbia Tri-Star and Game Show Network (among others) to present the ITV version of various game shows.

HYPERVIDEO AND SMIL

A term invented by a company called Veon, hypervideo is an Internet-only technology (for the moment) that enables producers to embed hotspots or links inside a streaming video. Using a combination of proprietary and standards-based protocols and display languages such as SMIL, Veon's authoring tool enables producers to create streaming interactive hypervideos in real-time. Hotspots can include links to the Web, links to open another movie within that streaming video, ATVEF data triggers, links to call up graphics and data elements on a Web page in synch with the streaming video, and so on. Although companies like SofTV and others also use SMIL to create synchronized streaming video applications on the Web, Veon's implementation truly reinvents the concept of television as interactive. (Watch Point Media is also exploring hypervideo technologies; Veon was purchased by Philips in June of 2001.)

The paradigm shift here is that producers can look at the hypervideo as an interactive video fabric rather than television programming supporting data overlays.

INDIVIDUALIZED TV—MULTIPLE CAMERA ANGLES

Used by ACTV to describe one of their proprietary ITV solutions for multicamera digital video switching, the term Individualized TV—more commonly known as multiple camera angle switching—is a technology that allows the viewer to choose from four or even five presented streaming video choices. In the case of sports and live events, the viewer can select from multiple camera views. In the case of targeted advertising or specially designed content, the viewer can select from four commercials, which the software in the box then remembers. ACTV formed a subsidiary called Digital ADCO to manage this software and business opportunity. Although NDS, a company owned by News Corp., does not use the term "individualized TV" to describe their software solution, it also offers multiple camera viewing technologies in a product called XTV. ACTV is currently testing in trial and will most likely deploy its product with AT&T Broadband in the future. NDS has made many deals with companies around the world (their biggest being BSkyB in the United States); Microsoft is also a big investor.

SYNCHRONIZED TV

Synchronized TV, essentially an Internet application that receives HTML data broadcasts synchronized with television programming, has, of late, become a hot area of growth. Why? Interactive syncTV apps can easily be downloaded over a Web site. Once the relevant TV show comes on, viewers can chat in real time about the characters or topics, read factoids, click on E-commerce links provided by advertisers, play along with game contestants, and so on. Essentially, anything one can do in an Internet environment, one can do in a syncTV app while the show is on. When broadband ITV platforms are available with Internet access over TV, syncTV apps will be available over that one screen display. Companies playing in this space include ACTV's HyperTV subsidiary with Liberty Media's Liberty Livewire, Spiderdance, and ABC. Many shows are getting this treatment—*History IQ* from Spiderdance, for The History Channel, a multiple choice game show; webRIOT, on MTV, also from Spiderdance; *Cyberbond* broadcasting on TBS; and even the movie, *The Sixth Sense* as it broadcasts on Starz Encore, the movie packager.

VIDEO-ON-DEMAND

Sure to become one of the most popular ITV solutions, video-on-demand (VOD) provides viewers with access to movies when they want them and a robust billing backend. A more advanced version of what the industry calls Near Video-on-Demand (NVOD or Pay-Per-View), VOD is enabled by digital video servers from companies such as DIVA, C-Cube, nCUBE, Concurrent, SeaChange, and several others. Cable operators are especially interested in buying these servers. Complete services such as DIVA and Intertainer, have partnered with Hollywood content providers provide an "end-to-end solution." Some cable companies like the idea of outsourcing that service without troubling themselves with licensing issues. Others that want a greater share of T-commerce revenues, are setting up their own VOD service within the "walled-garden."

Table 20-2 is a chart of various ITV solutions.

THE ISSUE OF PRIVACY

ITV presents new opportunities and risks for the producer and viewer, respectively, concerning the matter of privacy now that middleware, servers, and databases can be linked in a common digital environment. For example, it is easy to combine the data tracked from viewing habits (of an individual or members of an entire household) with the data collected from interactive clicks and T-commerce selections. This kind of data is very valuable, and advertisers are salivating at the potential. Is this what is actually driving development and investment? That is

Table 20-2 ITV Solutions

Solution	Platform Technology	Functionality	Sample Companies	Business Model
Digital Set-Top Box (DSTB)	Needs middleware/ operating system. Middleware: Liberate, OpenTV, PowerTV, Microsoft TV, Spyglass, MediaHighway, ConnecTV, Streamaster, @HomITV, JavaTV API.	High-speed data and video interactivity (some offer a smart card slot, additional ports, DVD, and other functionalities).	Boca Research, Mitsubishi, Motorola/ General, Instrument, Pace Micro, Scientific-Atlanta, Stellar One, Thomson, others.	Deals with network, cable subscription, advertising, T-commerce.
Digital Video Recorders (DVR) Also called Personalized TV and Timeshifting.	Digital hard-disk drive combined with data broadcasting service and box software.	Digital hard drive records programs and enables more control over selecting and recording programs.	Standalone boxes, DISHPlayer, ReplayTV, Showstopper, TiVo, Ultimate TV.	Per unit with free service. Per unit with monthly service.
Electronic Programming Guide	Cable and satellite set-top box systems. Eventually, all devices.	Interactive access to TV schedules and info about programs. Data sent through VBI currently.	Gemstar, TV Guide, and other companies that try to go beyond their patents.	Part of the service. Advertising.

TABLE 20-2 ITV Solutions (continued)

SOLUTION	PLATFORM TECHNOLOGY	FUNCTIONALITY	SAMPLE COMPANIES	BUSINESS MODEL
Enhanced TV	General term for broadcasting HTML data to ATVEF-compliant device.	Platforms interpret data sent through the VBI by broadcasters. Software interprets this data and presents enhancements on top of video like a Web page. Data is written in HTML. Down the road, the VBI won't be needed because of digital signal.	ATVEF boxes, e.g. using middleware from Liberate, Microsoft TV, Wink, WorldGate. Standalone boxes, like AOLTV, DISHPlayer, WebTV Windows 98, WebTV Plus.	Set-top box per unit with cable subscription. Advertising, T-commerce.
HyperVideo	Internet	Real-time authoring of streaming video featuring clickable hotspots.	Veon, Watch Point Media.	Development, backend tracking, production, and design services.
Individualized TV, also called Multiple Camera Angles	Software on set-tops—proprietary.	Viewers can choose from several camera angles and smart TV ads. One path multicast.	ACTV	Cable subscription, advertising, and T-commerce.
Synchronized TV	Proprietary broadcast application viewable on Internet browser.	Pushing data to Internet application in sync with TV programming.	ACTV's HyperTV with Liberty Livewire, Spiderdance.	Business development deal with broadcaster as well as T-commerce.
Video-on-Demand	Digital video servers.	Streams MPEG2 video in cable or DSL environment.	Concurrent DIVA, Intertainer, nCUBE, SeaChange.	Subscription, advertising, T-commerce.

debatable. For example: How will the producer provide the kind of application content with data feedback that advertisers want and at the same time ensure their viewers' privacy? Will existing cable regulations, which protect data collected from viewing habits from being distributed or sold, continue to apply? This last question has been the subject of recent attempts to extend this regulation to emerging video devices and delivery options currently not covered. California State Senator Debra Bowen, a Democrat from Redondo Beach, believes such regulation was needed. Over this last year, Bowen introduced a bill into the State Legislature. After passing several committees, the bill was defeated, most likely because of strong pressure from ISPs, Microsoft, and AOL lobbying against it, and the fact that it was trying to regulate an undeveloped and Internet-related industry on a statewide basis. This bill is in cold storage. Bowen has, however, reintroduced this bill, but limited language to just cable and satellite providers. It is passing committee after committee so far on its way up to the Legislature. Regardless of industry solutions that may emerge to solve this touchy situation, the need to ensure viewers' privacy is essential.

Open Access vs. Forced Access

An important issue still unresolved is called Open Access, or as opponents are calling it, Forced Access. In short, this controversial issue pits old foes (telecommunications and cable providers) against each other as each seeks to protect new markets important to it. The controversy stems from the fact that cable companies want the right to make back their money from the multimillions they invested in advanced digital cable networks built over public cable franchises in local communities. To do that, cable companies want to provide people with high-speed Internet access via cable modem, and they want to have the right to select a cable technology ISP partner to provide that service. The end result may influence the new media landscape for years to come. AT&T and AOL, the most prominent opponents in this struggle, have modulated their diametrically opposed positions over this last year—not surprisingly—to appease FCC inquiries during their respective mergers. Specifically, AT&T said it would open its networks to ISP Mindspring (bought by Earthlink), on a partnership basis and open their networks to others after they had concluded their arrangement with Excite@Home in 2002. AOL in a, postmerger announcement with cable operator, Time Warner, pulled back their position saying "let the market decide." Both cable MSOs are now partnering with multiple ISPs, but only very slowly.

CONTINUING CHALLENGES, PROBLEMS, AND RISKS

Regardless of the developments and improvements over the years, there are still many challenges and risks that plague the producers of ITV applications and content. Many issues need attention. Below is a review of current issues affecting ITV, culled from several professionals working in the industry.

- Content
  - There is lack of funding for ITV projects.
  - ITV services are not getting repeat viewers. Is ITV purely a novelty?
  - Not enough content is available.
  - Producers and T-commerce retailers must do more research on the kinds of products and services that work best in an ITV environment. (This is often not the same as on the Internet.)
  - Producers must do create ITV content that is more emotive, less purely like a Web page.
  - Broadcasters seem hesitant to launch programming.
- Production
  - Production processes must be worked out.
  - Standards are still unresolved.
  - Confusing whether to work with NTSC, ATSC, DVB, or…?
  - The ITV industry needs to become more "open source."
  - Industry needs a single middleware platform common to delivery platforms.

- There are needs for database technology that connects seamlessly to ITV broadcasts.
- Power and support is lacking in set-tops for important IP plug-ins.
- Producing ITV over VBI boxes still requires full knowledge of extensions per box.
- Business models are still unformed.
- Bandwidth problems persist, but are being resolved.

· Advertising
  - Advertisers have a lack of knowledge about what ITV is and what platforms are available.
  - Advertisers want to start testing but resist because of the low installed base of boxes.
  - Advertisers are unhappy with the L-shaped interface, downsized video, overlapping of commercials, and possible complaints from talent in video programming.
  - Advertisers and video producers, who spend millions on ads, are concerned their commercials will be "polluted" by enhancements. If interactive enhancements are designed to be extremely integrated into the commercial, will viewers know what to do?
  - Ad agencies are slow to experiment with new technologies, even though advertisers are interested.
  - Ad agencies must communicate more with their new-media divisions.
  - What does the ITV rate card look like?
  - Problems exist with time to act when rotating "i" is seen.
  - Lack of numbers persist regarding participation with ITV commercials.
  - Can lengthy ITV participatory advertising work or deliver branding effectively within :30 and :60 commercials? Will viewers get the same "feeling"?
  - DVR technologies will throw advertising models out the window.

- General
  - Not enough is being done to inform the consumer about ITV.
  - There is a lack of research data into what people want from ITV.
- Future
  - Wider deployments of interactive TV software and enabled set-top boxes are needed.
  - Better marketing of ITV to potential viewers and subscribers is needed to get the word out.
  - Adoption and advancement of the ATVEF specification is needed.
  - More relationships with collaborators and business partners to build access and revenue are needed.
  - Improvements to the development of successful business models are essential.
  - Tools and services for advertisers, content producers, and broadcasters are needed.
  - A lack of production methodology necessitates training and higher budgets.
  - Budgets for video and enhanced data programming may be bigger than expected.
  - The presentation of the show cannot be predicted until produced.
  - Challenges to create a billing system exist, with too much focus on set-top boxes.
  - Clicking to the Web takes viewers away from shows and advertising.
  - Producers must spend a disproportionate amount of time on technology tools.
  - Not enough people have all production skills necessary.
  - Users may still remain passive.
  - Some platforms still require big facility upgrades.
  - Overly complex enhancement programming may overwhelm the user.

– Size of video window and screen presentation is always a challenge.
– Too few options for interactivity exist at the moment.
– Bandwidth is still a problem.

Interactive enhanced television has been a long time coming and obviously has even further to go before it can realize its full potential (which varies depending on whom you talk to). Nevertheless, much has been accomplished, and the potential rewards of ITV are too compelling to dismiss for its unresolved issues. It is essential that digital content creators stay abreast of developments in this field, and anticipate what kinds of ITV programming demands the public will have. Working with viewers, content creation producers will, at the very least, reap rich rewards, and—quite possibly—redefine the television viewing experience as we know it in the coming decade.*

*Feedback on this paper may be sent to swedlow@itvt.com or to enhancedtv@afionline.org. For further information about the Workshop and the topic, go to http://www.afionline.org or *InteractiveTV Today [itvt]* at http://www.itvt.com.

CHAPTER TWENTY-ONE

DIGITAL CENTRALCASTING

LOWELL MOULTON
SENIOR TECHNOLOGY CONSULTANT
SYSTEMS INTEGRATION,
SONY ELECTRONICS INC.

Digital CentralCasting (DCC) is a business structure for on-air operations and distribution of content for geographically diverse television station groups using wide area network (WAN) technologies. DCC is the consolidation of several operations usually performed at the local television station to one or more centralized locations. Each station group must analyze current and projected workflow and decide the right amount of consolidation that will benefit their unique business. Some station groups may find a CentralCasting model works best, whereas others may find that regional ZoneCasting makes more sense. Some broadcasters may find that CentralCasting does not enhance their profitability at all.

Many station groups require a hub-type system of telecasting to their stations to distribute programming, promos, and advertising spots. Some groups have station concentrations that support regional news or sports networks. Digital CentralCasting may appeal to station groups that have multiple network affiliations, multiple syndication agreements, and diverse programming formats.

The unique nature of the individual broadcast groups is key to developing the proper architecture and approach to the DCC operation. Issues that drive the choice of a DCC model include the number of properties owned by the group, their geographic location, and the ratio between network and syndicated pass-through programming to local production. These factors can greatly affect the logical distribution of work.

The location and ownership of the Digital CentralCasting (DCC) facility is another factor to be considered. We have identified at least three possible scenarios of DCC location and ownership:

- *Station Group Model.* The station group owns the DCC facility, chooses the location for it, and possibly creates regional DCC hubs.
- *Hosting Service Model.* The DCC facility is operated by a host fiber carrier company and location is determined by the host service (Qwest, Williams, AT&T, etc.).
- *Independent Service Model.* The DCC operation is colocated with other DCCs as part of a for-hire service company, analogous to existing cable models. Location is determined by the service company.

The primary motivation for implementing DCC is financial, specifically by consolidating a number of tasks duplicated at each individual station, thus reducing the head count and the amount of capital equipment required at individual facilities. Operations that may be consolidated at the DCC include:

- *Administration.* Executive, sales and operations management, accounting/billing.
- *Programming.* Traffic, continuity, creative services (graphics), reconciliation of as-run logs, and archiving.
- *Technical Ops.* Master Control operations, logging and recording off-air signals, transmitter remote control.

Local station management may focus on local news production, local sales, day-to-day administration, community relations, maintenance, and station management. Live sports and community affairs production may also remain at the local station level.

Each business model must be carefully studied and customized for the individual station group. Groups that have already implemented Digital CentralCasting report reductions in operating costs attributed to staff reductions in master control, traffic, programming, accounting, administration, and production. One group reports that it reduced master control personnel by almost half and another group had a 70 percent error reduction rate in spot playback by moving to a DCC model.

FINANCIAL CONSIDERATIONS

The fundamental purpose for implementing CentralCasting is to increase profits. CentralCasting can offer many scenarios for increasing profit:

- Consolidation of management, administration, programming, and billing functions.
- CentralCasting Operations Center staff requirements can be offset by reductions of staff at local stations.
- DCC capital investment can be offset by long-term reductions in equipment investments at local stations. Consolidation of program and commercial storage and playout can offer synergies not available to multiple parallel master control efforts spread across geographical areas.
- Centralizing the technical staff can benefit MPEG-2 transport stream encoding, monitoring, and quality assurance. These operations require special knowledge and highly trained and skilled technicians, as well as expensive specialized test equipment.
- New functions are enhanced by having skilled technical staff in one location. Functions such as multichannel ATSC multiplex, PSIP and EPG insertion, data casting IP encapsula-

tion services, Internet video streaming, archive, media asset management, and Web repurposing for searchable video content can be added in a single standardized manner.

- Centralizing closed-caption encoding and EAS generation can create economies of scale. Both analog and digital signals can be generated from the DCC Operations Center during the ATSC transition period.
- Consolidating HDTV playout monitoring and storage facilities allows businesses to fund fewer high definition playout, archive, graphics, and master control switching and monitoring facilities for this specialized service.
- AC3 and Dolby E technical expertise, metadata generation, and multichannel audio monitoring can be consolidated at a central location.
- Repurposing and regionalizing content can be done with consistent aesthetic quality in centralized special facilities.
- Conversion from point-to-multipoint satellite distribution to a bidirectional network-centric distribution allows any station to send and receive video to and from any other station. This new program distribution functionality may provide new business opportunities that were not available in the point-to-multipoint topology of current network operations. For instance, once the full duplex data paths are common to all stations in a group, then fast-breaking news stories can quickly be fed from the local station to the CentralCasting Operations Center without having to rent ad hoc satellite uplink services and transponders.
- Broadcasters may even be motivated to trade higher network expenses and endure a slightly lower quality on air look if new business opportunities create enough new revenue. Or they may find that they achieve lower operating costs and maintain the current quality of on-air look. They may even achieve an improved on air-look while reducing costs.
- A digital end-to-end solution avoids the need for continual and repeated quality assurance checks at numerous points throughout the signal path. Quality assurance can take place during the analog-to-digital conversion and digital compres-

sion ingest phase. Once material is ingested it need not be measured again for technical quality.

- Increased automation integration at the DCC reduces errors in spot playback, thus resulting in fewer make-goods and opening more time slots for paid advertisements.
- The DCC can be located in areas with good quality of life and a low cost of living, thus reducing costs over operations centers located in higher cost geographical areas.
- Increased profits allow creative energies to be focused where they count most, such as in increased investment in local news production. Local news is one way local television stations can differentiate themselves from each other and cable television, Internet TV, and direct broadcast satellite.

NETWORK COSTS

Recurring telecommunications costs are an essential consideration when considering new program distribution topologies. Recurring telecommunications costs (fiber circuits, last-mile connectivity) will offset some of the savings listed in the previous section, but with careful planning the net result can still be a substantial increase in profit. Digital distribution via wide area network (WAN) technologies currently makes sense for groups of 10 to 25 stations, but for larger groups the costs of these telecommunications services becomes more significant and may become prohibitive. Larger station groups will want to explore hybrid satellite, VSAT, and terrestrial WAN technologies to keep monthly recurring costs to a minimum.

Broadband, last-mile local loops alone range from approximately $3K per month for DS3 to approximately $5K per month for OC-3c. Public Utilities Commission (PUC) tariffs are based on multiples of DS-0's, whereas tariffs for pure optical transport do not yet exist. This loophole may be worth researching by a broadcaster planning a large network. Monthly core network connectivity (Inter Exchange Carrier or IXC) fees can include vertical and horizontal (V & H) mileage charges that may be $20K per month or higher per node. We estimate that monthly connectivity costs for a group of about

15 stations can be on the order of $250K to $400K per month. Three- to five-year contracts can be negotiated that can substantially reduce these monthly costs.

The types of services and cost structures that a telecommunications provider such as Qwest, Williams, or AT&T can provide to support Digital CentralCasting for station groups must be researched. All of the fiber companies polled by Sony expressed interest in providing a wide range of services to broadcasters, including the hosting of servers and other on-air equipment and operations.

RISKS

Reliance on one or a few geographic sites for CentralCasting facilities can have downsides:

- Redundancy and back-up power systems will be essential for the CentralCasting technical operations center.
- Reliance on telecommunications vendors for real-time program distribution and guaranteed quality of service over WANs is similar to relying on satellite and telco vendors for traditional program distribution. What is new is the use of packet switching WAN technologies to distribute programs. Careful attention will be needed when negotiating Service Level Agreements (SLAs) with these WAN service providers. SLAs will need to specify maximum packet loss, packet jitter, and end-to-end latency requirements.
- Digital CentralCasting will create new operational structures and processes. Operations staff may be required to learn new ways of working and new skills. Change management will become very important.

ENABLING TECHNOLOGIES

Many technologies have converged in recent years to make Digital CentralCasting a viable business model for groups and networks; these include:

- MPEG-2 Data Compression
- Multimedia Content File Transfer
- Real Time Streaming via Wide Area Networking
- Media Asset Management
- Multichannel Automation Systems
- Lower cost servers, RAID, and archive tape formats

DATA COMPRESSION

Widespread adoption of MPEG-2 digital video and audio compression technology has revolutionized how video and audio are stored, allowing the migration from tape-based storage to server-based storage. MPEG-2 has also made program distribution via wide area networks (WANs) possible. Standards have developed to allow interoperability between equipment manufactured by multiple vendors and networks operated by different service providers. These standards enable the full benefit of working in the digital domain to be realized.

VIDEO, AUDIO, AND METADATA FILE TRANSFERS

Standards for video, audio, and metadata file formats allow interoperability between equipment provided by different manufacturers. These standards can be roughly divided into standards for files used for multimedia file transfer and standards for files for stream transfer. File transfers are generally guaranteed error free via TCP/IP and typically use bidirectional links. The choice of stream or file transfer is predicated on the type of interchange required, for example authoring or finished material interchange, content storage, emission, transmission or store and forward publication. (See the last section of this chapter for additional information on AAF and MXF file formats.)

STREAMING VIDEO THROUGH WIDE AREA NETWORKS

Real-time streaming of broadcast quality television programs across large geographic areas has traditionally been done using satellite and terrestrial microwave radio technologies. The quality of service and pricing structure of these legacy distribution systems are the benchmark by which broadcasters will measure new program distribution technologies.

The convergence of information and broadcasting technologies in the Digital CentralCasting model necessitates a multitiered signal distribution system. Baseband SDI, SDTI-CP, and DVB-ASI signals can be switched with traditional broadcast circuit switching equipment. MPEG-2 transport streams can be routed by utilizing IT packet switching technologies (Ethernet, ATM, and Fiber Channel).

Most broadcasters are comfortable with transferring non–real-time video files over WANs but these same broadcasters have more apprehension with the prospects of streaming real-time broadcast quality video over WANs. At least three factors contribute to this apprehension:

- Many broadcasters are unsure of WAN quality of service issues and how to negotiate Service Level Agreements (SLAs) with these service providers. Guaranteed quality of service prevents visible picture impairments and provides predictable end-to-end network latency. These are just two of the many issues that broadcasters must negotiate with their WAN service providers.
- Network access tariff and pricing structures for the high bandwidth WAN services required for real-time broadcast quality program distribution are difficult to negotiate because they cross multiple jurisdictions and use technology new to many telcos. For instance, some Local Exchange Carriers may not have experience in providing OC-3c local loop connectivity to their customers.
- Once broadcasters understand the network quality of service and pricing structures they must use this data to ascertain if there is a compelling business reason for changing their program distribution structure.

Stream transport over wide area networks (WANs) is now commonplace using ATM. Methods for guaranteed quality of service in Internet Protocol (IP) networks will soon be commonplace because of new protocols such as IntServ, DiffServ, and MPLS. (See the last section of this chapter for more details on guaranteed quality of service in streaming video through wide area networks.)

MEDIA ASSET MANAGEMENT

Information technologies such as the Internet, hierarchical storage, file servers, and relational database applications have made media asset management (MAM) and remote search, browse, and transfer from digital video archives a reality. Remote access and operations are made possible by low resolution proxy and streaming content and metadata applications such as MAM.

Interfaces to legacy systems, such as electronic newsroom systems, must be considered. Metadata compatibility between legacy systems is an issue. Use of metadata allows easy enterprisewide searching, thus making new resources available to producers.

TRAFFIC, BILLING, AND AUTOMATION

Traffic, billing, and automation software are key technologies required for multichannel playout facilities. Each station may use different station automation systems. It is important that the central traffic and automation systems seamlessly integrate with these legacy systems. Legacy systems may have trouble operating across multiple time zones, managing files across wide area networks, and supporting metadata.

Automation control of broadcast facilities allows the operation of multiple facilities from a central location. Automation software provides machine control to initiate and complete operations often performed by human operators. Modern traffic and billing software allows the scheduling and tracking of program content across multiple sites using wide area network technologies. Operations include device machine control and remote station monitoring. Traffic software manages and runs program schedules and logs and controls the automation software. Successful program playout initiates billing confirmation by the billing software, which is linked to the traffic and automation software. Traffic, billing, and automation software vendors require detailed operational scope to tailor their software feature sets to the requirements of the DCC operation.

Station groups may currently have software packages in inventory that can be upgraded or modified to add additional channels and include expanded functionality. Older software and hardware packages that operate in single, closed environments, however, may not adapt to a wide area network environment.

The software manufacturer often specifies the preferred computer hardware platform. Computer hardware costs, licensing fees, training fees, and on-going support program fees vary with the scope and complexity of the system.

LOW COST SERVERS AND ARCHIVE TECHNOLOGIES

The cost of hard disk systems continues to decrease. These reduced costs make long form program storage economically viable. Current MPEG-2 storage systems cost approximately $100 to $200 per hour for video servers, $50 to $100 per hour for RAID and about $1 per hour for DTF archive tape media. (See the end of this chapter for more information on DTF tape archive formats.)

DCC SYSTEM OVERVIEW*

The Digital CentralCasting system typically consists of a central digital routing switcher, server storage for program playout, near-line storage, archive and station automation software and hardware. This is coupled with state-of-the-art technologies for transferring material to and from the broadcast stations as data streams over broadband wide area networks. Figures 21.1 and 21.2 show block diagrams of typical DCC systems.

*This chapter discusses a CentralCasting approach that distributes real-time program streams to remotely located transmitters. Alternately, a store-and-forward approach could distribute files to local station servers, and local automation systems could play those files to air. For the sake of brevity these store-and-forward solutions are not discussed in this chapter. Store-and-forward should be considered a viable option when planning a CentralCasting system.

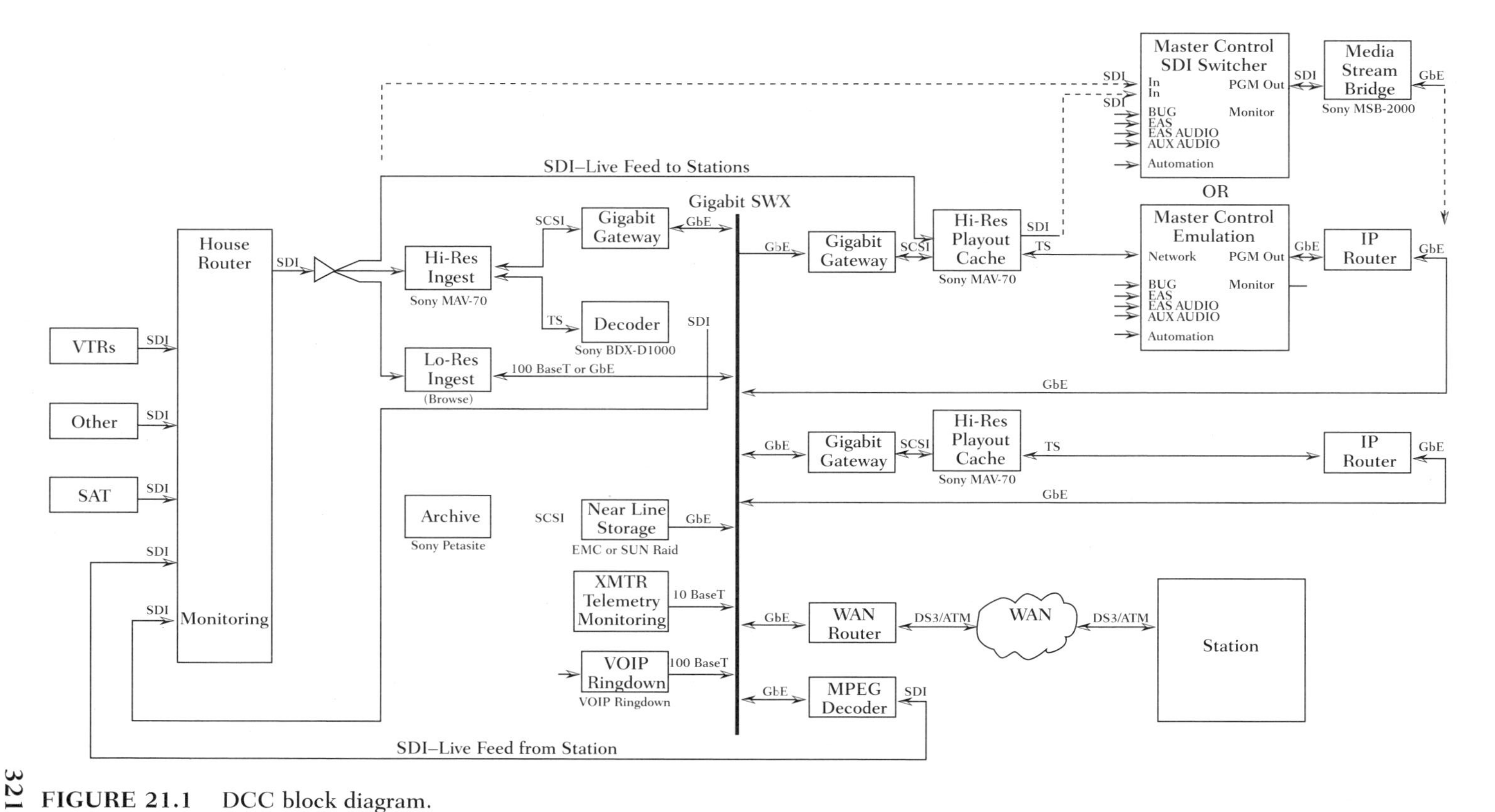

FIGURE 21.1 DCC block diagram.

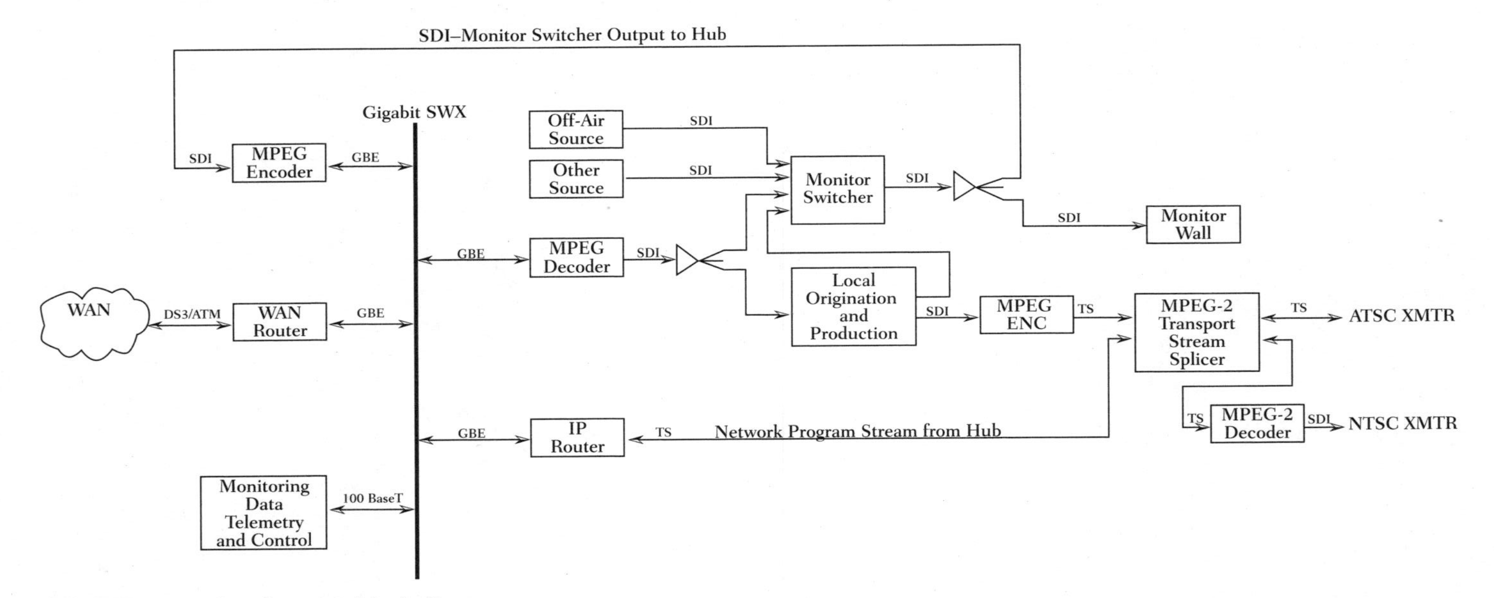

FIGURE 21.2 Local station block diagram.

The DCC operation center is also responsible for ingesting material from a variety of sources, including the local stations themselves, and queuing this material for playout across multiple time zones. It also receives monitoring feeds from the stations concerned, allowing the DCC Operations Center staff to fully monitor and log all functions of the station's transmission operation currently undertaken by MCR staff at each station individually. A DCC facility depends on the following key technologies:

- All digital, end-to-end
- MPEG-2 Compression
- Media Asset Management
- Hierarchical Storage Management models
- Broadband connectivity to wide area network (WAN)
- Upgrade paths for future technologies (e.g., 8-channel audio and HDTV)

The DCC is designed around the following operational processes:

- Digital end-to-end solutions reduce the need for continual and repeated quality assurance checks at numerous points throughout the signal path. Quality Assurance takes place during in the ingest phase. Once material is ingested it need not be measured again for technical quality.
- The traffic system prepares one or more playlists per local station. These multiple playlists are passed to the DCC central automation system, which controls both the ingest process and playout functions.
- The ingest phase converts the SDI or NTSC program signals from satellite, fiber, or VTR sources to a MPEG-2 4:2:2 Profile @ ML stream, which is then converted to a file in the ingest server. This ingest server functions as a cache, buffering the input stream until it becomes a complete file, which is then transferred to near-line and archive storage. The program

material remains in this file format until it is converted back to a stream at the playout server.

- Low resolution frame-accurate proxy files for browsing and metadata are also created during the ingest phase. Lower resolution Internet streaming files such as MS Media Player and Real Video formats and thumbnail storyboard keyframes can also be captured at ingest. These lower resolution versions of the content stored on the various servers are useful for viewing long-form content from remote locations.
- Locally originated programs, commercials, and promos are sent to the DCC OPS Center as MPEG-2 Transport Streams or as MXF files from the station via the WAN connection. These spots are stored as files and then played out or transferred to near-line and archive storage as directed by the automation system.
- The automation system directs files to be played out to be cached on the on-air and protect servers. The on-air servers feed a master control switcher and the composite program output of the master control switcher is then streamed to the WAN via network interfaces.
- Program streams from the DCC operations center are routed via WAN links directly into the local station's ATSC modulator input. The local station NTSC transmitter can be fed via a MPEG-2 decoder if the same program is transmitted on both channels. Locally produced content such as news and live sporting events can be encoded locally into an MPEG-2 transport stream and then spliced into and out of the program stream from the DCC. Alternately this local switching could be performed in the SDI domain by decoding the MPEG-2 stream from the DCC operations center.
- Local stations are equipped with remote control switchers for sending off-air monitor feeds and other baseband video feeds such as commercials, promos, and long-form content back to the DCC operations center.

FILE FORMATS FOR DIGITAL CENTRALCASTING

File transfer data format interoperability is extremely important. The EBU/SMPTE Task Force for Harmonized Standards for the Exchange of Program Material as Bitstreams conceptualized two basic types of files, Complex Content Packages and Simple Content Packages. Complex packages were envisioned for use in authoring applications applied to audio, video, and data content, whereas Simple packages were generally intended for streaming of interleaved audio, video, and data content. The Advanced Authoring Format (AAF) is an implementation of a Complex package and the Material eXchange File (MXF) format is an implementation of a Simple package.

ADVANCED AUTHORING FORMAT (AAF)

The Advanced Authoring Format (AAF) file format allows images, sound and metadata to be easily exchanged as files across multiple platforms and applications. AAF solves the problem of multivendor, cross-platform interoperability for computer-based digital production. AAF does a number of things:

- It allows complex relationships to be described in terms of an object model.
- It facilitates the interchange of metadata and/or program content.
- It provides a way to track the history of a piece of program content from its source elements through final production.
- It makes it possible to render downstream (with appropriate equipment).
- It provides a convenient way to "wrap" all elements of a project together for archiving.

By preserving comprehensive source referencing, and abstracting the creative decisions that are made, AAF improves workflow and simplifies project management.

The AAF Association is a nonprofit organization that works with standardization bodies to promote AAF technology internationally. AAF Association members include many equipment manufacturers, broadcasters, multimedia, and postproduction companies. With the accelerating progress of digital technologies, an open interchange standard is essential to enable the digital production facilities of the future. The lack of integration of multivendor products has proved to be a significant obstacle to the rapid acceptance of computer-based digital tools for professional production. AAF represents a broad industry initiative to remove those obstacles. AAF is a cross-platform Open Source application that is free for all to use on either Windows, Mac, or UNIX operating systems. See http://www.aafassociation.org for more information, including the AAF white paper titled, "AAF—An Industry-Driven Open Standard for Multimedia Authoring."

MATERIAL EXCHANGE FILE (MXF)

The Material eXchange File (MXF) is a file format designed for the exchange of files between file servers, tape streamers, and digital archives. Whereas AAF is primarily intended for authoring interchange, MXF is primarily intended for store-and-forward and broadcast playout interchange. MXF uses a fixed-byte header and a simplified metadata structure that is easy to implement in tape machines and servers and has low latency. MXF is based on the SMPTE Key-Length-Value coding format. The MXF file interchange format defines a standard wrapper for video, audio, and metadata. The wrapper is compression format independent and therefore can provide metadata transparency between systems that use different compression formats.

In MXF, the video, audio, and metadata are interleaved such that it is possible to start working on the program material before a file transfer is complete. This is particularly important when transferring large files over slow networks.

The MXF wrapper provides a means for clear identification of the all aspects of the payload. Different "templates" of increasing complexity are defined. The simplest template provides for the transfer of a single complete program. More complex templates provide the means to transfer a sequence of program segments with simple edit instructions, thus enabling the segments to be joined together. This is useful when distributing a single copy to multiple broadcasters, some of whom might not wish to show certain scenes.

MXF and AAF use the same data model. For example, AAF might be used in an online editing environment with the output moving via MXF to an on-air operation. Use of AAF in the editing environment and MXF in the playback environment provides a reliable essence and metadata pathway from content creation to play out. Both AAF and MXF enable the reliable interchange of essence of metadata within their respective application areas.

MXF is currently implemented as a container streaming format based on SDTI connections. MPEG-2 4:2:2P @ ML through SDTI-CP and DV DIF through SDTI are currently being implemented. MXF offers considerable extensibility to maintain format longevity and includes options to extend its template for the complex metadata support that may be required in libraries and data archives, as well as by future applications.

The MXF streaming file format interleaves video, audio, and data content on a field-by-field or frame-by-frame basis. This interleaving technique ensures a number of core user requirements are met, including recovery from transmission breaks, cuts-only editing, and retransmission before file transfer completion.

MXF and AAF both use the industry public SMPTE metadata dictionary, which allows systems to share metadata, and users and vendors to add features.

Sony is currently developing products that support AAF for production and MXF for streaming file distribution via FTP file transfer or other file transfer technologies. For additional information on MXF see http://www.pro-mpeg.org.

QUALITY OF SERVICE: ATM VS. IP

It is common to see regional program distribution using SMPTE 259M uncompressed serial digital video via private terrestrial fiber optic technologies. These technologies offer a quality of service that is straightforward and provides the highest quality video distribution available. But these systems can be extremely expensive to implement if dedicated dark fiber is not readily available at all of the intended nodes.

The following discussion compares ATM, which is the current first generation technology of choice to achieve guaranteed quality of service, to UDP/IP, which may be the predominant second generation technology for delivering real-time video over WANs in the future.

Asynchronous Transfer Mode (ATM) wide area network access via DS-3 or OC-3c local loops is widely available and is commonly used to distribute compressed real-time digital video across regional, national, and international networks. Guaranteed quality of service via ATM is currently a mature technology. Multicast via ATM using LAN Emulation (LANE) technology is still a new technology and should be thoroughly researched prior to implementation. Multicast via IP is a more mature technology than multicast over ATM. ATM is based on permanent virtual circuits (PVCs) and switched virtual circuits (SVCs), which create point-to-point connections through the WAN to achieve guaranteed quality of service.

Broadcasters should also consider Internet Protocol (IP) technologies when considering real-time compressed video transport via WAN technology. The Internet has the potential to become the de facto universal platform for global communications. Internet technologies will become part of the broadcast facility as packet switching technologies move into playout facilities from the distribution paths. It is conceivable that real-time signal distribution via Internet Protocol will become commonplace in all aspects of television production.

The network technology that comes into the broadcast facility will probably be IP and not ATM. Most television production and maintenance staff have a much better understanding of IP, and because IP is a much simpler network technology than ATM, and because often the same technical staff is responsible for engineering the Web and e-commerce facilities of their groups.

The original functionality requirement for the Internet was for minimal connectivity; therefore the Internet was designed to be a simple network with smart edge devices such as Personal Computers. By contrast, ATM was designed as a complex network with simple edge devices such as telephones.

Protocols for guaranteed quality of service in Internet Protocol (IP) networks are becoming commonplace. State-of-the-art network routers from all major backbone router vendors can classify, queue, and schedule (CQS) packets in accordance to their QOS class. These new routers address the limitations of existing best-effort shortest-path IP routing protocols and allow non–shortest path traffic engineering.

The Internet Engineering Task Force (IETF) has specified a relatively complex end-to-end model for QOS called Integrated Services (IntServ) and a much simpler edge-to-edge QOS model called Differentiated Services (DiffServ). DiffServ Expedited Forwarding Per Hop Behavior (EF-PHB) specifies that packets must depart a DiffServ node as fast or faster than they arrive at the node's input queue, which is analogous to the connection-oriented services provided by ATM networks. Multiprotocol Label Switching (MPLS) is a technology that specifies ways in which Layer 3 IP traffic can be mapped onto connection-oriented Layer 2 transports like ATM and frame relay.

Modern switching networks running these new QOS protocols can provide service between the United States and Europe with less than one 1,500-byte packet of jitter. Forward Error Correction (FEC) is necessary in all digital transmission systems, including satellite and terrestrial RF systems and WAN systems, to correct for packet loss caused by burst errors—which QOS cannot predict—such as static interference from spurious RF emissions. IETF RFC 2250 addresses payload formats for mapping MPEG-2 into Real Time Protocol (RTP) and IETF RFC 2733 addresses FEC for RTP payloads.

Once the full-duplex data paths provided by common DS3 and OC-3c WAN connections are enabled, fast-breaking news stories can easily be fed from the local station to the CentralCasting Operations Center.

SONY DIGITAL TAPE FORMAT (DTF)

Digital Tape Format (DTF) designates a Sony-trademarked technology that employs cassette-based, $^1/_2$-inch (or 12.65mm) magnetic tape and a specified format for the purpose of data storage in high-performance computing applications. The primary characteristics of DTF technology are high data capacity, high-speed data transfer, and high data reliability. In terms of Sony's current production-model DTF tape drive (GY-8240), these characteristics are expressed as follows:

- *High data capacity.* Up to 200 GB (native) data on a single large cassette.
- *High-speed data transfer.* Sustained transfer rate of 24 MB/sec (native) or burst rate of 40 MB/sec (native).
- *High data reliability.* Corrected bit error rate of less than 1 in 10e-17.

There are two models in the GY-8240 series. The GY-8240FC is equipped with two data ports for direct Fiber Channel Arbitrated Loop interface. The GY-8240UWD features two data ports for Ultra-Wide Differential SCSI interface. Both models provide two RS-232 serial communication ports for drive control and self-diagnostics. They also provide an Ethernet communication port for monitoring drive status and self-diagnostics, retrieving drive logs, and downloading upgrades to drive firmware. GY-8240 drives are equipped with the Sony Tele-File System, a newly developed application of noncontact flash memory in the form of a label attached to the cassette. The Tele-File memory stores information relating to tape and system management. Drive management routines use this information to dramatically reduce tape load and unload times and thus speed up access to user data. In addition DTF-2 drives support loading of tapes in the middle of the tape to significantly reduce load times. (See Table 21.1 for additional GY-8240F specifications.)

The first generation of DTF (DTF-1) was approved as an industry standard at the December 1996 general assembly of

TABLE 21.1 GY-8240 DTF-2 Tape Drive Specifications

Format	Sony DTF™ (Digital Tape Format) ECMA 248
Interface	Ultra SCSI-2 Fiber Channel
Data capacity	
L cassette	200 GB (formatted, uncompressed)
S cassette	60 GB (formatted, uncompressed)
Data transfer rate	24 MB/sec, sustained 40 MB/sec, burst w/ Ultra SCSI, 100 MB/sec burst w/ FC
Time-to-write 200 GB	2.3 Hours
Search speed	1.4 GB/sec
Bit error rate	Error rate < 1 in 1017 bits read: Reed-Solomon error correction code with read-after-write and automatic retry
Drive reliability	Average MTBF >200,000 hours; read/write passes >20,000
Head life	Typically greater than 5,000 hours
Dimension (W × H × D)	12.5 × 8.75 × 19.1 inches
Weight	55.1 lbs
Power consumption	< 300W
Environmental	
Operating temperature	5°C to 40°C (41°F to 104°F)
Nonoperating temperature	–20°C to 60°C (–4°F to 140°F)
Operating relative humidity (noncondensing)	20% to 80%

ECMA, the standard-setting body in Geneva. Its formal ECMA name is 12.65 mm Wide Magnetic Tape Cassette for Information Interchange–Helical Scan Recording-DTF-1. Sony intends to pursue adoption of DTF as an international standard through application to the International Standards Organization (ISO).

DTF is proving to be an extremely robust technology with an abundance of potential for greater speed and capacity. DTF-2 is a high-density, high-speed format enabling an uncompressed capacity of 200 GB per large cassette and a transfer rate of 24 MB/sec with a Ultra-SCSI or Fiber Channel interface. Native capacity and transfer speed is expected to double again to 400 GB per large cassette and 48 MB/s in DTF-3, which will also provide a fiber channel capability and is projected to be available in 2002. DTF-4 will see still another doubling of both density and speed. All future generations will assure full backward compatibility. This migration of the DTF technology also directly benefits the PetaSite mass storage robotics library by doubling and quadrupling storage capacity with each new version of DTF. Using backward compatibility, the PetaSite library will allow DTF-1,2,3, and even DTF-4 to reside in the same physical robotics. Currently, using DTF-2, the PetaSite library will store up to 11 Petabytes of native data storage.

DTF is a natural outgrowth of some 50 years of Sony technology and product development experience in magnetic tape storage. DTF's core technology is the Digital VTR, which Sony first introduced more than 10 years ago when it implemented the D-1 format for professional broadcast and other demanding industrial applications.

Digital VTR was a logical choice as the core technology for DTF when one considers the speed, capacity, reliability, and compatibility requirements it must satisfy in order to be effective in professional applications.

With its proven line of professional VTRs, Sony has established an indisputable reputation for delivering state-of-the-art data storage characterized by high-speed data transfer, high-density recording, and reliable drive mechanisms. Sony has shipped more than 450,000 professional Betacam series VTRs, including both analog and digital units. This number represents approximately 90 percent of the world market.

Sony has combined critical technologies from professional VTRs with advanced technologies for computer data storage to create DTF's high recording density, high-speed data transfer, and high data reliability characteristics. The DTF product line incorporates helical scanning, semiconductors, tape media, and library-level storage technologies as well as critical performance considerations based on in-depth analysis of the differences between helical and linear recording, the two major magnetic tape recording methodologies in existence today.

To meet the future requirements for increases in recording, density modifications to the track pitch must be made. Track pitch (track width) primarily refers to the recording density of tape devices, and recording density increases as track pitch decreases.

It should be emphasized that the current DTF generation (DTF-1) currently provides a track pitch of well under 40 microns, with a sufficient allowance for specifications that bring it into the evolving data storage arena. DTF-3, which employs four-fold recording density, is based on proven consumer and professional video recording technology. This approach is a milestone in the development of eight-fold recording density, a project that is already underway.

SUMMARY

Digital CentralCasting is the convergence of many key new technologies that enable new business models for television program playout. New creative possibilities can be realized and profitability productivity and on-air quality can be increased.

Groups that have already implemented Digital CentralCasting report that they were able to increase profits by making job cuts in master control, traffic, programming, and production; by reducing capital expenditures; and by reducing error rates in spot playback.

Because one model for Digital CentralCasting does not fit the requirements of all station groups, each group must decide which CentralCasting model works best for it.

CHAPTER TWENTY-TWO

PRODUCTION AND POSTPRODUCTION FOR 4:3 AND 16:9

JOHN RICE

As legend has it, George Eastman asked Thomas Edison what size motion picture film should be. Edison held his thumb and index finger about one inch apart and said, "This big." And so the 4-by-3 aspect ratio was born.

"Aspect ratio" refers to the relationship of the width to the height of a film or video image. According to *Mark Schubin's HDTV Glossary* (Union Square Press, 1990), it's "sometimes expressed as two numbers separated by a colon (e.g., 4:3) or sometimes as one number, with a colon and the number one implied (e.g., 1.85 is the same as 1.85:1). Aspect ratios of advanced television (ATV) systems are sometimes expressed relative to nine (e.g., 16:9, 15:9, 14:9, and 12:9)."

Although the Eastman–Edison legend is fun, it's probably more true that Eastman's assistant, William Dickson, established the frame size by slicing still photo film down the middle, creating the 35 mm size. There are two standard ways of identifying aspect ratios. One method is to base the height of the picture as "1" and the width as the relative size. For example, standard television is 1.33:1, widescreen is 1.78:1, Cinemascope is 2.55:1, and so on. The other method is based on picture elements, or squares. So, a TV image is 3 squares high and 4 squares wide. An HD or widescreen image is 9 squares high, 16 squares wide. We'll use the 4:3, 16:9 description here because it is easier to type.

Although feature films long ago broke the boundaries of the 4:3 box with everything from CinemaScope, VistaVision, and even Cinerama, that aspect ratio did become the foundation for television. Sure, there were issues to be dealt with when one of those wide feature films needed to be brought to the TV set. For the purists, *letterboxing* lets the full, wide image of the original director's vision be enjoyed. Viewers simply had to ignore the black bands above and below the images. Or, to fill the TV screen, the film would be *panned and scanned*, moving the original widescreen film image around to get the most important part of the shot to fill the TV screen from top to bottom. Sometimes this is done with great care and even the director's input. Sometime it isn't.

There didn't seem to be a lot of impetus to find any middle ground between the aspect ratios of feature films and television, until the coming of Digital Television (DTV) and its subset, High Definition Television (HDTV).

Over many years of studying and debating the various technical standards that would become HDTV, the issue of the shape of the screen brought out a vigilance on the part of many in the television and film community. The final determination was that the HDTV screen would have a 16:9 aspect ratio; one-third wider than Eastman's 4:3 (see Figure 22.1).

The DTV transmission standards, of which there are numerous options, allow a broadcaster to send out either 4:3 or 16:9 images. And, of course, as long as there are plain old TV sets out there and plain old TV signals being broadcast, there will always be a need for plain old 4:3 images regardless of what is being sent out over the DTV channels.

The availability of widescreen images in HD or standard definition (SD) provides the promise of "movielike" pictures on your TV. In fact, many early studies of HDTV found that people were more excited about the new, wider aspect ratio than the improved detail of the picture. But therein lies the rub. For the foreseeable future, anything that is going to be seen on

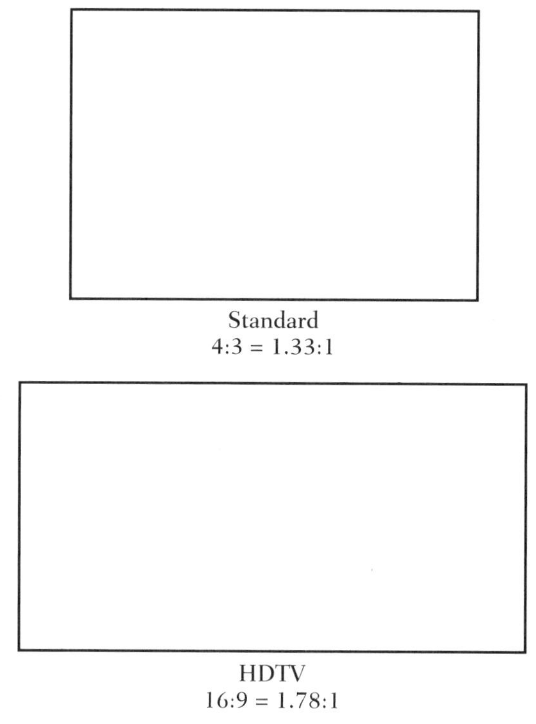

FIGURE 22.1 Television aspect ratios, shown with equal areas.

both these boxes must be created for both aspect ratios. And no matter how you cut it, one of those aspect ratios is going to have less of the picture in it than the other. We're not talking about image resolution, we're talking about picture content.

The most obvious, and still the most common answer, is to shoot in 16:9. When it comes time to create a version for 4:3, just cut off the sides. Seems simple enough. In fact, it's been going on for years. Many television shows shot on film have been shooting 16:9 images. The eyepiece of the camera is etched with the 4:3 aspect ratio. You don't have to worry about what is in those side panels because you're going to throw it away when you edit it for television anyway. And, with the anticipation of DTV, many of the shows we've been watching in recent years have actually been shot in 16:9 to protect the material for rerelease possibilities in the future. Here, you do pay attention to what is in the side panels, but you make sure that while it is pretty, or interesting, or that there isn't a crew member standing in the corner of the shot drinking a cup of coffee, you also make sure that there is nothing so important in the side panels areas that it will be missed when they're cropped from the picture (see Figure 22.2).

The vast majority of material being created for 4:3 and 16:9 distribution is being shot this way today. But this scenario is not without its challenges.

For example, if a shot is framed to take full advantage of the 16:9 frame, lopping off the sides can literally cut an actor (or part of the actor) from the scene (see Figure 22.3). Perhaps the classic example here is the face-to-face two-shot that we've all seen in more movies than we can remember. One actor is on the left, one on the right. There is a space in the middle. It is an effective and dramatic shot for the right piece of dialog; one shot with the two talking to each other. But, chop off the left and right sides, and you could be left with nothing more than a pair of noses and a lot of blank space in the center of the screen.

The solution in transferring a feature film to television is to take that single dramatic shot and edit it. Cut to actor one, full screen, for his line. Cut to actor two, full screen, for his line. Sure, it works, but you aren't seeing the shot the director first envisioned. And if these guys are in a rapid-fire chatty scene the editing can be dizzying.

And although that may be the best compromise for a feature, it is not going to happen for a television show. It takes time and costs money. In today's production economy, you want to shoot it once, use it often, and not have to run back to the edit room for every different distribution.

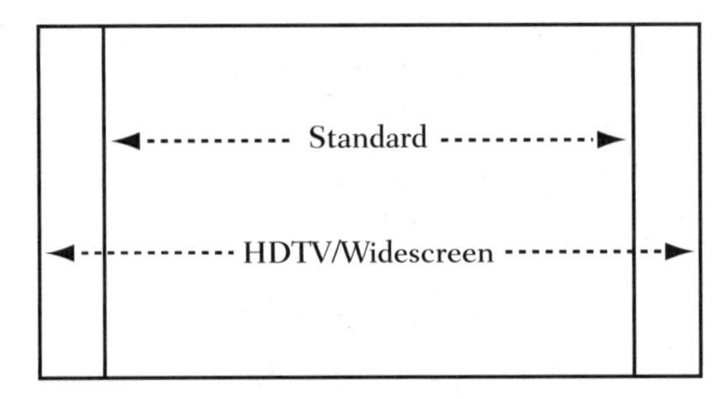

FIGURE 22.2 Common top and bottom.

FIGURE 22.3 A shot composed for 16:9 loses necessary story elements when shown as 4:3.

So, we are back to the option of shooting the 16:9 shot and keeping the sidebars safe. Your 4:3 version is protected, but at some point in the process you are going to feel like you are wasting all that side space or, worse yet, the composition of the wide shot will feel cramped in the middle and empty on the edges (see Figure 22.4).

Regardless of the problems, most dual-aspect ratio productions live with the best compromise their creative teams can find; frame the important stuff for the 4:3 screen and fill the rest of the screen with interesting—but not necessarily important—props, people, or scenery.

There are even larger issues here than the composition of a static shot. Remember, we are dealing with moving pictures, or perhaps better put, pictures in which things move. Things like cars, animals, and people. And they often move into the frame or out of the frame. The timing of an entrance or exit can be vital to the effect of scene's dramatic or comic content. But if you are shooting for both aspect ratios, one of your versions will be wrong (see Figures 22.5 and 22.6).

For the majority of people working in the world of dual aspect ratios, there really didn't seem to be any option other than the side-bar solution. But an interesting alternative has begun to emerge.

In the fall of 1999, CBS began airing shows both in standard NTSC and in HDTV. At that time, Paramount Pictures decided that all their shows, regardless of the network that airs them, should be shot in such a way that an HDTV version can be extracted in the future for reruns and ancillary markets.

A few years before the CBS announcement, Paramount examined the composition compromise adopted by other studios to accommodate the 4:3 aspect ratio of NTSC and the 16:9 of HDTV. But according to John Sprung, Director of Technology for Paramount Television Post Production, "our cinematographers informed us that that relationship of frames is difficult to work with, and produces severely impaired compositions in one or both versions.

"We did a lot of thinking and testing," says Sprung, "which resulted in the Common Sides process." (See Figure 22.7).

Simply stated, "Common Sides" means to shoot for a 4:3 frame, then after the show is finished (edited), a 16:9 HDTV or widescreen extraction can be made from within the 4:3 frame, maintaining the same left and right borders on both versions. Chances are a viewer will be less bothered by not seeing an actor's feet (as demonstrated in Figure 22.7) than by seeing half an actor (as in Figure 22.3).

FIGURE 22.4 A shot composed for 4:3 gets a loose, empty feeling when shown as 16:9. Note especially the upper left.

FIGURE 22.5 If a cut on an entrance is timed right for standard TV, then it will be late in HDTV. If a cut on an exit is timed right for standard TV, then it will be early in HDTV.

In this scenario, the eyepiece of the camera is now marked to identify the bottom portion of the picture that will be excluded from the 16:9 version (see Figure 22.8).

Is it a perfect solution? Well, not exactly. There are still framing issues. The shot that looks well balanced top-and-bottom in 4:3 may now seem to have too much head room in 16:9. There are alternative ways of extracting the 16:9 image that include shaving a little off the top and a little off the bottom. Or there is the option of changing the areas eliminated on a scene-by-scene or even shot-by-shot basis (see Figure 22.9).* (This is possible as well in the "side panel" scenario). But just as in a pan-and-scan transfer, changing the area of extraction within a program will take more time and, yes, more money.

*Thanks to John Sprung, Paramount Pictures Television, for diagrams.

FIGURE 22.6 If a cut on an entrance is timed right for HDTV, then it will be early in standard TV. If a cut on an exit is timed right for HDTV, then it will be late in standard TV.

In actual practice, Paramount never uses a fixed common top line, rather doing a tilt-and-scan. This takes much less time and money than a pan-and-scan because a position can be established that takes some off the top and more off the bottom of an image that works for the vast majority of the show. The whole show is converted in this position, and time codes are noted for the shots that need a different positioning. A show can be tilt-and-scanned in roughly twice the nominal air time, i.e two hours for a one hour show. Pan-and-scan would require about 4-6 times the time and expense of tilt-and-scan.

The reason for the fixed common top ground glass markings is that there's no way to make variable ground glass markings, and you have to have some reference to how big the HD frame will be.

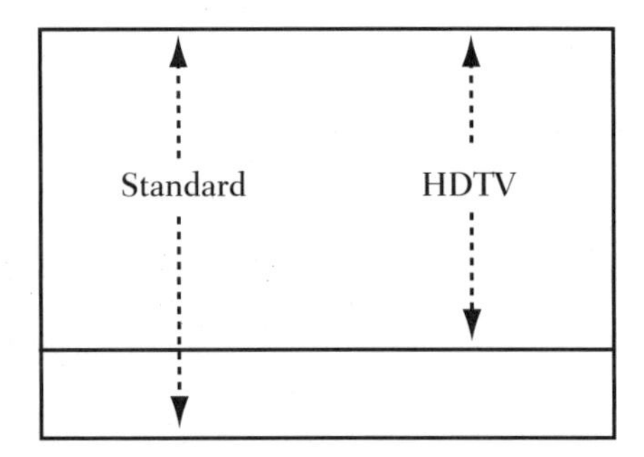

FIGURE 22.7 Common top and sides.

In transferring feature films to TV, the anamorphic 2.35:1 shows are forced to do a pan and scan, because there's nothing available on the film outside that frame. The flat 1.85:1 pictures (60–80% of all movies) have more image above and below the theatrical frame, and they use it for television transfers.

There is also the issue of image resolution. For productions shooting in 35 mm film, as are most of Paramount Television's current fare, this is a less important matter. But what may be perceived as a technical compromise in terms of video-format resolutions may be worthwhile for better shot composition.

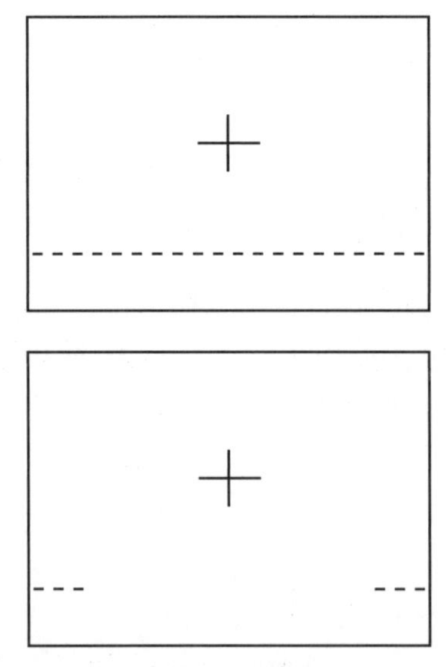

FIGURE 22.8 Suggested ground glass markings.

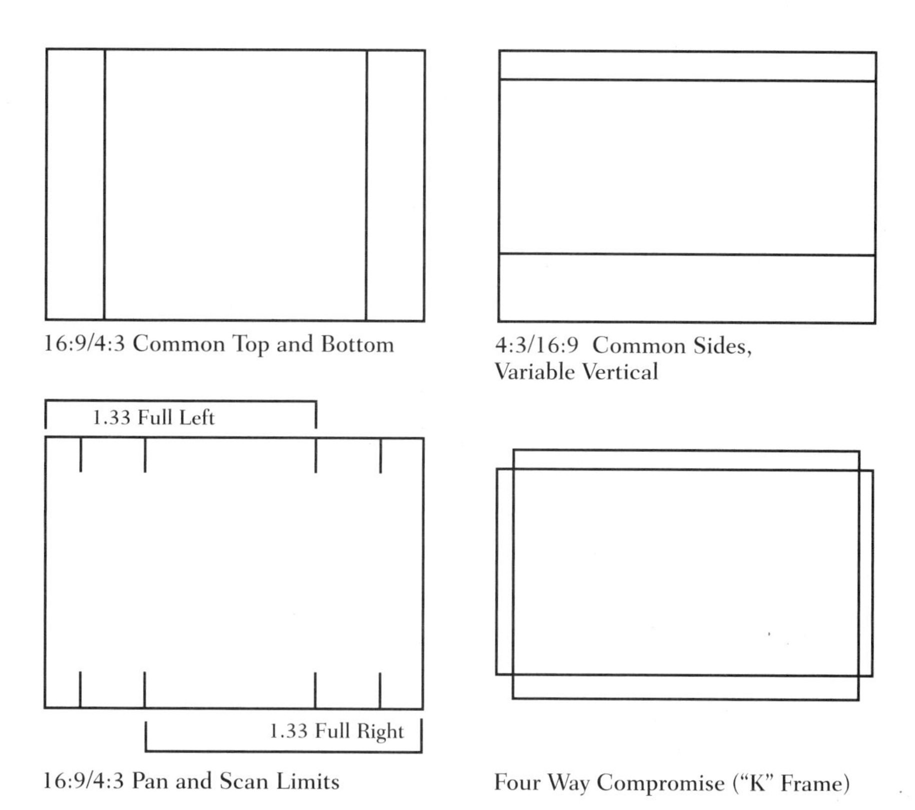

FIGURE 22.9 Composition compromises for NTSC/HDTV simulcast.

In today's production environment, television shows and feature films are being shot on everything from 35 mm film and HD video to DVCAM and even DV. Don't discard the idea because you are afraid you are giving up image quality. At least, don't discard it until you've done a test or two to see how appreciable the quality difference is to the viewer's eye.

The key for any content creator looking to maximize the distribution of his content is to find the best way to do it once. In the film and television industry, there are vehement advocates for both "side panel" and "common sides" production. And chances are, when you watch a show or movie from either side of the debate, you can't immediately tell which scenario was used.

What you will see, hopefully, is a good product, made by talented, creative people, who have found the most effective application of dual aspect-ratio shooting.

CHAPTER TWENTY-THREE

TELEVISION: THE "HIGH DEFINITION"

BOB ALLEN
DUNCAN KENNEDY

It's new, enriched, wide, liberating, and it sounds as good as it looks. All that's left is for content creators to explore its many possibilities. High-definition television (HDTV) is, from the perspective of those engaged in creating content of one kind or another, simply a next-generation tool. It is a deepening of the palette, or another gag in the production bag of tricks, depending on your point of view. As with any other tool, the art is in how you use it.

The truth is that the number of pixels presented; the flatness, brightness, and clarity of a screen; the dimensionality of the way the audience experiences the sound; and even the opportunity to "talk back" to the TV set (which the digitization of television offers) is really not so compelling. The miracle is that there *actually is television*. The fact that—at an ever-increasing rate—images and information are transmitted through the air, down a wire, or out into space, and arrive in one piece in the family living room is really still—after all this time—magic. The promise of this new technology is the breadth of this new creative palette as it refines an existing art form.

One of our clients recently commented after a pitch on the potential of a 500-channel digital network, "Just because we have 500 channels doesn't mean we should use 500 channels." She was right, and the wisdom certainly applies to the HD environment also. What should be used (and this is not always easy to do in the midst of the "thrill-wave" of a new technology) is just the right package of media necessary for the production to be as impactful as possible.

With this in mind, HD can do a few important things for the viewing audience. First, it is novel. If we make the assumption that by 2005 there will be sufficient deployment of HD displays and receivers to provide a broad audience, then there are several years of green field to present content that looks "new and improved." If you are reading this, you have probably seen a 1080I picture and might even take it for granted that you can see the hair on an antelope's nose from 30 feet away, but most people haven't yet experienced it. The novelty of the, "Oh my God, what is that?" quality of HD is a seductive tool to get a well-placed message under the radar of critical audiences. It is the visual equivalent of surround sound. It is a loud, brass fanfare for the eyes. The opportunity is in the content to which that fanfare draws the viewers' attention.

Next, HDTV is enriched. HD done well (and that's not a trivial stipulation) packs more story nutrition per bite than we could have ever hoped for in the Old World of NTSC. As opposed to the familiar 4:3 aspect ratio of video, 16:9 is a true organic aspect ratio. It is the way our eyes are aimed, it is the way our mind processes the world. The nature of this new frame means we can compose visual statements that contain more humor, more action, and more pure information than ever. That information is compounded by stunningly vivid image quality, and its value is exponentially increased with the addition of 5.1 surround sound.

Certainly, the "pipe" to the home or the desktop has been widened by the advent of broadband delivery. But the real frontier isn't the famous "last mile" to the home; it's the final 30 feet to the eyes and ears of the audience. HD has wildly increased that throughput rate, and the story potential is stunning. The art of directing in this new medium is a compound task, reminiscent of the live stage and the big screen. It is leaving the soft-edged, 4–by–3 world—where the close-up was king—a lonely afterthought.

Finally, HDTV is liberating. At least 18 distinct formats, and an endless variety of ways to squeeze, slice, dice, and package material all open the door for the use of television as a true medium for storytelling in all its richness. The NTSC world has been a cloistered, hidebound environment beset by rules. It is in many ways reminiscent of the current issues surrounding the Internet: limited bandwidth, limited color reproduction, limited audience. HDTV may finally make soccer and hockey the successful televised sports they have tried for years to become because they are *wide* events and this is a wide medium. It has the potential to revitalize news (actual news as well as "infotainment"), because it can manifest the reality of a situation much more faithfully. It will, and already has, engendered new forms and certainly will enable more valuable nonentertainment uses of the medium, including simulation, training, telemedicine, and even art. It is an aesthetically pleasing experience to see paintings on that screen; not the same as seeing them in a museum, but equally valuable and far more accessible.

This wide, dense, almost noise-free stage is fertile ground for a new generation of storytellers who have been raised with ubiquitous, multimodal, image-based information as common as tap water. The Beatles changed popular music forever when they started to play with possibilities of the new stereophonic sound. CNN changed the evening news forever when it delved into the realm of 24/7 satellite access. The changes that will be wrought by HDTV will be profound and the rewards great for those content creators bold enough to push ahead and explore the frontiers of what this new creative palette can accomplish.

CHAPTER
TWENTY-FOUR

IN SEARCH OF THE NEW VIEWING EXPERIENCE

JERRY WHITAKER

High-definition television has improved on earlier techniques primarily by calling more fully upon the resources of human vision. The primary objective of HDTV has been to enlarge the visual field occupied by the video image, an attribute that can be used to benefit interactive TV applications. This attribute has called for larger, wider pictures that are intended to be viewed more closely than conventional video. To satisfy the viewer upon this closer inspection, the HDSTV image must possess proportionately finer detail and sharper outlines.

CRITICAL IMPLICATIONS FOR THE VIEWER AND PROGRAM PRODUCER

In the search for a "new viewing experience," early experimenters conducted an extensive psychophysical research program in which a large number of attributes were studied. Viewers with nontechnical backgrounds were exposed to a variety of electronic images, whose many parameters were then varied over a wide range. A definition of those imaging parameters was sought, the aggregate of which would satisfy the average viewer that the TV image portrayal produced an emotional stimulation similar to that of the large-screen film experience.

Central to this effort was the pivotal fact that the image portrayed would be large—considerably larger than current NTSC television receivers. Some of the key definitions sought by researchers were precisely how large, how wide, how much resolution, and how far the optimum viewing distance of this new video image.

A substantial body of research gathered over the years has established that the average U.S. consumer views the TV receiver from a distance of approximately seven picture heights. This translates to a 27-inch NTSC screen viewed from a distance of about 10 feet. At this viewing distance, most of the NTSC artifacts are essentially invisible, with the exception of cross color. Certainly the scanning lines are invisible. The luminance resolution is satisfactory on camera close-ups. A facial close-up on a modern high-performance 525-line NTSC receiver, viewed from a distance of 10 feet, is quite a realistic and pleasing portrayal. But the system quickly fails on many counts when dealing with more complex scene content.

Wide-angle shots (such as jersey numbers on football players) are one simple and familiar example. TV camera operators, however, have adapted to this inherent restriction of 525-line NTSC, as witnessed by the continual zooming in for close-ups during most sporting events. The camera operator accommodates for the technical shortcomings of the conventional television system and delivers an image that meets the capabilities of NTSC, PAL, and SECAM quite reasonably. There is a penalty, however, as illustrated in Figure 24.1. The average home viewer is presented with a very narrow angle of view—on the order of 10 degrees. The video image has been rendered "clean" of many inherent disturbances by the 10-foot viewing distance and made adequate in resolution by the action of the camera operator; but, in the process, the scene has become a small window. The now "acceptable" television image pales in comparison with the sometimes awesome visual stimulation of the cinema. The primary limitation of conventional TV systems is, therefore, image size. A direct consequence is further limi-

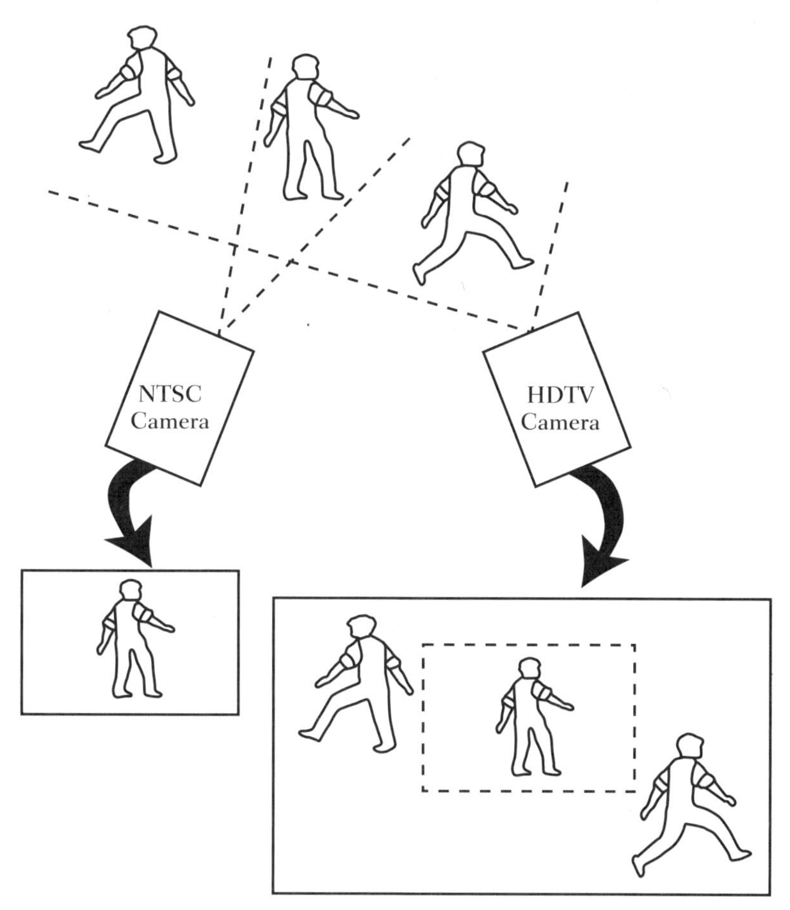

FIGURE 24.1 An illustration of the differences in the scene capture capabilities of conventional video and HDTV.

tation of image content; the angle of view is constricted constantly by the need to provide adequate resolution. There is significant, necessary, and unseen intervention by the TV program director in the establishment of image content that can be passed on to the home viewer with acceptable resolution.

Compared with the 525-line NTSC signal (or the marginally better PAL and SECAM systems), the ATSC DTV system and the North American HDTV studio standard (SMPTE* 240M) and its digital representation (SMPTE 274M) offer a

*Society of Motion Picture and Television Engineers, the leading video production standards organization.

vast increase in total information contained within the visual image. If all this information is portrayed on an appropriate HDTV studio monitor (commonly available in 19-, 28-, and 38-inch diagonal sizes), the dramatic technical superiority of HDTV over conventional technology can be seen easily. The additional visual information, coupled with the elimination of composite video artifacts, portrays an image almost totally free (subjectively) of visible distortions, even when viewed at a close distance.

IMAGE SIZE

If HDTV is to find a home with the consumer, it will find it in the living room. If consumers are to retain the average viewing distance of 10 feet, then the minimum image size required for an HDTV screen for the average definition of a totally new viewing experience is about a 75-inch diagonal. This represents an image area considerably in excess of present "large" 27-inch NTSC (and PAL/SECAM) TV receivers. In fact, as indicated in Figure 24.2, the viewing geometry translates into a viewing angle close to 30 degrees and a distance of only three picture heights between the viewer and the GDTV screen.

Compare this with the 10-degree viewing angle for conventional systems, as shown in Figure 24.3.

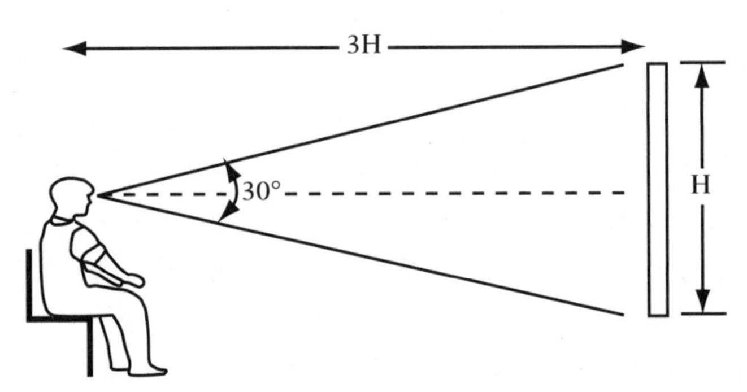

FIGURE 24.2 Viewing angle as a function of screen distance for HDTV.

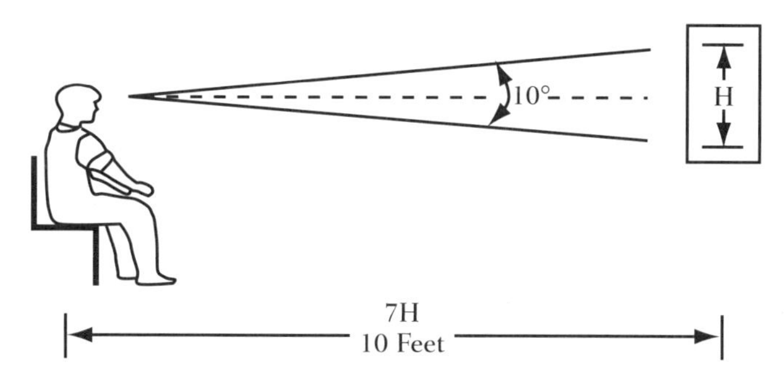

FIGURE 24.3 Viewing angle as a function of screen distance for conventional video systems.

HDTV IMAGE CONTENT

There is more to the enhanced viewing experience than merely increasing picture size.[1] Unfortunately, this fundamental premise has been ignored in some audience surveys. The large artifact-free imaging capability of HDTV allows a new image portrayal that capitalizes on the attributes of the larger screen. As mentioned previously, as long as the camera operator appropriately fills the 525 (or 625) scanning system, the resulting image (from a resolution viewpoint) is actually quite satisfactory on conventional systems. If, however, the same football game is shot with an HDTV camera, the angle of view of the lens is adjusted to portray the same resolution (in the picture center) as the 525 camera when capturing a close-up of a player on its 525 screen. A vital difference between the two pictures emerges: The larger HDTV image contains considerably more information, as illustrated in Figure 24.1.

The HDTV picture shows more of the football field—more players, more of the total action. Thus, the HDTV image is radically different from the NTSC portrayal. The individual players are portrayed with the same resolution on the retina—at the same viewing distance—but a totally different viewing experience is provided for the consumer. The essence of HDTV imaging is this greater sensation of reality.

The real, dramatic impact of HDTV on the consumer will be realized only when two key ingredients are included:

- Presentation of an image size of approximately 75 inches diagonal (minimum).
- Presentation of image content that capitalizes on new camera freedom in formatting larger, wider, and more true-to-life angles of view.

FORMAT DEVELOPMENT

Established procedures in the program production community provide for the 4:3 aspect ratio of video productions and motion picture films shot specifically for video distribution. The format convention has, by and large, been adopted by the computer industry for desktop computer systems.

In the staging of motion picture films intended for theatrical distribution, no provision generally is made for the limitations of conventional video displays. Instead, the full screen, in wide aspect ratios—such as CinemaScope—is used by directors for maximum dramatic and sensory impact. Consequently, cropping of essential information may be encountered more often than not on the video screen. This problem is particularly acute in wide-screen features when it is necessary to crop the sides of the film frame to produce a print for video transmission. This objective is met in one of the following ways:

- Letterbox transmission uses blank areas above and below the wide-screen frame. Audiences in North America and Japan have not generally accepted this presentation format, primarily because of the reduced size of the picture images and the aesthetic distraction of the blank screen areas.
- Printing the full frame height and cropping equal portions of the left and right sides to provide a 4:3 aspect ratio. This process frequently is less than ideal because, depending on the scene, important visual elements may be eliminated.
- Programming the horizontal placement of a 4:3 aperture to follow the essential picture information. Called pan-and-scan, this process is used in producing a print or in making a film-to-tape transfer for video viewing. Editorial judgment is

required for determining the scanning cues for horizontal positioning and, if panning is used, the rate of horizontal movement. This is an expensive and laborious procedure and, at best, it compromises the artistic judgments made by the director and the cinematographer in staging and shooting, and by the film editor in postproduction.

These considerations are also of importance to the computer industry, which is keenly interested in multimedia technology.

One of the reasons for moving to a 16:9 format is to take advantage of consumer acceptance of the 16:9 aspect ratio commonly found in motion picture films. Actually, however, motion pictures are produced in several formats, including:

- 4:3 (1.33)
- 2.35 (used for 35 mm anamorphic CinemaScope film)
- 2.2 in a 70 mm format

Still, the 16:9 aspect ratio generally is supported by the motion picture industry. Figure 24.4 illustrates some of the more common aspect ratios.

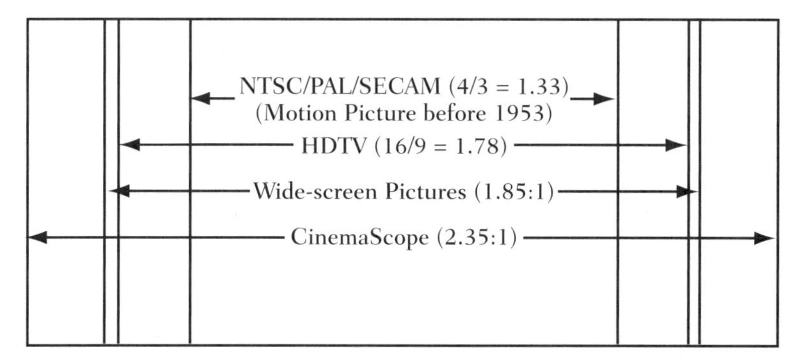

FIGURE 24.4 Comparison of the aspect ratios of television and motion pictures.

AURAL COMPONENT OF VISUAL REALISM

The realism of a video presentation depends to a great degree on the realism of the accompanying sounds. This important point should not be lost on the designers of interactive TV systems. Particularly in the close viewing of HDTV images, if the audio system is monophonic, the sounds seem to be confined to the center of the screen. The result is that the visual and aural senses convey conflicting information. From the beginning of HDTV system design, it has been clear that stereophonic sound must be used. The generally accepted quality standard for high-fidelity audio has been set by the digital compact disc (CD). This medium covers audio frequencies from below 30 Hz to above 20 kHz, with a dynamic range of 90 dB or greater.

Sound is an important element in the viewing environment. To provide the greatest realism for the viewer, the picture and the sound should be complementary, both technically and editorially. The sound system should match the picture in terms of positional information and offer the producer the opportunity to use the spatial field creatively. The sound field can be used effectively to enlarge the picture. A surround sound system can further enhance the viewing experience.

HEARING PERCEPTION*

Human hearing is quite sensitive in some respects and less so in others. In the past few decades, psychoacoustic research has discovered and explored several areas in which human hearing exhibits such reduced sensitivity. This knowledge has been applied to digital encoding systems to achieve greater coding efficiencies for audio signals. Unlike standard file-based data compression systems that analyze a bit stream for statistical redundancies and remove them in a fully retroactive way (so-called *lossless* coding, such as that used in various "zip" utilities), so-called *perceptual coding* systems analyze digital audio signals and re-encode them at lower bit rates with generally

*This section contributed by Skip Pizzi, Microsoft.

inaudible results, because they exploit the areas of human hearing that are relatively nondiscriminating. This encoding is lossy in that once it is performed the audio content cannot be fully reconstructed to its original form, but ideally, the listener cannot tell the difference.

The use of perceptual coding has become commonplace in today's digital audio system. An example is the popular MP3 format, named after a perceptual coding format for digital audio signals (MPEG-1 Audio Layer 3). To understand the value of these systems, consider the fundamental uncompressed format for digital audio coding, linear *pulse code modulation* (PCM). This format necessarily observes the Nyquist value for sampling of the digital signal (i.e., sampling at twice the frequency of the highest audio frequency intended to be reproduced on the system, or higher). For the full-range audio used in broadcast television, this implies an audio bandwidth of approximately 20 kHz, requiring a sampling rate of at least 40 kHz. Typical sampling frequencies used are 44.1 kHz (the audio CD sampling rate) or 48 kHz (common in digital television systems). For high-fidelity dynamic range, a resolution of 16 to 24 bits must be applied to each sample. This results in a digital audio coding that produces approximately 1 Mbits/s of data per audio channel. For the multichannel audio systems used in DTV or cinematic productions, this can amount to over 5 Mb/s or nearly 40 MB per hour of audio.

Perceptual coding systems can provide substantial reduction in this data rate—on the order of 10—without significant audio quality loss. This requires substantially less bandwidth for transmission, and correspondingly less storage space, hence the popularity of such data compression systems. This reduction in data rate is performed by adaptively reducing the resolution of groups of audio samples, while leaving the sampling rate unmodified. In this way, the audio bandwidth remains unchanged, but the dynamic range is reduced (i.e., the signal becomes temporarily noisier).

The perceptual coding systems can get away with this added noise because their operation exploits the *masking* properties of human hearing. One form of this phenomenon, *spectral masking*, dictates that in the presence of a prominent tone, other slightly lower level tones at nearby frequencies are rendered inaudible. This means that the loudest tones in a given moment of sound produce a selective desensitizing of the hearing sense to other nearby frequencies. Another form of the process, called *temporal masking*, acknowledges that a loud aural event desensitizes the hearing sense to other, quieter sounds immediately after (and even slightly before) the loud sound.

Perceptual coding systems use these *selective desensitivities* to their advantage by placing the noise products that result from their resolution reductions into the desensitized zones, thereby making the added noise inaudible. Through the use of these techniques, relatively high-fidelity audio is possible at data rates of less than 50 Kbits/s per channel. This data rate would produce a dynamic range of about 10 dB (an unlistenably noisy signal) using linear PCM coding, whereas with perceptual coding this same number of bits can offer the equivalent of a 90+ dB dynamic range (very high fidelity).

In multichannel systems, further reductions in data rate can be taken because similar audio signals often occupy more than one of the audio channels at any given moment. In other words, it is extremely unlikely that all six channels of a surround sound signal will contain completely discrete signals having no common information. When such common information does occur, it need not be uniquely coded for each channel. This technique is often referred to as *joint coding*, and its efficiency is utilized by the systems employed in today's digital multichannel sound-for-picture formats, such as Dolby AC-3 and DTS.

The Aural Image

There is a large body of scientific knowledge on how humans localize sound. Most of the research has been conducted to study lateralization with subjects using earphones to listen to monophonic signals. *Localization* in stereophonic listening with loudspeakers is not as well understood, but the research shows the dominant influence of two factors: *Interaural amplitude* differences and *interaural time delay*. Of these two properties, time delay is the more influential factor. Over intervals related to the time it takes for a sound wave to travel around the head from one ear to the other, interaural time clues determine where a listener will perceive the location of sounds. Interaural amplitude differences have a lesser influence. An amplitude effect is simulated in stereo music systems by the action of the stereo balance control, which adjusts the relative gain of the left and right channels. It is also possible to implement stereo balance controls based on time delays, but the required circuitry is more complex.

A listener positioned along the line of symmetry between two loudspeakers will hear the center audio as a phantom or *virtual image* at the center of the stereo stage. Under such conditions, sounds—dialog, for example—will be spatially coincident with the on-screen image. Unfortunately, this coincidence is lost if the listener is not positioned properly with respect to the loudspeakers. Figure 24.5 illustrates the sensitivity of listener positioning to aural image shift. As illustrated, if the loudspeakers are placed 6 feet apart with the listener positioned 10 feet back from the speakers, an image shift will occur if the listener changes position (relative to the centerline of the speakers) in just 16 inches. The data shown in the figure is approximate and will yield different results for different types and sizes of speakers. Also, the effects of room reverberation are not factored into the data. Still, the sensitivity of listener positioning can be seen clearly. Listener positioning is most critical when the loudspeakers are spaced widely, and less critical when they are spaced closely. To limit loudspeaker spacing, however, runs counter to the purpose of wide-screen displays. The best solution is to add a third audio channel dedicated exclusively to the transmission of center-channel signals for reproduction by a center loudspeaker positioned at the video display, and to place left and right speakers apart from the display to emphasize the wide-screen effect. The addition of surround-sound speakers further improves the realism of the aural component of the production.

Matching Audio to Video

It has been demonstrated that even with no picture to provide visual cues, the ear–brain combination is sensitive to the

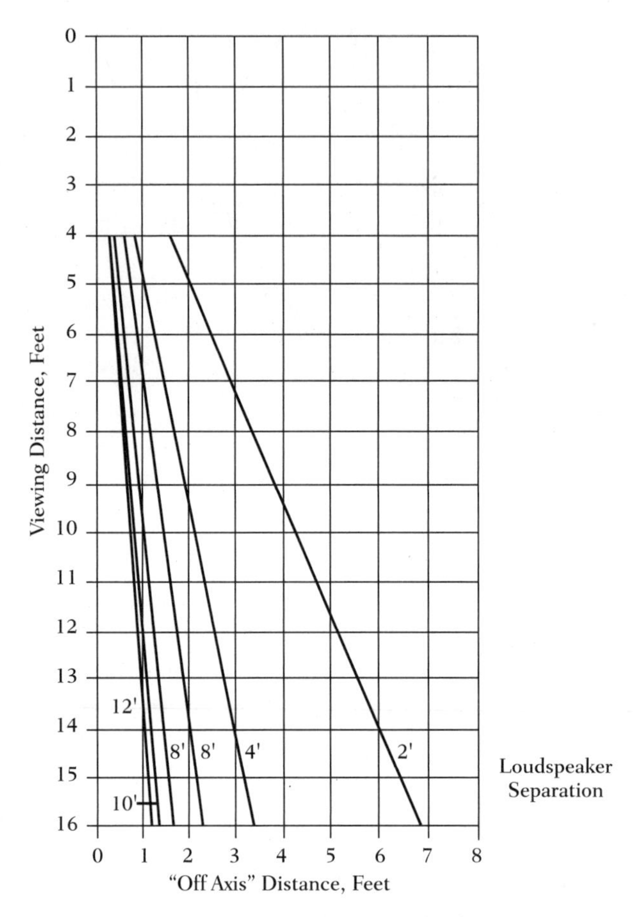

FIGURE 24.5 The effects of listener positioning on center image shift.

direction of sound, particularly in an arc in front of and immediately in back of the listener. Even at the sides, listeners are able to locate direction cues with reasonable accuracy. With a large-screen display, visual cues make the accuracy of sound positioning even more important.

If the number of frontal loudspeakers and the associated channels is increased, the acceptable viewing–listening area can be enlarged. Three-channel frontal sound using three loudspeakers provides good stereo listening for three or four viewers, and a four-channel presentation increases the area even more. The addition of one or more rear channels permits surround sound effects.

Surround sound presentations, when done correctly, significantly improve the viewing experience. For example, consider the presentation of a concert or similar performance in a public hall. Members of the audience, in addition to hearing the direct performance sound from the stage, also receive reflected sound, usually delayed slightly and perhaps diffused, from the building surfaces. These acoustic elements give a hall its tonal quality. If the spatial quality of the reflected sound can be made available to the home viewer, the experience will be enhanced greatly. The home viewer will see the stage performance in high definition and hear both the direct and indirect sound, all of which will add to the feeling of being present at the performance.

In sports coverage, much use can be made of positional information. In a tennis match, for example, the umpire's voice would be located in the center sound field—in line with her observed position—and crowd and ambient sounds would emanate from left and right.

Several methods have been used to successfully convey the surround sound channel(s) in conventional NTSC broadcasts. The Dolby AC-3 sound system used in the ATSC DTV system offers 5.1 channels of audio information to accompany the HDTV image.

Making the Most of Audio

In any video production, there is a great deal of sensitivity to the power of the visual image portrayed through elements that build the scene such as special effects, acting, and directing. All too often, however, audio tends to become separated from the visual element. Achieving a good audio product is difficult because of its subjective content. There are subtleties in the visual area, understood and manipulated by video specialists, that an audio specialist might not be aware of. By the same token, there are psychoacoustic subtleties relating to how humans hear and experience the world around them that audio specialists can manipulate to their advantage.

Reverb, for example, is poorly understood; it is more than just echo. This tool can be used creatively to trigger certain psychoacoustic responses in an audience. The brain will perceive a voice containing some reverb to be louder. Echo has been used for years to effectively change positions and dimensions in audio mixes.

To use such psychoacoustic tools is to work in a delicate and specialized area, and audio is a subjective discipline that is short on absolute answers. One of the reasons it is difficult to achieve good quality sound is because it is hard to define what that is. It is usually easier to qualify video than audio. Most people, given the same video image, come away with the same perception of it. With audio, however, accord is not so easy to come by. Musical instruments, for example, are harmonically rich and distinctive devices. A violin is not a pure tone; it is a complex balance of textures and harmonics. Audio offers an incredible palette, and it is acceptable to be different. Most video images have any number of absolute references by which images can be judged. These references, by and large, do not exist in audio.

When an audience is experiencing a program—be it a television show or an aircraft simulator computer game—there is a balance of aural and visual cues. If the production is done right, the audience is drawn into the program, putting themselves into the events occurring on the screen. This *suspension of disbelief* is the key to effectively reaching the audience.

IDEAL SOUND SYSTEM

Based on the experience of the film industry, it is clear that for the greatest impact, HDTV sound should incorporate, at minimum, a four-channel system with a center channel and surround sound. Figure 24.6 illustrates the optimum speaker placement for enhancement of the viewing experience. This view-point was taken into consideration by the ATSC in its study of the Grand Alliance system.

DOLBY AC-3

Under the ATSC DTV sound system, complete audio programs are assembled at the user's receiver from various services sent

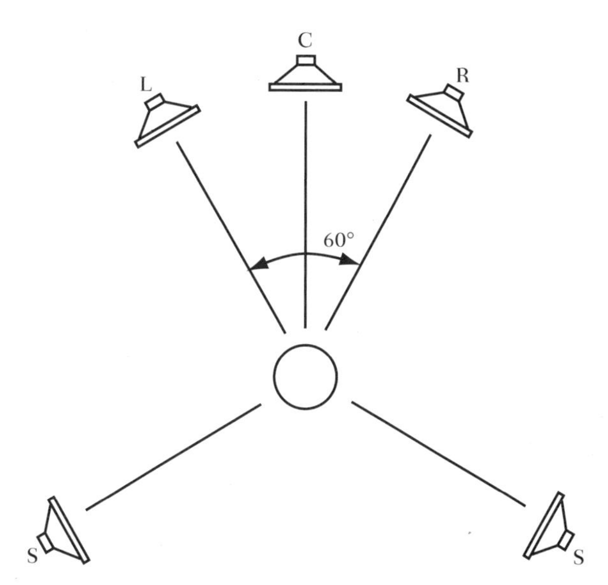

FIGURE 24.6 Optimum system speaker placement for HDTV viewing.

by the broadcaster. The concept of assembling services at the user's end was intended to provide for greater flexibility, including various-language multichannel principal programs supplemented with optional services for the hearing impaired and visually impaired.

A variety of multichannel formats for main audio services also is provided, adapting program by program to the best stereo presentation for a particular offering. An important idea that emerged during the process of writing the standard was that the principal sound for a program should take up only the digital bit space required by that program. The idea was born that programs fall into production categories and may be classified by the utilization of loudspeaker channels. The categories include:

- *1/0—one front center channel, no surround.* 1/0 is most likely to be used in programs such as news, which have exceedingly strict production time requirements. The advantage in having a distinct monaural mode is that those end users having a center channel loudspeaker will hear the presentation over only that one loudspeaker, with an attendant improvement over hearing mono presented over two loudspeakers.

- *2/0—conventional two-channel stereo.* 2/0 is intended principally for pre-existing two-channel program material. It is also useful for film production recorded in the Dolby Stereo or Ultra Stereo formats with a 4:2:4 amplitude-phase matrix (for which there is an accompanying indicator flag to toggle surround decoding on at the receiver).
- *3/0—left, center, and right front channels.* 3/0 is expected to be used for programs in which stereo is useful but surround sound effects are not, such as an interview program with a panel of experts.
- *3/2/.1—left, center, right front, left and right surround, and a low-frequency effects channel.* 3/2/.1 is expected to be used primarily for films and entertainment programming, matching the current motion picture production practice.

MONITORING

Aural monitoring of program production is a critical element in the production chain. Although the monitor system—with its equalizers, power amplifiers, and loudspeakers—is not in the signal path, monitoring under highly standardized conditions has helped the film industry to make an extremely interchangeable product for many years. With strict monitor standards, there is less variation in program production, and the differences that remain are the result of the director's creative intent. Monitor standards must address the following criteria:

- Room acoustics for monitor spaces
- Physical loudspeaker arrangement
- Loudspeaker and electronics requirements
- Electroacoustics performance

Such attention to detail invariably improves the viewing experience.

C H A P T E R
T W E N T Y - F I V E

COMING TO A THEATRE NEAR YOU: DIGITAL CINEMA

BRIAN MCKERNAN

The cinema is the most dramatic form of moving-image content; now it's going digital, with important implications for filmmakers and exhibitors alike. Given all the technological progress of the twentieth century, it's hard to believe that the photography and exhibition of theatrical motion pictures remains essentially unchanged since the days of Thomas Edison. Cameras, lenses, and film stocks have all improved, of course. And innovations such as sound, color, and wide screen pictures have vastly transformed the filmgoing experience since the early silent days. But the basic technological concept for the cinema is the same as it was more than 100 years ago: A long, flexible ribbon coated with photographic emulsion and perforated on either side with sprocket holes is mechanically pulled through a shutter, where it is intermittently exposed to light for the recording of a continuous series of still pictures; then this media is chemically processed and printed, and—after editing—duplicated and shipped to theatres, where it is projected by another mechanical device.

There are many good reasons why this cinema model has been in place for so long, the principal one being that it works extremely well. The amount of picture information that can be recorded within a 35 mm film frame exceeds 2,000 lines of resolution (ordinary NTSC television provides 480 lines) and when projected at 24 frames per second, film provides the richest palette of expression available for moving-image communication. Film images are robust enough to be projected and displayed onto theatre screens at hundreds of times their original 35 mm-wide size to provide a high-quality, larger-than-life viewing experience. And, depending on the creativity of filmmakers involved, the resulting entertainment experience can potentially earn many times more money than was spent actually creating it.

A century of 35 mm filmmaking around the world has made it the only truly global imaging standard. Film is also an excellent production format for many kinds of television programming, outliving dozens of video formats that have appeared since the introduction of commercial video recording in 1956. The technology of the cinema is mature, and has benefited from years of continuing innovations, including improved optics, emulsion advances, lighting inventions, video assist, and multichannel Dolby sound. Now, however, as movies enter their second century, the biggest innovation of all is the eventual elimination of photographic film itself. As with every other form of content creation, the cinema is going digital.

A BETTER WAY TO GET THE JOB DONE

By its simplest definition, *digital* refers to a method of representing information (e.g., sound and still or moving pictures) as a series of 0s and 1s. The origins of this technology are decades old, but it's really only in the past 20 years that inexpensive digital recording became practical. This was a consequence of the computer revolution and the mass manufacturing of digital microprocessors, storage disks, and other components, which tend to increase in power and affordability as time goes on (see reference to Moore's Law, page 116). The advantages of digital recording over earlier analog technologies are many. Digital data can be manipulated, edited, and processed far more extensively and easily than analog information. Because this data is really only 0s and 1s, it's far more immune to the pitfalls of analog recording. Overdub an analog tape several times and audible hiss builds up; not so with digital. Combine several layers of film to photograph a special-effects shot and visible grain darkens the picture. Digital, meanwhile, has much greater resistance to multigenerational degradation. Sounds and images can be overdubbed, edited, composited together, and otherwise manipulated for an infinite range of creative options.

One by one, starting in the late 1970s, enterprising technology companies developed digital systems to synthesize, manipulate, and edit digital sound, and—later—motion images. Simple datatypes came first: audio, then still imagery, then video, and later, high definition. Digital recorders, mixers, and editing systems not only began to replace their analog counterparts, they could outperform anything analog systems could previously do. At the same time, other computer technologies were adapted for television and motion-picture production tasks, particularly in the area of postproduction, or the editing of picture and sound, special effects, and all the other tasks that need to take place after actual filming—or production—to achieve a finished program.

Capitalizing on the increasing capacities and declining prices of disk-based computer storage, companies such as Avid and Lightworks developed *nonlinear editing* systems during the late 1980s; these were nonlinear because they stored moving images and sound on disk drives. With no reels of film, video, or audio to shuttle through, picture and sound information could be accessed far more quickly than ever possible, saving time, and—more important—money. This image and sound information is digitally compressed to enable it to fit on the disk drives; working in this way, creative film and video editors can "cut and paste" the sequence of shots and scenes in a TV show or movie, and then cut the original video or film footage accordingly. Today nearly every theatrical movie's sound and picture are edited with a digital nonlinear system.

Computer graphic imaging (CGI) also began to be used increasingly to generate imagery for theatrical motion pictures, starting in the early 1980s. Science-fiction films, especially, capitalized on the creative latitude of CGI, which, in talented hands, can be used to make whatever pictures a screenwriter or director's mind can conceive. CGI images, digital in nature and usually recorded on some form of data-tape cartridge, could be printed to 35 mm film on systems such as the Celco film recorder, which was originally developed to make photographs of digital data streams beamed down from orbiting satellites. Then the CGI footage could be intercut with live-action film to accomplish seamless cinematic illusions. *Tron* (1982), *The Last Starfighter* (1984), and *Young Sherlock Holmes* (1985) are early examples of live-action movies incorporating CGI. In 1995 *Toy Story* premiered, which consisted totally of CGI. Meanwhile, as this digital filmmaking evolution progressed, producers and directors increasingly realized that digital editing and postproduction tools could bring increased efficiency and creative expression to the entire filmmaking process. Today, digital postproduction is the norm for feature filmmaking (not to mention television production).

Postproduction is but one area of cinema undergoing a digital transition; the others include cinematography, distribution, and exhibition. Each of these areas is advancing into the digital domain at its own pace, and each is not necessarily dependent on the progress being made by the others. That being said, however, success in one area tends to encourage growth in the others. Director George Lucas has indicated that his satisfaction with digital postproduction technologies has encouraged him to explore digital cinematography. Certain studios have indicated that if other filmmakers follow suit, it makes better economic and creative sense to stay totally within the digital domain from production right on through to exhibition. One hour's worth of 35 mm film costs thousands of dollars. One hour's worth of digital HD tape costs less than $100.

Currently, most of the movies exhibited digitally are principally shot on film. Conversely, the vast majority of people who have seen the all-CGI movies *Toy Story*, *Shrek*, or *Final Fantasy* saw them projected off of film. Definitions of what constitutes *digital cinema* vary; to some it just means projecting a film digitally. For certain high-end Hollywood postproduction companies it means using a telecine or DataCine system to transfer a movie shot on film, and mastering it to a digital recording format. To certain other filmmakers or movie studios digital cinema means being digital from the first shot to the last exhibition. All of these definitions are valid. However you define it, the digital cinema transition is an inexorable reality. Film's more than 100-year foothold will probably ensure its existence far into this century as well. But in the same way that horses and sailboats have survived automobiles and outboard motors, film will gradually yield increasing territory to digital imaging and projection technologies. Whether or not filmmakers will evidence greater or lesser talent and creativity in the age of digital cinema is another matter entirely.

CINEMATOGRAPHY

As mentioned, video recording arrived on the scene in the late 1950s, and despite occasional attempts to use it as a substitute for film in theatrical moviemaking, its shortcomings in terms of resolution and other visual attributes prevented its widespread adoption. Video did, however, prove valuable to filmmaking process. Video assist, in which a video camera is attached to a film camera to simultaneously record the same image, emerged in the late 1960s, pioneered by such directors as Jerry Lewis. Video assist enables directors to instantly play back shots and see if they got what they wanted, without waiting for film developing. When high-definition television (HDTV) recording appeared in the late 1980s, attempts to shoot theatrical movies on videotape still suffered in terms of picture quality (*Julia and Julia*, 1988).

Nevertheless, technology continued (and continues) to improve. Manufacturers such as Sony and Panasonic spent heavily on R&D for HD cameras and recording systems in anticipation of the U.S. government's mandated digital conversion of the nation's television system. That conversion is proving to be a longer process than originally anticipated for numerous reasons, chief among them being the shortcomings of the over-the-air digital television (DTV) transmission standard chosen for the United States But new high-quality digital HD (minus the "TV") cameras and recording systems are proving to be a boon to the world of television—and cinema—production. Even non-HD digital formats developed for making TV shows (such as Digital Betacam, DVCAM, DVCPRO, and D-9) or home videos (such as Mini-DV) can produce images of sufficiently high quality to satisfy filmmakers—especially if they're working with a limited budget. As with so many other innovations, however, certain filmmakers tend to be first in seizing the initiative and getting the proverbial ball rolling.

Many years before the coming of the digital age, Walt Disney set an example in Hollywood as a filmmaker willing to invest in new technologies (Technicolor, RCA Photophone sound, etc.) to improve his studio's films. Others would later follow his lead. More recently, *Star Wars* director George Lucas pioneered computerized motion-control photography, EditDroid nonlinear editing, and the THX theatrical sound system.

In 1996 Lucas wrote a letter to Sony's research center, in Atsugi, Japan, and asked if an HD camera could be developed that recorded digital images at the same speed that film cameras use: 24 frames per second. Video cameras—digital or analog—traditionally record images at 30 frames per second, with each frame repeated twice in a process known as *interlace scanning*. Converting video into film footage requires a mathematical compensation known as 3/2 *pulldown*, which produces motion aberrations that many filmmakers find objectionable.

Sony's response was a 24-frame digital HD camera known as the CineAlta, which employs the *progressive* picture scanning used in CGI, as opposed to the interlaced images usually found in video and TV. Lucas promptly put CineAlta work in shooting his *Star Wars Episode II* (2002) film after successfully testing Sony's Digital Betacam format for selected shots in his *Star Wars Episode I: The Phantom Menace* (1999). Lucas' satisfaction with the CineAlta system was summed up during Sony's press conference at the 2001 National Association of Broadcasters' convention, when he commented, "I think that I can safely say that I will probably never shoot another film on film." Lucas also commented that he saved $2 million using digital tape instead of film, and that he was able to work much faster.

Making theatrical movies on digital videotape isn't limited to George Lucas, or to the 24-frame CineAlta HD format. Director

Spike Lee used the Sony DVCAM digital video format designed for TV news on his movie *The Original Kings of Comedy* (2000) and the Mini-DV format intended for consumers for his film *Bamboozled* (2000). Both movies were transferred to 35 mm film for widespread theatrical exhibition. Low-cost independent films benefit from the use of digital video because of the economy of these formats and the integrity of digital images. Independent directors such as James Mathers and Harry Shearer report satisfaction with Panasonic's digital DVCPRO TV news format for movies, whereas others have used JVC's D-9 digital camcorder format.

The cost savings and good performance of digital video are attracting an increasing number of directors, which range from the Swedish "Dogma" school of filmmaking (which prefers cheaper formats and minimal lighting) to big-budget A-list directors such as James Cameron and Robert Rodrigues, who are using CineAlta and all the craft and lighting usually lavished on 35 mm film to make big-budget movies. Meanwhile, the distribution and exhibition side of cinema is also transitioning to digital.

DISTRIBUTION

The transmission of video was one of the earliest applications for communications satellites, which were originally proposed in 1945 by futurist (and *2001: A Space Odessy* author) Arthur C. Clarke. Since the mid-1960s, satellite-service providers and television broadcasters have greatly advanced this technology, and today live, high-quality global telecasts of soccer, football, and other sporting events are an immensely profitable industry. In 2000, Boeing Satellite Systems was formed with the purchase of Hughes Space and Communications Company, a longtime leader in satellite innovations. Boeing soon set about identifying new markets and quickly saw that satellite distribution of motion pictures was one of them.

As mentioned earlier in this chapter, the distribution of theatrical movies really hasn't changed in 100 years. After postproduction, films are duplicated and shipped to theatres in motor vehicles owned by courier services. A typical feature-length movie comprises several fairly heavy reels of 35 mm film, which are shipped in large metal cases; these film copies scratch easily, degrade each time they are projected, and are easily pirated. A single release print can cost anywhere from $1,000 to $3,000, and every theatre showing a particular title has to have at least one copy, which wears out after several weeks of exhibition. Satellite delivery of motion pictures to theatres, however, could reduce distribution costs dramatically. The U.K.-based market-research study *Electronic Cinema: The Big Screen Goes Digital* (Screen Digest Ltd., 2000) estimates that the worldwide cost of making release prints is $5 billion, and that digital cinema distribution could eliminate 90 percent of that expense.

An added benefit to distributing movies digitally to theatres is security and encryption. Although motion-picture content beamed down from space can theoretically be received by anyone with the proper antenna and equipment, digital cinema distribution providers such as Hughes (now Boeing) have been transmitting high-quality video for many years for NASA and military clients, whose security needs far outweigh those of Hollywood. Most movie piracy, in fact, is accomplished using hand-held camcorders to shoot movies right off the screen (which is why most bootleg videos look so bad).

Satellite transmission isn't the only means by which movies can be digitally distributed to theatres. In June 2000, 20th Century Fox and Cisco Systems transmitted the animated feature *Titan A.E.* from Burbank, California to Atlanta's Woodruff Arts Center. Qwest Communications' Burbank-based CyberCenter streamed the movie via the Cisco 7140 Virtual Private Network, which includes firewalls, security routers, and encryption.

Denver oil billionaire Philip Anschutz, a part owner of Qwest, has been purchasing bankrupt theatre chains with the reported intention of using that company's fiber optic network as a means of distributing movies. Other organizations with networking technology, such as Global Crossing and Pixel Systems, have also been active in purchasing or acquiring theatres. And once a theatre is converted for digital projection, other forms of entertainment—such as live sports, concerts, or business meetings—can be presented on screen in digital HD as an alternative to feature films. Suddenly there are more reasons for going to "the movies," and overbuilt megaplexes have more ways to attract paying customers to big-screen experiences.

Currently there are just over 30 digital cinemas worldwide, but the number is growing. Many of these cinemas have motion-picture content delivered to them on multiple DVD-ROM discs prepared by Technicolor, a leading Hollywood film lab. Playout to screen is accomplished using a server, which incorporates numerous computer hard drives and compression technology (usually either Wavelet or MPEG) to retain the high quality of cinematic images while also fitting them onto relatively limited storage space. Unlike projected film, which degrades and picks up dirt with each showing, movies played off servers look the same first time, every time, and thus maintain consistent, high-quality images.

PROJECTION

Although servers take the place of spooling reels of film in a digital cinema, a projector is still required to "throw" the image at a movie screen. For projection of cinema-quality images, two major technologies are currently employed, with more on the way. Texas Instruments' Digital Light Processing (DLP) Cinema technology employs a digital micromirror "black chip" to display film frames one at a time at high speed, as with film. JVC's D-ILA uses a liquid crystal display and an electron gun. These technologies have been licensed to several manufacturers (including Barco, Christie, Digital Projection, Kodak, and Sony), which today are building cinema-grade digital projectors. Although there are some Hollywood directors who feel that digital projection technology does not yet render motion-picture images with the same color fidelity as film, others are satisfied, and—of course—technology continues to improve. Even some of the digital projectors made for business presentations are finding their way into cinema use, especially in small theatres showing independent movies made with non-HD digital video formats. Younger, independent producers and directors find the quality of these affordable digital cameras, edit systems, and projectors perfectly adequate, especially when budgets are limited, and making non-"Hollywood" movies is their prime goal.

SUMMARY

The digital cinema revolution is underway. Film and digital will coexist for many years, and some aspects of filmmaking (such as postproduction) have already transitioned to digital. Important questions have yet to be answered, most notably that of who will pay the $125,000 per-screen price tag for theatres to convert to digital (especially when many theatre chains are bankrupt). An even larger question is "Why fix what isn't broken?" Film has worked well for more than a century; why not stick with the status quo? The answer, as with any business, will no doubt boil down to money.

Digital technologies offer an advantage here, both from a creative and an exhibition angle. Filmmakers routinely turn to digital imaging for everything from mundane tasks (change the color of an actress' dress) to creating otherworldly science-fiction landscapes. Technicolor Digital Cinema, a joint company of Technicolor and Qualcomm, announced plans in March of 2001 to outfit 1,000 theatres with digital servers and projectors in the near future. Independent studios, such as Madstone films, have indicated plans to establish a chain of digital cinemas dedicated to independent digital productions.

This chapter provides but an overview of the technology of digital cinema; entire books have been written on the subject. One of the best of these is Scott Billups' *Digital Moviemaking* (Michael Wiese Productions, 2000; ISBN 0-941188-30-2). In the meantime, it's important to bear in mind that the digital cinema transition will be a long one, with potentially rewarding opportunities for a wide range of interests—filmmakers, equipment providers, theatre installers, telecommunications concerns, and the entire motion-picture industry. Most important, however, the reduced costs represented by digital cinema have the potential to democratize filmmaking, which has seen its costs skyrocketing in recent years. Given the power of this most dramatic form of moving-image communication, digital cinema is a very good thing indeed.

CHAPTER TWENTY-SIX

USING 1080/24P OR 1080/60I FOR DIVERSE PRODUCTION APPLICATIONS

LAURENCE THORPE

Just about every cinema in the world runs motion-picture film through their projectors at 24 frames-per-second (fps). To avoid protracted periods of darkness and to avoid flicker as successive film frames are projected each film frame is typically flashed twice on the cinema screen—at the rate of 48 fps—using a twin-bladed mechanical shutter. Thanks to the *persistence of vision* of the human visual system, these separate frames of film are perceived as acceptably merged moving pictures and flicker is reduced to a more reasonable level.

The film speed of 24 fps has been with us a very long time. The 24 fps film-camera-capture rate was born in the late 1920s with the advent of sound for movies. It was the lowest frame rate that could still satisfactorily reproduce the new phenomenon of audio on the film's optical track (24 fps was higher than the film speed that preceded it in the era of silent movies). This is a very important historical fact: 24 fps is actually a product of sound rather than imagery! The fact that it rendered motion reproduction somewhat more accurate in the pictures themselves was a happy and fortuitous coincidence.

Over the next five or six decades, there was little impetus to raise the capture rate any further because of the cost increases associated with the attendant elevated consumption of 35 mm film. "Good enough"—as applied to displayed motion film imagery—was a criterion established long before the advent of television.

Indeed, much later, in the 1980s, The Society of Motion Picture and Television Engineers (SMPTE) was to conduct a long study on the merits of raising the film rate to 30 fps, and in 1988 they published a famous report on its considerable visual advantages. It was, however, totally ignored by the industry. Evidently 24 fps film had become deeply rooted in the psyche of most human beings, and few were complaining. Meanwhile, film stock costs have continued to rise. All of this has mitigated against any industry move to an enhanced frame rate (other than the use of high-frame-rate photography for slow motion). Thus, 24 fps remains universally accepted as a quite adequate frame rate for capturing moving images. Thousands of movies are presented to hundreds of millions of people every year in cinemas all over the world at 24 fps.

THE TECHNICAL ASPECTS OF 24-FRAME CAPTURE

Despite its widespread acceptance, 24 fps film does have technical limitations. Most of the populace has, at one time or another, probably wondered why the wagon wheels reverse themselves in their favorite Westerns. Many have invariably endured the eye-strain associated with sitting close to a gigantic movie screen, when fast camera pans, or fast-moving scene content, assail the eye and brain with a visual staccato judder that almost hurts. Ever notice the "stuttering" of the windows of buildings as the film camera pans across a dark cityscape? Yet, if familiarity does not quite breed contempt, it most certainly has bred a noncritical acceptance.

In purely technical terms, 24-frame image capture represents gross temporal subsampling of the moving picture. Technically, it works as well as it does only because a great deal of real-world motion is actually not all that fast. Indeed, the greater portion of today's average movie constitutes a large number of skillfully edited sequences of scenes that in themselves are largely still or quite slow moving (including the camera zooms, pans, and tilts), and are thus captured really quite well. Over the decades, the refinement of camera dollies, jib arms, and Steadicam, coupled with the expertise of film crews, has greatly contributed to the precision and smoothness of film camera moves, which in turn ameliorates the motion judder associated with 24-frame capture.

Aside from the technical aspects of 24-frame capture are the very important creative dynamics. Passionate advocates of film have long differentiated it from video on the very basis that its more subtle picture portrayal is the very essence of effective storytelling. Many largely attribute this to the way that film cameras are shuttered and operate at the low 24-frame rate. Over time, this unique frame rate has been most skillfully exploited by legions of cinematographers to refine an entire art form in storytelling. This is a core creative issue in any decision on choice of picture-capture rate.

TECHNICAL ASPECTS OF 60 HZ CAPTURE

Sixty years ago an electronic communication phenomenon burst onto the consumer scene and was forever to change the world. Television brought real-time program creation (absolutely inherent to a medium initially born without a concept of recording) and real-time motion picture display and sound directly into the living room.

American television was born and raised on the basis of 60 Hz picture capture and 60 Hz picture display. (Hertz [Hz], represents cycles per second.) This provides moving images comprising 30 separate frames per second, with each of those frames displayed twice as a field of odd- and even-numbered scan lines. These scans are known as *fields*, which are *interlaced* together to make a whole image. The choice of 60 Hz was made because there were technical advantages back then to locking the picture rate to the already entrenched electric power frequency of 60 Hz. Sixty years later, American NTSC television is still 60 Hz (it is actually 59.94 Hz—for reasons we need not examine here). Many other regions of the world base their television systems on a 50 Hz rate (again, because of the frequency of local electrical power). Either of the two—50 Hz or 60 Hz—is now a hugely established norm in virtually every living room in every world.

Over the ensuing six decades, the refinements to both television technology and television production have been unceasing. These encompass both the sheer breadth of program genres and the associated techniques of program creation. Very early in television's history the high picture-capture-rate quickly endowed television program directors and camera operators with a creative impetus that sought to push the medium to its limits. Live television created its own imaging imperatives. Integral to these were the speed with which the television cameras themselves could be moved—in terms of dollying, panning, and tilting. Later, the speed with which the focal length of the zoom lenses could be exercised would become an integral creative aspect of picture capture. The combination of camera motion and lens zooming became a distinctly recognizable "footprint" of television imagery, creating pictures that would uniquely set it apart from those of the cinema portrayal. Indeed, many pictures simply cannot be satisfactorily reproduced at all in 24-frame capture.

CREATIVE ASPECTS OF 60 HZ CAPTURE

Certainly, the 60 Hz television system is capturing considerably more information in the temporal domain. On moving pictures this surely imparts a greater sensation of reality. On fast moving pictures—such as those encountered in sports, wildlife shows, and certain drama segments—it captures and portrays quite an authentic and most compelling reproduction of the fluidity of fast motion. Of special importance, it has allowed the television display to progressively increase in brightness over the years without an attendant disturbing increase in flicker. Even on the still image, there are more images per second being assimilated by the eye and brain, and although difficult to quantify, this too is believed to add a psychophysical picture "presence" that is not obtained with shuttered 24-frame film imagery. Traditional artifacts (picture defects) associated with the interlaced 60 Hz (and 50 Hz) scanning are seemingly largely unnoticed by consumers the world over. Their effect on television engineers is quite another matter!

Sports productions especially capitalize on the high picture-capture rate of television. Fast-moving action in the scenes themselves, in combination with vigorous movement of cameras attempting to rapidly scan that action over significant distances (e.g., on a ball field) all contribute to the "look" of live television (which can be further affected by the rapid and long zooms so familiar in sports coverage). Other technical innovations, such as slow-motion playback of recorded material and electronically shuttered CCD cameras, also add to the creative exploitation of that higher 60 Hz picture-capture rate.

The television documentary comes in many forms: wildlife and natural history, biography, investigative reports, recapture of historic events and places, mountain-climbing and underwater sagas, human drama stories related to medical, scientific, and societal challenges—the range of topics covered only increases every year. The very nature of the different documentaries themselves calls for a wide variation in picture aesthetics, creative shooting techniques, and editing. Certainly the capture of birds in flight, racing cheetahs, and other specialized wildlife photography demand the highest picture capture rate. Many documentaries, by their very nature, specifically seek a stark realism that is better imaged with the 60 Hz system. Others call for a heightened sense of drama, and indeed might specifically seek to replicate the special visual practices of movie-making. In such a case the 24-frame capture might well be more attractive to a producer.

THE 24-FRAME FILM AND 60 HZ VIDEO DEBATES

Over time, the emerging dialectic between the proponents of film imaging and television imaging would speak of the subtle storytelling nature of the former in contrast to the startling visual presence of the latter. Ardent supporters of the 24-frame rate of film speak of a video look that is too real, and they largely attribute this to the much higher picture capture and display rate (with "television" lighting, and other practices, also substantially contributing to this look). Within the enduringly futile debates between "video" and "film," the superior temporal resolution latent within television portrayal would be hopelessly pitted against the superior narrative capability of shuttered 24-frame film. In attempting any creative comparisons, quite different languages were used by the two sides. All of the disconnects that endure in communication to this day between advocates of motion picture film imaging and those of television imaging stem—first and foremost—from an inability to objectively recognize and adequately describe the relative merits of both picture capture rates. Yet, they are both very important.

THE "FILM LOOK"—A CLOSER LOOK

In this new era of compelling choices among production media, it has become especially important to better describe all that constitutes the treasured "film look" sought by many producers and directors of photography (DPs). It is far too simplistic to attribute the entirety of that look to the 24 fps capture rate. Key elements of that "film look" can, in fact, be achieved with either 24-frame or 60-field high definition (HD) video.

Primary contributors to the many dimensions of imagery that collectively constitute the "film look" are: tonal reproduction, color reproduction, exposure latitude, and picture sharpness. *Exposure index* (operational sensitivity of a specific camera system) refers to the capture capability under specific levels of scene illumination.

Today, digital HD can rank with the best of 35 mm motion-picture film in each of these dimensions of the picture. Images shot in 24-frame HD, however, retain one singular advantage over their 60 Hz counterparts; they use *progressive scanning* (as opposed to television's interlaced scanning), which presents each picture in its entirety, one at a time. This increases the vertical resolution and reduces some associated fast-motion edge-artifacts. This may or may not be important, depending on the particular degree of picture sharpness sought by the producer (after all, filtration is commonly employed to curtail excessive picture sharpness for certain scenes). Otherwise, digital HD and 35 mm film are essentially on a par in terms of the remaining three key picture attributes.

Other critical dimensions of the image relate to the specific craftsmanship of cinematographers as they exercise key aspects of camera optics (see Figure 26.1). Certainly, traditional film-style lighting, optical filtration, and use of matte boxes can all be applied equally to a digital HD camera (24 Hz or 60 Hz) as

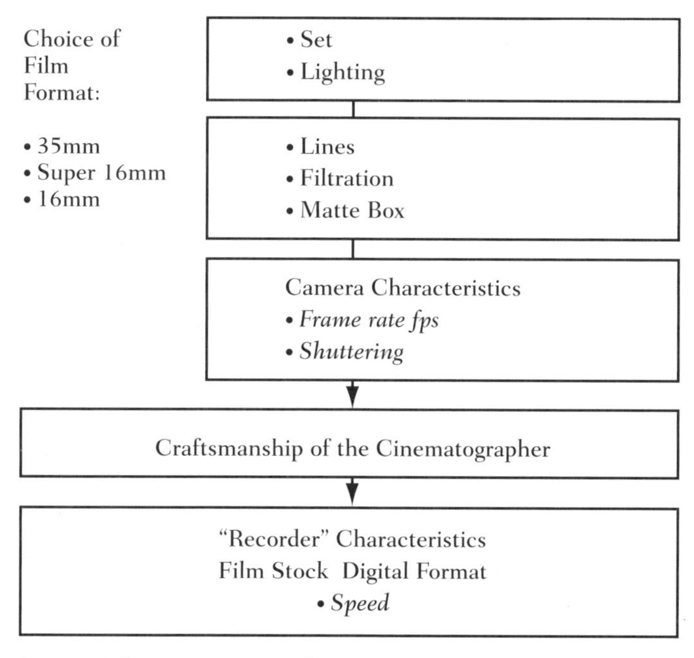

FIGURE 26.1 The separate contributors to the final "Look" of captured imagery.

to a film camera. A most crucial other dimension of the picture—one that can be skillfully used to impart a very special attribute to the look of the scene—is that of *depth of field*. Here again, there remain significant optical differences between the 35 mm optical film format and the 2/3-inch CCD chip that captures light for HD photography. The great advantage in exposure index (sensitivity) of the contemporary HD camera will, however, allow for a higher degree of neutral density filtering and associated lens-aperature setting. That, in turn, facilitates the use of shallow depths of field in the HD camera (which can come remarkably close to the images created with a 35 mm film camera) for what is generally sought in subtle image portrayal. In the case of Super 16 mm film, parity in depth of field can readily be achieved with the 2/3-inch HD camera.

Thus, the creative discussion ultimately reduces to the one final picture dimension—one that's associated with motion capture and portrayal—and that's the issue of 24 versus 60 pictures per second. That, in turn, depends on the creative aspirations of the producer and the DP for each particular production. If 24 fps is important to add some final "dimension" of the film look, then that is what should be employed. If smooth motion rendition is more desirable, then the 60 Hz capture makes more sense.

The choice of either 24 or 60 pictures per second is readily available. The choice of what is ultimately used on a given shoot is dependent upon program genre.

CREATIVE IMPACT

The inescapable larger reality is that, from a creative viewpoint, both 24-frame and 60-field digital HD imagery are equally valid. Scripts, storyboards, and the specific aesthetic aspirations of the production team dictate the particular choice used for a given program.

Consumers the world over are very accepting of both 50 and 60-Hz television and 24-frame film. All these capture and display rates are well ingrained in the minds of virtually all. They accept either, and thus the program-origination choice reduces to that of the program produced.

When fluid, smooth motion reproduction is important, 60-Hz capture will always retain a distinct advantage. Arguably, this applies to the greater degree of television production. Such images are also a major consideration if the program is to be distributed in multiple formats: digital HDTV, digital standard-definition television (SDTV), and analog NTSC or PAL (a likely scenario for many years to come).

TECHNICAL IMPACT

HD video shot at 24 frames has the advantages of progressive (as opposed to 60 Hz's interlaced) scanning; these advantages include improved vertical resolution and lower scanning artifacts. This may or may not be important, depending upon the picture content—again, this comes back to program genre.

The U.S. DTV transmission standard (see Chapter 17, *The DTV Transition*) is unique in that it allows transmission of a wide range of frame or field rates (with different digital sampling structures). Broadcasters might elect to transmit 1080/24P for some programs—to gain transmission data rate efficiencies—and, in doing so, they can technically capitalize on this new digital transmission freedom. Here, 24P origination might become a major consideration. (See Figure 26.2.)

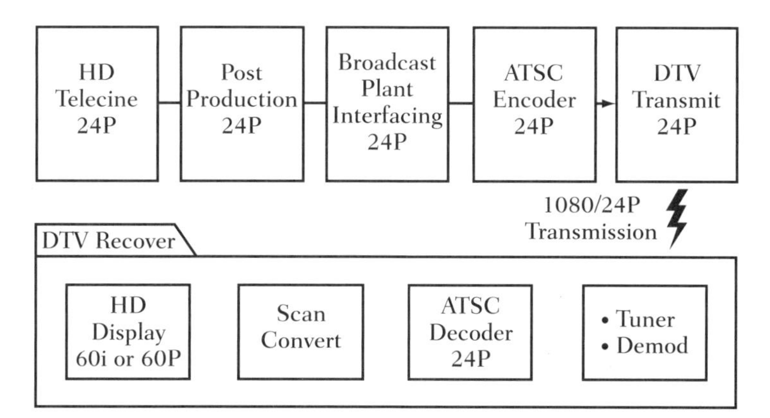

FIGURE 26.2 DTV Broadcasting allows tranmission of 24P directly to the DTV receiver.

A great deal of current 525-line television programming does in fact intermix both 24-frame film-originated material and direct "60I" (30 frames displayed twice in interlaced fashion) video. In the digital world, new choices for mixing media will be available: shoot both film and HD at 24P; shoot both film and HD at 30P; shoot film at 24 or 30 fps and HD at 60I. Although any decision might be swayed by which media dominates the mix, the technical decision as to which is the optimum mix of media and picture-capture rates should probably be determined by the final primary distribution format. For example, if the final distribution is downconverted NTSC or 60I widescreen SDTV there would be much to favor the HD being shot at 60I (then, only the film material would have to contend with the frame-rate-conversion in postproduction). See Figure 26.3.

APPLICATIONS OF 24 PROGRESSIVE AND 60 INTERLACE HDTV

PRIME-TIME TELEVISION PRODUCTION

The greater percentage of total prime-time programming is originated on the well-established 24-frame film. More recently, this has increased—a testament to a universal desire to pro-

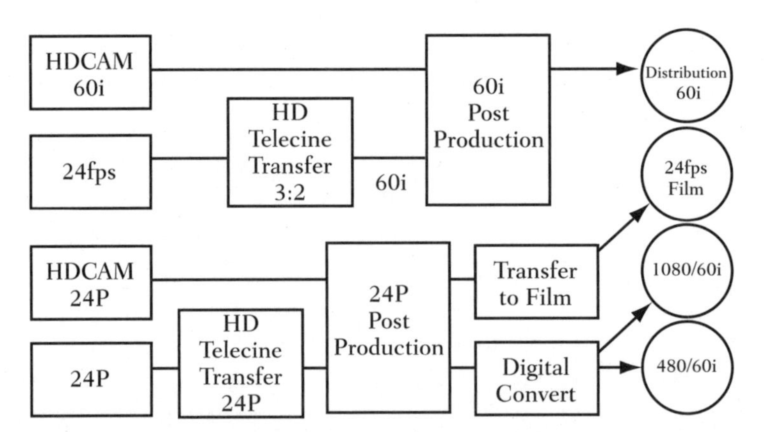

FIGURE 26.3 The preferred picture capture rate may be influenced by the dominant distribution format.

tect shelf life of program assets in the new era of DTV. This long-established tradition had the singular advantage of facilitating high-quality telecine transfer to any of the international television formats (NTSC in the 60/59/94 Hz regions and PAL/SECAM in the remaining 50 Hz regions).

Production costs for prime-time television programming have risen sharply over the past half-decade. A lot of U.S. shows are going abroad in pursuit of lower costs. HDTV production offers attractive possibilities for curtailment of costs.

Certainly the recording media itself is substantially lower in cost than motion picture film stock (in addition, it eliminates the attendant film processing and transfer costs).

The ability to immediately screen a given take in full HD quality offers opportunities to strike a set on the same day of the shoot—with a new set in place early next morning. This can contribute to cost savings—assuming, of course, that such set turnover is important to a specific show.

Live blue/green screen work in HD is far more efficient than the traditional film/optical methodology—it can be seen in real-time on the set (facilitating, among other things, framing and optimization of lighting and electronic controls). This is important for producing certain kinds of programs.

Certain labor agreements and/or workplace traditions may possibly provide for cost savings when producing on digital rather than film.

Those shows presently being shot on 24-frame film are prime candidates for serious consideration of shooting in 24P HD. The experienced DP can quickly gain the expertise to create imagery very close to what they had achieved on 35 mm film, but at a lower cost. As the DP gains in skill, new creative opportunities will arise that exploit their own innate experience and some of the truly unique creative capabilities of the digital HD camcorder or camera.

The remaining prime-time programs presently being shot on 60 Hz video (with few exceptions, these are shot in standard 4:3 aspect ratio) are certainly prime candidates for consideration for being shot on digital HD. With the advent of multiformat widescreen DTV, their shelf life will be augmented. Here 1,080-line, 60I HD is probably the better choice, because it will retain many of the characteristics of the former 60I-based standard-definition origination. In the new global DTV environment, the widescreen higher resolution HD origination will also better support high-quality distribution masters of both 525- and 625-line programs.

TELEVISION COMMERCIAL PRODUCTION

A great deal of national television spots are shot on film, mostly 35 mm. Although the majority of these are shot on the traditional 24-frame film, there is also a wide use of 30-frame film because of its friendlier relationship to the American 60-field television system (no 3:2 pulldown issues). Super 16 mm and even 16 mm film are used to produce many lower budget, local commercials.

Those commercials presently being shot on 24-frame (or 30-frame) film are prime candidates for serious consideration of shooting via 24P (or 30P) HD. Again, in the hands of an experienced DP, the picture produced will have most of the characteristic attributes of motion picture film familiar to the DP. There are other commercials, however, that will be well-served by the higher 60-picture capture rate. Fluid, fast motion is a key element of many commercials. Others utilize slow motion techniques—a key element, for example, in many commercials promoting hair treatments, detergents, the pouring of liquids, etc. Traditionally, a high-speed 35 mm film (frame rates varying up to 120 fps are not uncommon) is used. Recent tests have shown that electronic shuttering in the digital 60I HD camera, in combination with slow-motion digital recorder playback, can superbly simulate what is typically achieved with film running as high as 90 fps.

INDEPENDENT FILM COMMUNITY

This constituency has grown dramatically over the past 20 years. It currently encompasses a vast potpourri of creative program makers within all facets of visual and aural communication—low-budget movie-making, natural history, documentary, corporate, business and industrial, commercial, television, college student, and film schools. This is a community that largely shoots on 16 mm film because of budget constraints. Yet, often their work must be "blown up" to 35 mm for theatrical release—a process that adds to productions costs and entails a further loss in image quality. Digital origination has already been discovered by many in the independent community.

All of the major film festivals, both domestic and international, have recently shown a quite extraordinary number of award-winning works that were shot with digital cameras and subsequently transferred to 35 mm film for theatrical showing. What is particularly astounding is that many of these have originated not on professional digital camcorders, but on the consumer Mini-DV! Accordingly, many have been shot with either 60-field or 50-field digital cameras. Clearly, the old barriers between 24-frame and 60-pictures-per-second are being lowered. Storytelling is not being impaired. Instead, creative bridges are being built between digital video and film.

MAJOR MOVIE PRODUCTION

Movie production in the United States is a highly refined engine of theatrical entertainment program production. It is supported by a vast "cottage industry" infrastructure of talents and facilities that can, with extraordinary efficiency, be drawn upon to assemble a well-oiled machine that can create everything from sophisticated block-buster movies to countless low-budget films for an ever-hungry audience.

This is an industry that does not in any way consider its mechanisms of operation to be broken. On the contrary, their pride in a decades-old *modus operandi* is one of a creative community that believes every new movie is a refinement on the last (in terms of creativity, innovation, and hopefully, profit-making!). The very concept of 24P electronic cinematography is, accordingly, anathema to some veterans of an industry that intractably perceives "video" to be radically inferior to film. To them, digital HD is an upstart electronic technology daring to encroach on an art form they perceive to be unique. Indeed, they further maintain that anything that smacks of "television" should have no place in theatrical entertainment.

Many others, however, recognize that the world is changing and that technology is not all evil. They have closely followed the truly astonishing developments in digital HD video and they wisely appreciate that 24P HD, especially, is far too important to be dismissed out of hand.

It is within the context of these two worlds of video and motion picture film that a new convergence of digital HD and film is taking place. HD is taking a giant leap across to the domain of motion picture film by implementing the emulation of one of its most prominent characteristics—the 24-frame rate. A crucial bridge has been built. A common picture capture rate platform expedites higher quality transfers between the two media.

This move has created a great deal of industry "buzz" and many conclusions are being drawn—some of them prematurely. HD at 24P is not intended to totally displace film, nor to displace 60I HD. What it does provide is a seamless transfer between motion picture film and HD—and that opens up new creative options.

For some producers, 24 fps film remains their preferred acquisition format, but 24P HD allows for effective management of postproduction for multiformat DTV on the basis of a single 24P master. For others, 24P acquisition provides new creative and budgetary flexibility.

The most exciting thing about 24P today is how it's catching on in so many areas: commercials, sitcoms, prime-time dramas, major feature films, and independent features—and the list is getting longer every day.

Those most visible of all forms of content creation—movies and TV—are being energized by this new digital format. One can only wonder what creative minds will do with this technology as the cinema advances into its second century.

CHAPTER TWENTY-SEVEN

LOW-REZ

SCOTT BILLUPS

"Empty pockets never held anyone back, only empty hearts and empty heads."

—Dr. Norman Vincent Peale

You've made up your mind to shoot a digital movie for under $10,000. You're going the low-cost route and using the Mini-DV format. Okay, let's work with it. There are essentially so many variations that it would be impossible to cover them all, so let's pick the worst-case scenario and then you can make adjustments with regards to your own production.

Situation: You've got a barn-burner of a script and a new credit card with a $10,000 limit. That's it.

Figure out a production time-line and then stick to it. One of the most important elements of success is to have a set of specific goals that you can articulate. That way, everyone who joins your little crew knows where they're going and when they need to be there. Goals give focus and deadlines give the whole production its pace. In a large production this is a natural byproduct of the budget and scheduling, but smaller productions all to often wither away from "tomorrow-itis." Create an intelligent schedule and then stick to it because a goal without a deadline is only a wish.

Get a recycler magazine from the closest metropolitan city and do the bulk of your shopping from it. (You can also check out my online persona at www.PixelMonger.com for deals on used equipment and production packages.) You should be able to pick up a good used, FireWire (IEEE 1394)-equipped Mini-DV camcorder for $500. The important thing here is to shoot everything in 16:9 aspect ratio, so you're going to look for a camera that actually uses a 16:9 imaging chip. Because your choices here are severely limited, your next best option would be to get a camera that stretches the conventional 4:3 (Panasonic, Canon) image rather than one that crops the top and bottom of the 4:3 (JVC, Sony). Total: $500

For a simple option you could check out one of the several reputable used equipment dealers such as B+H Photo in New York City (www.bhphotovideo.com). They are more expensive than buying out of the recycler, but then you've got a camera that is backed by a substantial and reputable company.

Next look around for a used DV iMac or, better yet, a used G3 or G4 with at least 256 MB of RAM and the largest monitor you can find. The Mac should cost under $600 and hopefully comes with Apple's very hip editing software package, FinalCut Pro, or Adobe Premiere 5.0 or higher. If you can score a copy of FinalCut Pro do it.

Because you're going to need to stay in the 16:9 aspect ratio without reprocessing your image, consider buying the DV ToolKit from ProMax (www.promax.com) for $200. This software plug-in allows you to edit in FinalCut in native 16:9 aspect without further degradation to the signal's integrity. Total: $800.

FireWire is going to give you about four-and-a-half minutes of audio and video per gigabyte, so figure on picking up around 80 GB of FireWire drives, which will give you room to move around. Sometimes you can find really good deals on drives, but you never know where they've been so it's not a bad idea just to bite the bullet and get some new ones. ProMax sells a really good 37 GB for around $400 each (get two) and they have the most knowledgeable sales staff in the industry. Total: $800.

Pick up the best, used, minishotgun microphone you can find for around $100, and you've just slammed together your entire digital production environment for a tad over $2,000.

The bad news is that because this system is built around the 4:1:1 Mini-DV format, you're starting out at a 5:1 compression ratio with greatly reduced colorspace. The good news is that it won't get any worse. The FireWire accommodates the DV's 3.6 Mb second data rate with room to spare and the DV-friendly FinalCut edit software is transparent to the final image.

EQUIPMENT

Foam-core reflectors from the art store and a white army-surplus parachute are the two most important elements of your outdoor shooting package. For interior shots consider bouncing sunlight into the room from reflectors (foam core covered with aluminum foil) located outside in direct sunlight. Once you start with lights you've gotten into an area that costs money.

Many rental houses carry inexpensive lighting kits, such as the Lowel DP kit, which are quite versatile. Just remember that you're trying to achieve a film look, so you're using the ND filters, right? Total: $100 per week rental.

Almost as important as the lights themselves are the tools to bounce, diffuse, and cut the actual beams. Again, foam core with reflective surfaces as well as black. A good hardware store will have spring clamps that look somewhat like large clothespins. Get a dozen of these as well as a few rolls of double-sided foam tape and several rolls of duct tape. Because you're a filmmaker, you should always call it "grip" tape once you're out of the hardware store. Total: $100.

PRODUCTION

Drive your production or it will drive you. Hopefully you've gathered a few friends around to help with the actual shoot, and you've even got a few actors from the local school or workshop to work for free. If they are professional actors you've made a deal to pay them a deferred SAG salary if the film gets distribution. The important thing to work on here is not so much the technical aspects of the production but rather the social skills involved in keeping everyone happy and focused. More than any other aspect of no-budget production, teamwork and a congenial environment will lead to a successful shoot. Total: Tape stock $100; food $100; slush $100.

Now shoot your film. As to how to do that, that's the subject for a very long book in itself, and there are thousands of them out there. I'm partial to my own, *Digital Moviemaking* (ISBN 0-941188-30-2; see www.mwp.com).

People have been directing films for over a century, and good and bad filmmaking is still very much in evidence. Look at the films you admire, turn down the sound, and watch them over and over. What can you learn from what the director did? With VHS and DVDs so accessible, it's never been easier to learn from the masters. Your digital movie is your shot at recognition and greatness; go for it, and remember that less can often be more.

Often overlooked in microproductions is the obligatory wrap party. I know that you're short on funds but there is simply no better way to say thank you to all the people who put up with your sniveling insecurities and unreasonable demands. Beer, wine, pizza, and a bottle of good tequila. Total: $250.

EDITING

Apple's FinalCut Pro works equally well with PAL or NTSC and is a little easier to use than Adobe Premier.

Because DV only needs about 18 GB for 90 minutes, and you've got room to spare, you can forget offline editing all together. Just sit your butt down and get it done. Either you've got the ability, or you'd better know someone who does.

If you plan to transfer your final, edited product to motion-picture film, one of the seemingly endless bits that you must keep track of is the length of an actual reel of film. Unless you want the audience to know when the projectionist changes a reel, you'll need to adapt a well-worn strategy: Divide your movie up into 20-minute segments (19 minutes if you're editing in PAL) so that at the end of your first reel you have a scene cut. This will keep the unavoidable differences between the various reels from becoming apparent. Even though the reels are all developed by the same lab, they most likely won't be developed in order. Chemicals in the development process are constantly changing and often times produce noticeably different colors and densities between the head and tails.

DATA TRANSFER

The least expensive way to transfer your data is to simply clone the segments back out to your camera via FireWire. Keep in mind that the duration of a camera is more limited than tape decks and unless your DV camcorder is true 16:9 native, which it probably won't be, there could be significant degradation to the image.

A better method is to clone your movie to another digital format such as Digital Betacam or D-1, or get yourself a Sony DSR-30 DV record deck (about $3,000 new). This deck records in 16:9 aspect ratio and accepts the larger and longer three-hour DV tapes. Because it is an IEEE-1394 native system, there is essentially no distortion of your delicate signal. You might luck out and find a used one for $900 or less but plan on doing a rather aggressive search. Total: $900.

PRINTING TO FILM

The Electron Beam Recorder (EBR) is out. Maybe even the $6,000 to $12,000 for a decent kinescope. Really, the first order of business is to evaluate the reason behind your need for a film transfer in the first place. Is it for submission at a film festival or a screening at a theater? Does it need to be 35 mm, or will 16 mm suffice? How many copies are you going to need? There are so many variations that it would be impossible to cover them all, so let's stay with this worst-case scenario that you only want a single 16 mm print for that special film festival (or maybe you don't need to go out to film at all!).

But assuming you do, you will eventually find yourself with your project edited and scored. Oh yeah, throw in $200 for a canned sound track. You've been meticulously careful with your data stream and it is essentially in its pristine condition as it sits on the computer's disk. Add: $200.

Kinescope, the process of filming a monitor, is by far the least expensive method for transferring your movie. Companies like Ringer Video, in Burbank, California, use a high quality monitor and a special camera running at 23.976 fps to achieve a flicker-free transfer. They generally charge around $50 for a one-minute, 16 mm test and $80 per minute for actual transfer. Go for the test. This will give you a good reference of what you can expect from your inherent resolution. Total: $50.

Hopefully there aren't any unpleasant surprises and your meticulously cared-for image still looks good when projected. Heck, if you can afford it, have them do the whole thing.

CAMERA

There is quite a selection of 16 mm motion-picture cameras out there and if you don't already have a personal favorite you might want to consider renting an Arri-S, 16 mm MOS (without sound) camera from someone like Alan Gordon Enterprises for a week. You'll need it long enough to do a test, get it developed, and then check out your footage. Total: $225.

FILM STOCK SELECTION

The inherent sensitivity of a film to light is expressed in terms of its American Standards Association (ASA) number. A fast film stock needs less light to register an image than a slow stock does. Rule of thumb is that a film stock rated ASA 50 or less is considered slow, whereas an ASA of 200 or above is considered fast. The ASA 50 film needs twice the amount of light or an increase of two stops on the lens aperture to achieve the same exposure as the ASA 200 film stock.

Of course you've got to make your own decisions, but I like the grain and color space of the 16 mm, Kodak 7277, 320 ASA, tungsten negative film stock for this process. Keep in mind that this is a twin-sprocket stock, which will give you far less gate weave but a slightly smaller image area than the single-sprocket variety. Because you're eventually going to want a 1.85:1 protection mask, the slight side cropping from 16:9 to the narrower 1.85:1 won't be a big deal.

A 1,200-foot reel will cost you in the neighborhood of $400. Normally I'd recommend going to a company that sells short ends for the ultimate in economy, but that would entail a substantial amount of splicing and editing, something you don't want to get involved in. By recording your movie on three 1,200 foot reels you've got a cross-projection package that just about any festival can deal with. Total: $1,200.

CONTROLLING CONTRAST

One of the biggest problems inherent in the kinescope process is contrast. Professional labs that do this have made accommodations to their CRT. They know how different films handle and have gamma settings that correlate to the nuance of each stock. You, on the other hand, must use another method to reduce the contrast or gamma.

The simplest solution is to add a low-contrast filter, but they have a tendency to reduce the color saturation in this environment. *Test!*

There are several devices, such as the Arriflex VariCon, which mount to the front of the lens and shine a low-level light into the front element. This is a great tool for conventional studio and location work because it allows you to actually see the amount of contrast that you're controlling. The biggest drawback with using a VariCon-type device in kinescope is that the light has a tendency to reflect off the front surface of the monitor, giving you a hot spot in the center of your picture. Besides, this is way out of your budget.

Perhaps the most widely used method of lowering the gamma (and reducing grain in many stocks) is to underdevelop in processing. This is called *pulling*, and any good lab can do this although they do charge a bit more. Essentially, the film is overexposed when you shoot, generally by a stop; in processing you ask them to "pull a stop."

If you're using an ASA 100 film you'd meter it for 50 ASA. I personally like the results of pulling, it adds an inherent richness to the scene and you're going to need all the richness you can get. Again: *Test!*

Another method is to preflash the film. To do this you simply run the film through the camera in a dimly lit room. What it ends up doing is increasing the exposure of the shadows while not affecting the brighter parts of the shot.

Let's say that you've got a 12 × 12-foot room. You'd put a white card on the wall and then place the camera about nine feet from the card, making sure that the card's image fills the view. Then crank the focus in as close to the lens as it will go and stop down to *f*8. You want to use a combined light source that is about 10-candle power with a slightly warm color temperature. Small Christmas lights work quite well.

Mount one of these tiny lights on either side of the camera about 2 feet out from either side of the camera body. Make sure the lights are well behind the lens. (This is based on a 180° shutter and an ASA of 200.) Make accommodations for shutter angle based on the fact that light falls off at a square of the distance. Narrower shutters need more light or wider aperture. *Test!*

THE SHOOT

Liquid-plasma displays (LPD) are similar to the liquid crystal displays (LCD) on laptop computers, except they are much larger and far brighter. Like LCDs they don't flicker when filmed, so you don't need to make accommodations for frame rate. They are, however, expensive. If you don't personally know someone who actually owns one of these marvels of modern technology, go to a store that sells them and plead with a manager to let you run your video through the system while you record it with a film camera. If whining and groveling doesn't work, offer money. A couple crisp Benjamin Franklins should do the trick quite nicely and assure you of an uninterrupted, after-hours "kinnie" (kinescope) session. Total $200.

Rig a 1.85:1 mask onto the front of the camera. Turn off all lights or—better yet—surround the path between the LPD and the camera with duveteen or black cloth. Run an 85 percent gray signal through the monitor and use a light meter to get the appropriate setting.

TEST

Hopefully you've made a deal that allows you to run a test so that you merely bracket the exposure using successively larger and smaller apertures. You need to keep meticulous notes as to which *f* stop you are using. The best thing to do would be to cut several seconds of various scenes with dissimilar lighting together and then superimpose the successive *f* stop numbers over the various clip combinations. When the clips are played through the monitor you simply set the *f* stop on the lens to match the *f* stop that is superimposed on the shots: There won't be any confusion when you're sitting in a dark screening room trying to make your critical decision.

FESTIVAL TIME

So now you've got your print and there's a festival that's got you in its cross hairs. Now is the time to really turn up the effort, and don't be timid about blatant self-promotion. Make a concerted effort and maximize every opportunity to establish or increase your Internet presence. Remember that a strong Internet awareness is both self-propagating and compounding.

Always remember that "Advertising is what you say about yourself, public relations is what others say about you." An ounce of PR is worth a pound of advertising, and the nice thing about PR is that it's practically free. The popular misconception is that PR means press releases. As annoying as these terse missives were in the early days of faxes and desktop publishing, the Internet has allowed them to breed like mosquitoes in the pixelated pool of online, interconnectedness. You must create a unique voice that will break through the media clutter and enhance the perceived value of your project.

Good PR is inseparable from public opinion. It should always be based on your project's strongest points and echo the timbre and voice of the film. Keep driving these same points over and over, regardless of how repetitive you may feel they are. Actively seek out interviews about the project and line up interviews for your actors as well. Send out the best VHS dubs you can generate to people who write articles for film and entertainment magazines. A little personal attention from you can generate enormous returns in PR.

Unless you really hit one out of the ballpark, PR is going to give you the highest return on your investment. You've got dupe and shipping charges for 30 dubs with really nice labels. Total $300.

You've now got a little over $2,000 left from your original $10,000. Consider this your transportation, housing, and schmooze fund. Whither goest thy film, so too shall ye goest. And when you get there, don't hesitate to buy the next round. Always drink less than half of what you're pouring down the throats of potential patrons and never stop schmoozing your ass off.

Don't get involved in relationships that aren't business-related. A lot of deals have been lost because a newbie film-maker was busy hustling up company for some late-night tryst rather than sniffing out business and promoting the film. Oh, and have fun. People see you having fun, they'll figure that you're a fun guy or gal and that you've made a really fun movie. And don't forget to put me on your guest list.

"The first 90 percent of a project takes 90 percent of the time. The last 10 percent takes the other 90 percent."

—Peter Marx

CHAPTER TWENTY-EIGHT

INTERVIEW WITH GEORGE LUCAS: BREAKING THE CELLULOID CEILING

JOHN RICE
BRIAN MCKERNAN

Film director George Lucas has been a driving force in advancing content-creation technology for nearly a quarter century. His Oscar-winning 1977 film *Star Wars* pioneered the use of computerized motion-control model photography. Four years later, Lucasfilm's Computer Division began developing EditDroid, one of the very first nonlinear editing systems. Lucas was an original founder of Pixar, which pioneered feature-length movies made with computergraphic imaging (CGI). His Industrial Light & Magic (LM) subsidiary of Lucasfilm Ltd., north of San Francisco, has won 14 Oscars for its digital special effects work on some of the most successful movies of all time. Lucasfilm has also applied its digital expertise to television projects such as *The Young Indiana Jones Chronicles* and to an educational software division. Lucasfilm's THX initiative, meanwhile, has a long history of working with theatres and home-video distributors to improve sound reproduction. Lucas' latest technological innovation, however, is his boldest yet: He has photographed his next *Star Wars* movie (*Episode II*, to be released during the summer of 2002) on digital tape instead of film. Lucas had contacted Sony in 1996 to ask if they could develop a digital a 24P high-definition camera, one that could record images at the same speed as film (24 frames per second) and do so using the progressive picture scanning found in CGI. The result was a new camera system known as the *CineAlta*.

"We shot *Star Wars Episode II* in 61 days in five countries in the rain and desert heat, averaging 36 setups per day without a single camera problem," Lucas stated in an ad in the trade magazine *The Hollywood Reporter* on October 25, 2000. "We have found the picture quality of the 24P Digital HD system to be indistinguishable from film.

"Thank you, Sony and Panavision, for helping us meet our goal of making an entirely digital motion picture."

Six months later, during a press conference at the annual National Association of Broadcasters convention, in Las Vegas, Lucas elaborated on his experience with digital filmmaking, and made the historic statement, "I think that I can safely say that I will probably never shoot another film on film. It's the same issue with digital editing. I've been editing digitally for over 15 years now, and I can't imagine working on a Moviola again."

We spoke to Lucas about his transition to digital filmmaking on two separate occasions during March and April of 2001. What follows is a combined and edited transcript of those conversations.

Creating Digital Content: Can you give us a sense of where you think the entire entertainment community is right now, in terms of digital?

George Lucas: Obviously, the video industry has gone digital faster than anybody. The difference between tape and digital tape is very subtle. I think that is going to transition, reasonably fast, especially on the production side. And in film, it's going to be a slow process. We went through this when we developed nonlinear editing and everybody was poo-pooing it and fighting it, and they did it for ten years. And then, after about ten years, we sold it to Avid and it was still another three or four years before people started using the system.

CDC: What about digital as a creative tool; will digital image acquisition or digital postproduction change the way people create movies?

Lucas: Well, it definitely changes the way you create on a lot of different levels. I think it's as profound a change as going from silent to talkies and going from black and white to color. It gives the artist a whole range of possibilities that they never really had before.

In this particular case, the possibilities center around malleability. They [digital tools] are much more flexible, in terms of what you can do with the image, how you can change the image and work with the image. So, I think that that's the biggest issue that the filmmaker will be faced with. The equipment is easier to use and at the same time, it allows you to get more angles, and do more things than you'd normally be able to do. And then once you've captured the image, the digital technology allows you to do an unlimited amount of changes and work within a lot of different parameters that just were not possible with the photochemical process.

CDC: Will having used digital make your final product a different film than it would have been if it were shot on 35 mm?

Lucas: The medium that I'm in and that most people are working on at this point is the storytelling medium. For those of us who are trying to tell stories, the agenda is exactly the same. It doesn't make any difference what technology you shoot it on. It's all the same problem. It's just that telling stories using moving images has constraints. What this does is remove a large number of those constraints.

CDC: Constraints are creative, time, financial?

Lucas: There are all kinds. Yes, the constraints are time and money, but most importantly, creative. Now we can do things that we just couldn't do before in the special-effects business.

When we moved into digital, about 20 years ago now, we kept pushing things forward and were able to do things like the morphing sequence in *Willow*, the water creature in *The Abyss*, and—finally—the dinosaurs in *Jurassic Park*. You just couldn't do those kinds of things in an analog medium. It just was not possible, no matter how you thought of doing it.

CDC: Is the 1080-line 24P HD you used good enough, or would you like to see more resolution?

Lucas: Well, we always want to see more resolution. You've got to think of this as the movie business in 1901. Go back and look at the films made in 1901, and say, "Gee, they had a long way to go in terms of resolution. They had a long way to go in terms of the technology." That's where we are right now, and it's great because just think of what's going to happen in the next 100 years. I mean it's going to go way beyond anything we can imagine now.

With film, quite frankly, as a medium, the physics of it can't get much further than it is right now. We don't live in the nineteenth century, we don't live in the sprockets and gears and celluloid and photochemical processes like we did then. We've advanced beyond that. We are in the Digital Age, and [as far as] all the arguments about "It changes how you do it," of course it changes how you do things. Digital makes it easier for you to edit. Film is...a hard medium to work in. You've got to push real hard to get it to go anywhere. But once it's digital, it's...light as a feather.

CDC: What role did Sony and Panavision play?

Lucas: When we first starting pushing Sony, they had the camera. We just said, "You've got to make it available." And the first roadblock we came to was, "What about the lens?" So we went around to all the lens manufacturers [and] had a very hard time getting anybody to develop what we needed. Finally we connected with Panavision, which was very enthusiastic about it. We brought Panavision and Sony together, and they were both willing to make huge financial commitments in this area. We consulted with both of them and sort of became the matchmaker.

CDC: The Internet has been full of rumors that you shot digital and film side by side. Was there any film used to shoot *Star Wars Episode II*?

Lucas: There was absolutely no film at all used in the shooting of *Episode II*. We did have probably a case of film and a VistaVision camera with us, but to be honest the reason was more of an insurance issue than that we ever planned to use it. The insurance company kind of didn't "get it's head around digital" as fast as we'd hoped, and it said, "We will lower your rates if there's a film camera on a truck somewhere that you could get hold of if you had a problem." But we never had to use it.

We never had a problem with the digital camera, and we shot under unbelievably difficult conditions. We shot in the desert. We shot in the rain. Sony cameras have been used all over the world for newsgathering under...battle conditions. And they survive, they get the news. And this [CineAlta] is now just a hybrid of that kind of thing. There's no reason why it can't work in the jungle, the desert, the rain. We had no problems.

You can have rumors forever. But it doesn't make any difference because we did shoot the whole thing on digital. We had no problems whatsoever with the cameras or with the medium or anything. The film looks absolutely gorgeous. You can say whatever you want. Eventually the film will be out there in the theatres and you can judge for yourself, but I guarantee that 99 percent of the people that see the film will simply not know that it was shot digitally. Then there's the 1 percent who are the technophiles or the people who are the die-hard film people, as opposed to cinema people, and they'll say it's not real film. And you know, it's not real film; what we're talking about is cinema, which is the art of the moving image. It's very similar to film, but it's not the same thing.

CDC: Why do you think so many negative rumors circulated about your use of digital cinematography?

Lucas: I don't think anyone likes change. The same thing happened when we created EditDroid. When we created EditDroid nobody wanted anything to do with it. There were all these rumors that directors would take forever because they had so many more choices. It breaks down all the time. There were all kinds of reasons about why it wouldn't work. Well, it does work. You can argue it to death, but it works and it's better, and eventually people will use it because why would you work with a push mower when you can have a power mower? It just doesn't make sense.

CDC: Let's talk about digital delivery; there's been talk that you're hoping a good number of theatres are ready to do digital projection.

Lucas: Well, we're hoping. When we released *Phantom Menace* we were able to release it in four theatres digitally. Then our other San Francisco company, Pixar, pushed it to almost 20 theatres on *Toy Story 2*.

CDC: To what end; what's better about digital projection?

Lucas: The quality is so much better. I've spent a lot of my time and energy trying to get the presentation [of movies] in the theatre up to the quality that we have here at the Ranch or what you see in private screening rooms. That's why THX was invented and that's why the TAP Program was invented. We've worked very hard to try and improve the quality of presentation at the level of the audience actually seeing the movie. It's a rather thankless job, but it's something that does need to be done. And we've been pushing digital projection for years and years now, and working with all the vendors to try to improve it and try to get it working in the marketplace. And now the business models are coming together to try to actually take the next step, which is "How do you pay to get the thing into the theatres?" And THX is very active in that whole area.

Digital far and away improves that presentation. You don't get weave, you don't get scratchy prints, you don't get faded prints, you don't get tears—a lot of the things that really destroy the experience for people in the movie theatre are eliminated.

CDC: Is digital cinema projection technology there?

Lucas: I think the technology is definitely there and we projected it. I think it is very hard to tell a film that is projected digitally from a film that is projected on film. If you have brand-new prints, after a few weeks, it's very easy to tell which is which.

CDC: Do you foresee your content being repurposed repeatedly for other-than-large screens?

Lucas: I think ultimately, when broadband comes, that there's going to be a lot of entertainment that will be run through the Internet, and I think that's going to be a big factor in the near future.

CDC: Do you have that in mind when you're creating films, or are you still producing for the big screen?

Lucas: I'm producing for the big screen. Whether broadband Internet actually happens is still up in the air. We haven't gotten that far yet. I think content that will be produced for the Internet will have to be of a different caliber, because the kind of a revenue stream that's going to be involved is still up in the air. There could be Napsters out there, and if so it's going to be very hard to create for the Internet because you're going to have to do it for ten cents. You'll be basically giving your content away because there's no revenue stream coming in, and all you can do is hope to sell t-shirts or something to make money. So, you might be able to charge a little bit, but that whole issue of how you finance the product on the Internet is still completely unknown.

CDC: Are you investigating that?

Lucas: We have a Web site. We have documentaries on the Web site. We manage to make it work economically because we have a very few people who work for us.

CDC: Earlier you mentioned that the transition to digital creation of motion-picture content is comparable to the transition of silent to sound films. Is there an overall digital distribution revolution going on? If we came back here in 10 years, would we be talking about an entirely different world of entertainment?

Lucas: I think what's going to happen is that the invention of digital, in terms of distribution, is going to open up the marketplace to a lot more people. It's going to be much more like painting—or writing books, and it will take cinema in that direction, which means it's accessible to anybody. That will change a lot about how we view the entertainment business. Right now it's very tightly controlled by a very small group of individuals who make all the decisions about what is good and what is bad and what people should see and all that sort of thing. This is going to pretty much knock that away so that filmmakers from anywhere and everywhere will have access to the distribution medium.

CDC: What's that do to the broadcaster?

Lucas: I have no idea how that's going to affect the broadcaster because we don't really know where the Internet is going or how it's going to finance itself. But at the same time, there is no question that it's going to draw viewers away from the broadcast medium. The advertising dollars are going to possibly be spent on the Internet. If they are, then that will definitely shift that whole medium around quite dramatically.

CDC: Let's talk about the same thing from the consumer point of view. When I was growing up, entertainment options were television and movies. Now there are many more; I can even watch a feature film on my laptop on an airplane. We're very close to the ability of watching video on a PDA. What do you think that does to the accessibility or the quality of the content that's going to be available to the consumer?

Lucas: The technical quality will continue to improve no matter what medium is being used, and especially now, with digital. We're sort of at the very bottom of the medium. In terms of digital, we're working in very crude terms with what we're going to be able to do in the next 100 years. But in terms of the creative quality of the product, you have two forces at work. One is that you're going to infuse a lot of fresh blood and fresh ideas and exciting new people into the business. But on the other hand, you're going to have far fewer resources to create what you're going to create. So those may even themselves out. All in all, I think that visual entertainment, movies, and that sort of thing will pretty much be the way they always have been, which is about 10 percent is really good, 20 percent is sort of okay, and the rest of it is terrible.

CDC: Is there a technology or product missing right now that you wish were out there? Is there something you've dreamed of?

Lucas: No. In terms of what I am really dreaming about right now, I want to get a totally integrated system where it's all digital and we're all working in one medium. The biggest problem we've been having in the last 15 years is the fact that we have bits and pieces from different kinds of mediums, and we're trying to *kludge* them all together into one product. I think things will be infinitely easier and cheaper and much more fun for the creative person when we're working in one medium.

CDC: So, summer of 2002, going to the theatre and watching *Episode II*; am I going to know, assuming I haven't watched the trades, that this has been a digitally created film?

Lucas: No, I don't think anybody will ever know that the new film has been digitally created. The last film was digitally created. Most people didn't notice. We even had some test shots that we captured and inserted digitally and nobody could tell which were which. So I don't think anybody is going to be able to know the difference. I actually think *Episode II* looks better than *Episode I* in terms of the technical quality of the image and that sort of thing. So I'm assuming that people will never know. I think that in those theatres where it is shown digitally, it will be markedly better than a projected image. We've already seen our finals on *Episode II* projected digitally and also projected on film, and there's just no comparison between the two. The film images are very, very high quality but the digital is just noticeably better.

CDC: So what's my reaction, what's my son's reaction going to be, when we walk out of that cinema?

Lucas: Well, if you see it at a digital theatre, I hope you'll like it. I hope you'll say "Wow, that looked great!" If everything works well, you're not even going to think about that part. You'll just say "that was a great movie and it worked well." But you will definitely not say "Gee, it looked kind of crummy; it looked grainy; the blacks weren't right." You won't see any of those things. I guarantee that. I'll say that if you don't know it's digital, if somebody hasn't told you, or if you haven't read it somewhere, you won't have a clue that it's digital. You'll just say that was a nice, beautiful movie.

CDC: Are there any words of encouragement or warning that you think need to be delivered to the content-creation marketplace?

Lucas: Well, the one thing that we have documented without any doubt in the film business is that people like quality. They like quality sound and they like quality images. They like to look at something that really has a technical polish to it. They respond to that. We did it when we released *Star Wars* in 70 mm. We did it when we started THX theatres. It's been very clear that the people want to have a good image and good sound. The one advantage that traditional broadcasters have is that they have access to digital television. They can create a better, high-quality product. Now the compatible hardware is coming out, and I think that's going to be an important element in selling broadcast television in the next 10 to 20 years.

CDC: Digital television or HDTV?

Lucas: I would say digital television, but any kind of higher-quality image than what people are getting is important to them. That's always been the selling pitch for cable. The whole point to cable is that you don't get fuzzy images, and the primary reason people buy cable is to get better image quality.

CDC: Is there a definitive explanation—definition—of the digital world from content organization to content distribution? Is there something we can say to wrap our arms around where we are right now and where we are going?

Lucas: That's a hard question. I don't even know how to answer that one. Digital is just a different storage medium, a different capture medium. It doesn't change what I do at all. You know, it's lighter cameras, easier to use, and a more malleable image. I can change it and do things that I want to do, but they're all technical things, so it's just a more facile medium than the photochemical medium. I think that is the all-encompassing, overriding factor in all of this.

CDC: And in terms of digital distribution to theatres and digital projection at those theatres…?

Lucas: It's the same thing. You get a higher quality image and at the same time you get a system that is easier to use and, in the long run, is much cheaper.

CDC: Could you elaborate on the aspect of digital allowing you to be more creative?

Lucas: The example I always use came from my friend, [sound designer] Walter Murch. He described the period we're going through as being similar to the period in the Renaissance when artists went from fresco painting to oil painting. When you had to do frescoes you could do only a very small portion at a time, the plaster had to be mixed by a craftsman. The colors had to be mixed by other craftsmen, and the science of mixing those colors so that they would dry just the right color was all very important. And then everything had to match. Each section that you did had to match with the next section because you were doing what was, in essence, a large mosaic. You had to paint very fast because you had to paint before the plaster dried, and you pretty much had to work inside a cathedral or something, a dark space lit by candles.

And it was all science. It was technology at its height in those days because the plaster had to be mixed just right. Florence competed with Rome; it was, "Ooh, we've got a red that you can't believe." If you study the history of colors, going from red and blue to mauve was a huge advance in art—in technology—every time they came up with a new color.

Today in film it's really the same kinds of issues. With oil painting, suddenly you could go outside and work in sunlight, work with paint that didn't dry immediately, continue to work with it. Come back the next day, paint over it, and that was a huge advantage to the artist in terms of his ability to change his mind and work in a medium that was much more malleable and much more forgiving. One person could do it. You mixed your own colors and pretty much that's the way the color dried. You didn't need a bunch of experts back there doing it for you. You didn't have to have the 15, 20, or 30 guys mixing paint, mixing plaster, building scaffolds, doing things that you would normally have to do if you were an artist.

And oil let you interpret the way light played on things. You would never have ended up with the Impressionists if it hadn't been for oil, because they couldn't do that kind of art in fresco. And so did it change the medium? Yes. Did it change it for the better or worse? That's a personal opinion. But at the same time it allowed the artist to free up his thinking and to say, "I want to go to a different place than I've been able to go to before." And, you know, I greatly admire the fresco painters and I still think it's an incredible art form, but people choose their medium, and they choose it for the kind of work they do.

So that is the best description of digital. It's really going from the photochemical, fresco medium to the oil painting, digital era.

CDC: Did digital technology have an effect on location shooting?

Lucas: The on-location process when you're actually shooting digital is just fantastic because there are a lot of important things that happen. These things may seem very small to people that are not involved in the film business, but to those who are, these are very big issues. You don't have the situation of running out of film every 10 minutes. This means that you don't get to a point in the performance when everybody is ready to do that really perfect take, and somebody yells "Roll out!" and you have to wait 10 minutes for a film reload while the whole mood falls apart. And then the actors have to go and build themselves back up again. That's very debilitating, that kind of thing. It seems small, but it's very big.

The issue of being able to move the cameras around very fast, the issue of being able to watch what's going on. You see your dailies on the set as you're shooting, with the departments watching. So if the hair is messed up, if the wig shows, if the prop isn't there, all those kind of things that you'd normally see the next day in dailies, you're looking at them right now. You can make changes immediately. You don't have to go back and reshoot something. That makes a big difference. On *Episode II*, I had the same schedule and the same crew as *Episode I*, and I was able to go from 26 set-ups a day to 37 set-ups a day. I was able to shoot under schedule by three days, and a lot of this had to do with not having that extra hour at night watching dailies. Not having to do reloading and that sort of thing really does speed up the process in terms of the amount of work you can do in a day.

CDC: I'm not trying to get your endorsement here, but what did you think of the Sony CineAlta digital camera?

Lucas: We had a great time with our cameras. We never had any problems with them on the film. They were always there, always working, beautiful images; light to use, and I'm completely sold on digital. I can't imagine ever going back.

CDC: How about the operational aspects? Everyone on the crew is used to certain ergonomics that film cameras have.

Lucas: Well, again, most of the people who capture images are capturing them electronically. It is a very small group of people that actually still use film. Because most of the images that are captured are captured for the broadcast medium and most of the cameramen are using hand-held Betacams or some such camera. They are very used to the way it works. The only people that are having difficulty are the people who are used to big, heavy cameras with giant magazines on them. But once they get used to the nice, light, little cameras, they'll embrace them because it's not that difficult.

CDC: So, five years from now, we would not even be discussing how movies are made—digitally or on film?

Lucas: I don't think so. Ten years ago we went through this exact same controversy with EditDroid, and there was a huge backlash and a lot of fighting, a lot of rumors, a lot of craziness: It'll never happen, we'll never use it. Well, now it's hard to find somebody who still cuts on a Moviola. Digital nonlinear editing is so accepted that the idea of not using a nonlinear editing system is unthinkable, and nobody talks about it now. Nobody asks, "Did you cut this on a nonlinear editing machine, or did you cut this on a Moviola?" People don't even ask those questions anymore because it's obvious. Ninety-nine percent of the people are using nonlinear editing.

CDC: What do your peers think of you shooting movies digitally?

Lucas: I have very close friends who say they will never, ever go digital until the last piece of film is used up and the last lab is closed. Then there are others—Jim Cameron, and a lot of the other directors—who have come through here to see what we're doing. Francis Coppola is also very enthusiastic about what we're doing and is amazed at what we've been able to accomplish in terms of images. I think, eventually, most people will follow in the next few years. I have a very long post on my picture, over two years. When I started, obviously I was the first one to ever use the 24-frame digital camera, and I have a feeling that there will be at least a half-dozen to a dozen films, which will be released before my film, that have been shot digitally.

CDC: How does it feel to be a pioneer?

Lucas: I don't think of myself as a pioneer. I've gotten asked these kinds of questions early on in my career because I've spent a great deal of money on research and development, and I've tried to push the medium forward. It's really because I want to make the process of making films easier for the artist, because I want to make it easier for myself to cut the films, to shoot the films. And because I am an independent filmmaker, I have to make sure that every resource I have is used in the best possible and most efficient way. Sometimes advancing the technology a little here, a little there, does just that. It cuts your costs and it also makes it easier for the artists to realize their vision. I can think of a lot of crazy movies, but up to this point, I've always been hitting the concrete ceiling—or celluloid ceiling, I guess we should say—that has not allowed me to fulfill my imagination.

The main thing is I'm very happy now. I have always been pushing the envelope of the medium because primarily I want to get the best possible image and the best possible way of telling my stories. And I've always found myself bumping up against that celluloid ceiling, the technology that says, "You can't go here, you can't go there, you can't do that." Or the economic resources, such that "You can get there, but it's going to cost you an arm and a leg to do it." And my feeling is that the artist really needs to be free to not have to think about how he's going to accomplish something, or if he can afford to accomplish something. He should just be able to let his imagination run wild without a lot of constraints. And that's really what digital is allowing us to do.

CHAPTER TWENTY-NINE

INTERVIEW WITH JAMES CAMERON: 3D DIGITAL HD

BRIAN MCKERNAN, EDITOR

In the recent history of motion pictures, director/producer/screenwriter James Cameron is, to quote a line from his 1997 *Titanic* "The King of the World." *Titanic* won 11 Oscars, including best picture and best director, and has earned more than $1 billion worldwide. A technical innovator in addition to being a master storyteller, James Cameron's long string of hit movies (including *The Terminator, The Abyss, Terminator 2: Judgment Day*, and *True Lies*) all feature unforgettable visuals produced with pioneering digital effects and computer graphic imagery (CGI) spearheaded by the director's powerful will and creative vision.

Always on the lookout for the newest filmmaking technology to add to his arsenal of storytelling tools, Cameron's latest interest centers around the high definition (HD) 24-frame, progressive-scan (or 24P HD) Sony CineAlta digital video format. In this interview, conducted in April 2001, Cameron describes his impressions of this new content-creation technology and how he's using it for 3-D and underwater filmmaking projects.

Creating Digital Content: How would you describe the difference between film and digital HD?

James Cameron: HD is much more immediate. You don't have dailies, so dailies are a thing of the past. You look at the monitor, and what you see is what you're getting. So you can fine-tune the lighting and make all your decisions in the moment. This, in a way, changes the normal paradigm of photography because now the director is directly involved in the cinematography, which some cinematographers may welcome, and others may not like because there's always the possibility of recriminations the next day at dailies: "Hey, how come you didn't have more fill light? I can't see her eyes, etc."

HD is an opportunity to avoid that unpleasant aspect of the process. Of course there are a few cinematographers who are so good that it's always perfect the next day. But if you're not in that 1 percent I think it's good to have a dialog with the director or the producer—whoever the other creative force is if it's television or if it's film—and have that dialog at the time when you're looking at the monitor, and get exactly what you want before you walk away from that set-up.

CDC: What do you perceive the differences will be between HD and film? What benefits do you think HD will have?

Cameron: The way I see it, if you're shooting in a regular studio environment—with cameras on tripods, or on dollies, or on cranes—reloading is not that big a deal. It's a little bit of an encumbrance; you'll do four or five takes and then you'll have to reload, and it can break your energy. Obviously with HD, using a camcorder, it's almost an hour between reloads—with a VTR it's two hours. You're not breaking your energy, you're just flowing with it, and that's nice. You have the immediacy of actually analyzing the image that you're creating at the time. You can experiment with filters, with lighting, you see it in a very immediate sense, and you don't have to wait until the next day to see how it came out, which I like a lot because I like the ability—the freedom—to experiment without having to be too conservative.

And I think another of the really big advantages with HD is one that really hasn't been explored very much, but we're looking into quite a bit, which is the fact that with film cameras you've got to have a magazine. And the magazine is your film, the medium that you're recording to, that has to be physically adjacent to the lens and to the movement that drives the film. It's an ancient system; it's a hundred-year-old paradigm.

With HD—now that it's a good replacement for film in terms of image quality—you have an imager block about this big [Cameron holds up his hands to indicate] and you have a lens. That's all you really need at the place where the rubber meets the road, where the actors are, or where the scene is happening. The rest can be put further away. So now the opportunity for lightweight stabilization systems, lightweight body-mounted camera systems, aerial photography, car-mounted photography—all these things, I think will be increased dramatically. We should be able to do a lot more with a very flexible, fluid moving camera than we've been able to do in the past because the physical size of the camera is very small.

With the 3D system that we're developing we have two HD cameras side-by-side, and—with lenses, and servo motors, and everything else—the entire package can go on my shoulder, weighs 19.5 pounds, and is solid. And it can actually replace a 3D IMAX camera, which weighs 329 pounds. We've proven this by taking the output and blowing it up to IMAX, projecting it in a theatre with the glasses on, and having a really beautiful image experience.

CDC: You've spent a lot of time getting information from other folks.

Cameron: We spent a year working with Sony developing our specialized 3D camera system, in the process of which we've done a lot of testing with the 24P 900-series cameras, both the HDW-F900—the camcorder—and the C950—the studio model. And we've looked at the results on tape, projected through HD projection, and filmed out to 35 mm film and projected, and filmed out to 1570 IMAX format 3D and projected, and I'm completely satisfied with the results.

I think the question that's been bothering everybody for the last ten years is "When is this going to be mature? When is HD going to replace film?" The argument has been "When is it good enough to replace 35 mm?"

I think we've soundly and quietly passed that mark, and it's so far in our rear-view mirror that it's not really an issue of whether it's the equivalent to 35 mm. It's an issue of "Is it the equivalent to 65 original negative, or beyond that?" because that's the kind of data that you're getting.

Most people tend not to see the data because they're looking at a monitor. The monitor, despite how good the monitors are, doesn't show you everything that the camera's capturing. You have to go out to film to see that. And there's a tremendous amount of data there, almost an uncomfortable amount. But it's always good to have too much resolution and back off with filters and selective focus and so on, and with the various digital arts in postproduction, than to have too little. You can't put focus back once you don't have it, and you can't put resolution back when you don't have it.

CDC: You're mainstream, but you seem to have adapted to it quite well.

Cameron: I think it has great potential, I think there will be some short-term transition issues for people who won't be as willing to change the way they work, I think that trying to spend a lot of time getting HD to exactly duplicate a film look is, in a sense, almost counterproductive, because it produces its own aesthetic. And I think it's possible for one to embrace that aesthetic and say "This is a good thing that we're able to see this clearly."

We used to fight for this kind of resolution. We used to lug around these giant 65 mm cameras, and only in the last couple of decades have there been a couple of films made—actually shot in 70 mm. People did it for a reason. They did it because they love the clarity. There's a certain type of film that can benefit from that a great deal. And so that, I think, is exciting; that with these cameras we can essentially replace 65 mm cameras.

I'm past the whole 35 mm thing. These HD cameras get more than 35 mm. Now, there are certain things that still need to be worked out with respect to offspeed shooting, high-speed photography, and things like that, so there's still going to continue to have to be a hybrid between the two types of filmmaking, at least for a transition period. But eventually all this will get worked out, and HD will be able to do the high-speed photography and underspeed photography, and so on. I think that's probably only a year or so away, if that.

CDC: Do you feel like an HD pioneer?

Cameron: Yeah, I'm sort of joining a group of pioneers. I'm not saying that I'm the tip of the spear by any stretch; George Lucas has been talking about this for years, saying that we're going to be shooting on HD. And he's gone out and actually put his money where his mouth is, and has actually shot a feature on it, which I have not done yet. And I'd love to have this same interview a year from now, having shot a feature in HD and be able to give you a lot more of the very specific day-to-day, day-in, day-out aspects of how it works. But in terms of the testing that we've done over the last year, I'm pretty satisfied.

CDC: Can you talk about your relationship with the folks who are helping you out technically?

Cameron: Well, it's been great. Sony has been extremely forthcoming in helping to work out this new 3D camera system. In fact, they designed a special version of their 950 studio camera, which they call the 3D-T camera, meaning it's for 3D purposes and it's a "T" camera, which in their terminology means a telescoped camera, or one that's separated; the imager head is separated from the camera electronics.

And so we've been able to take that separately packaged imager head—it's been specially made for us to be very narrow. We can put two of them side-by-side and have it be less than or equal to the interocular distance of human vision. That means that we can shoot parallel 3D. And to my knowledge nobody's done this at film-level resolution before, other than with the IMAX 3D camera.

Like I said, the big 329 pound. Solido IMAX camera is a parallel system—two lenses side by side. But for 35 mm equivalent or large format I think this is a very, very revolutionary camera. At 19.5 pounds you can hand-hold it, you can put it on a Steadicam, you can use it like a conventional camera, and yet the end result is large-format 3D. So it's very revolutionary, and Sony has been very helpful with that. I've had to fly to Tokyo, they've flown to meet me here, we've flown halfway and met in Hawaii. We've done what was necessary to integrate very well. I've found them to be extremely responsive to our technical requirements; at first perhaps a little skeptical that this would be a good thing to do, and now very excited that it is a good thing to be doing.

CDC: Did you have your own requirements or specs that you wanted, and did Sony adapt to those?

Cameron: I think what happened was that Sony leaped so far ahead of the pack in developing the 24P technology that they hadn't had a chance to really figure out what it's meant to the world yet. They've been deluged by requests and by new concepts and new ways to use their technology to the point that it's hard for them to keep up. Of course, I'm grateful that they've been able to take my suggestions and my input and actually turn them into something. And I believe that they've actually productized the telescope version of the 950 camera as well, so that other people will be able to work with these very tiny, lightweight imaging heads.

If you're not doing HD for 3D, if you're just doing it for normal film production, you're talking about something like an 8-pound camera. That's revolutionary. The lightest nonsound camera—an MOS camera—is in the 20-pound range. And for a sound camera, you're talking about the 40- to 50-pound range with bells and whistles. I'm talking about an 8-pound camera with a ten-to-one zoom lens on. Pretty amazing.

So you start saying, "Well, I could put some gyros on that and hand-hold the whole system." Or "I could stick it on a Steadicam and be able to run with it at high speed, trailing a fiber-optic behind so you can get the image to the recorders." That's a pretty revolutionary concept.

CDC: Please explain your HD 3D concept.

Cameron: Well, 3D basically works by taking two separate images—a left image and a right image—and they're basically identical except they're taken from slightly different vantage points: One is here and one is here, both the same distance apart as our eyes are. And then we superimpose those two images onto the same screen at the same time, and we use polarized glasses, so that the left eye only sees the image it's supposed to see, and the right eye of the viewer only sees the image that it's supposed to see. And we actually reproduce it on the screen the way human beings see. If you have enough resolution, as you would with IMAX or with HD, and if you have stereo vision—we would call that stereoptic vision—it's the equivalent of the way we see.

If we do a shot with these cameras—walking into a room or being in some exotic environment, it will look the way it would look to you if you were really there. So I think it's a heightened kind of experience, far beyond a normal filmmaking experience.

CDC: What have you done as far as testing?

Cameron: We've looked at the way the cameras perform with lenses from different manufacturers. We've looked at the color space that's created by the cameras. We've looked at the dynamic range, the gamma, and so on, to make sure that the blacks are reproducing properly. And we've looked at the way in which you expose the image, what type of light levels are required, and so on. And we've found the cameras to be as sensitive as the high-speed Kodak negative rated kind in the 320 to 400 ASA range. I prefer closer to 320 for HD. But you're getting more—you're getting more grain and more detail than you would with Kodak negative of that same sensitivity.

The lenses take a little getting used to because the focal lengths are a whole bunch shorter than you're used to. And that brings with it this issue of increased depth of field. It's important to try to work with the lens close to wide open, near wide open, as much as possible, so that you have some selective focus, which does reproduce a kind of a film look. And it's not really going for a film look as a thing in itself, but really I think that's the way the mind interprets images in the real world. Our eyes focus on a certain thing and everything else goes soft, and so it's what you want to do with photography.

One of the criticisms of HD in video in general is that it carries too much depth of field. Panavision, especially, has been working to create lenses that are very fast, so that you can work at an *f* 1.6, or an *f* 1.8, or an *f* 2, and get the kind of selective depth of field that you want.

CDC: For you to jump into this [3D] marketplace is almost expected, having done the Universal Studios Florida *T3 3D* film.

Cameron: I've done one 3D project in the past, which was *T2 3D*, and we worked with very big, cumbersome camera rigs, and it really slowed us down. It required an awful lot of light. We had to light the entire set to an *f* 5.6, which any cinematographer will tell you if you're doing a night exterior, trying to light it to an *f* 5.6 is almost impossible. I think we had every Musco light in Hollywood rented to do these big exterior night shoots. In fact I don't know how anybody else was making a movie while we were doing *T2 3D* because we had the world lit up with these things. So of course with the HD cameras we don't have the same light requirements, our rig is very small, very mobile. So I think it's a way of making 3D production much more like normal production. I think it could attract the types of filmmakers who might be otherwise put off by these big monster cameras and way too much science going into every single shot.

CDC: Where do you see 3D applications being used?

Cameron: I certainly think 3D can be used to tell a dramatic story; there's no inherent reason why you couldn't do a feature-length dramatic fictional story in the 3D medium. It certainly is very good for natural history subjects, especially underwater subjects, because of the way in which objects float in front of the camera. It's exceptionally good for venue-based attractions in theme parks, and so on. And with some new technologies, some of which Sony is developing, and some other people, you'll be able to watch it at home on your monitor, as well, with glasses. I think that eventually you'll see it coming in as an alternative to one-eye vision, which we've all been used to for 100 years of filmmaking—it doesn't necessarily have to be the only way to watch a movie.

CDC: You see it [3D] as becoming much more mainstream.

Cameron: Yes, eventually I'll shoot a feature film with stars and the whole thing, and special effects, in 3D, and presumably release it in some 3D theatres, and some 2D theatres, and the consumer can make the choice. Some people are uncomfortable with 3D, about 10 percent of people can't see it or are not interested in it. For the other 90 percent we'll offer them a choice: You can see it in a 3D theatre or you can see it in a 2D theatre. It's like choosing between a THX theatre or a non-THX theatre. Or in the old days you used to choose between a 70 mm theatre and a local theatre. Consumers are smart; they know how to pick out and seek out the things that they want.

I think that it's critical for us as entertainers to find ways to give people special experiences in the theatre these days because home video systems are becoming so good. I think that people today are seeing and hearing so well in their home systems that a small mall cinema is only marginally better, aside from the social or group experience, or a date. But from a technical standpoint that gap is getting pretty narrow these days. We have to keep leapfrogging ahead and try to think of what is it that can make the experience special and magical again.

CDC: Will you be using HD in the future?

Cameron: Yes, I plan to shoot my next film in HD, absolutely. And we're looking right now at shooting our next season of *Dark Angel*, our TV show, in HD as well. Not only because of what it allows us to do in terms of the immediacy of lighting and so on, but also in terms of being able to run a production that's being shot in Vancouver, from Los Angeles. We can have immediate access to our dailies within minutes of when they're shot using a broadband network that can stream the HD to us. So by going all-digital, all electronic, there are a lot of advantages for TV production as well. We're looking into all of this. I can't guarantee it's all going to work out quite as rosy as it sounds, but it sounds pretty good.

CDC: If you're doing acquisition in Vancouver, how are you going to get it to L.A.?

Cameron: Panavision is offering a broadband production network that can stream HD over that network, point to point. They will have a point set up in Vancouver, and we'll set one up here at Lightstorm, in Santa Monica. We'll be able to get our material as it's being shot, or within minutes or hours of when it's shot. And that's not a surrogate; it's not film photochemically processed and then telecined to cheesy video and then hand-carried down 24 to 36 hours later.

We don't have to wait for photochemical processing, we don't have to wait for it to be hand-carried through customs, we don't have to worry about the negative being scratched or damaged, we don't have to take time for telecine. We can just get it directly here, and work with it here. Our postproduction is all done here, all our editing is done here. It's immediate—right now—the actual shot, all the data, so you could literally cut it in and air it. So, in theory, if we were under the gun in an episode, and we had a second unit shooting some inserts or some close-ups, or a new scene, or a change the network requested, and we literally had already mixed the whole piece and had just slugged in for those things, we could drop them in and air that day or the next day. We don't want to have to cut it that close, but it does give you certain options.

CDC: You could be in L.A. watching them shoot in Vancouver.

Cameron: I'm not sure that we'd want to maintain an open link, given the data costs. We'll certainly use it to send the actual finished shot to us. We may maintain an open link for standard-def video, but we could certainly have an HD open link for discussion of, say, lighting, or makeup, or some things where you need to see the actual image as it appears to the camera. We could do that.

CDC: Did you think 24P HD could surpass IMAX?

Cameron: I was skeptical for a long time that HD would ever equal 35 mm film as I knew it. And I think that's behind us at this point. Then the question became "Well, what is its potential? What is it the equivalent to?"

And it took a while to really find out. We had to keep filming it out to successively larger formats. We did 70 mm, and then we did IMAX. I think where we are right now with the 2/3-inch chip, the 2.2 million-pixel chip that is the heart of the Sony 900 cameras, is that we're not truly the equivalent of IMAX 2D, but we are a very, very good equivalent to IMAX 3D.

I say that very specifically, because if you have two cameras shooting side by side, each one gathering 2.2 million pixels of data, then what you're really displaying to the human visual cortex is 4.4 million pixels of data, and that's plenty, that's more than you need. And it produces a very, very good analog for human vision, or a surrogate for human vision.

CDC: What are the financial advantages of using digital 24P HD instead of film?

Cameron: Well, I think there's different ways to look at HD from a financial perspective. Whether HD can produce a huge savings in the cost of standard production right now, I'm not sure. We haven't really run the numbers in a side-by-side comparison for a feature motion picture with a full package of, let's say, Panaflex cameras versus a full package of, say, F900 cameras, or F950 cameras. My guess is they're going to be fairly equivalent.

That's not really where your savings are; I think it's really more in the manner in which you shoot, the speed of your photography, the speed of your lighting set-ups, the fact that you don't have to go back and shoot anything again because you've got it. You don't have to worry about film getting lost or damaged, or processed incorrectly, scratched, all that sort of thing. You don't have to worry about the lighting, the exposure not being correct because you've seen it. So I think that probably over the length of a movie I would guess there might be an incremental speed-up in the overall production process. Whether that's 5 percent of 15 percent I can't say right now, not having gone through that entire cycle.

But I think there's another way to look at it, too, which is that the things that make the most sense on films, financially, are those things that allow the film to make the most money. Because it doesn't matter if you have a dog movie that costs 5 percent less. What really matters is that you have a movie that people want to see. So whatever enhances your artistic process as a filmmaker is an enhancer to the bottom line of what the film will earn—because you'll make a better film. People will like it more, they'll go see it again—or more people will go see it—and you'll make more money. And that's really the way this needs to be looked at: Anything that enhances the creativity of the filmmaking process is going to be financially beneficial to the filmmaking business, even if the cameras cost the same.

CDC: And you think HD can help do that.

Cameron: I think it will. I think it's going to unlock a lot of new talent, I think it's going to make the filmmaking process more hands-on creatively for directors, and it's going to allow their creative relationship with cinematographers to be better, to be closer, to be less wasteful, and to be more creative. "More creative" in the sense of giving oneself permission to try things without having to worry that you've lost a day's shooting, when you go to dailies the next day and cringe, and hope that it worked. Because nobody wants to lose a day's shooting at $50,000 a day or $100,000 a day, whatever you're spending.

CDC: Will this revolutionize the "ESP" and "trust factors" between directors and DPs?

Cameron: We're going to need less ESP. It's not going to be sub-vocal, subconscious communication. It's going to be very conscious: "There's the image; what do we want to do? What do we want to change? Let's make it a little more low-key. Let's make it spookier. Let's play with the fill levels. Let's play with the back light. Let's add more smoke."

I think the director and the DP are going to be able to work together more closely because they're going to be standing at a monitor instead of having a "post-game discussion," "Monday-morning quarterbacking" what they shot the day before in dailies the next day, and hoping they can apply some lessons learned to the next material shot. Even though they may be a little dissatisfied with what they just saw, they won't have to put anything in their film that they're dissatisfied with. You're going to see it on a monitor, having just done a take or having done a rehearsal, and you're going to make the adjustments you need, and then you're going to shoot it. And then you're going to walk away knowing that you've done it. You may even tell the script girl which takes you like, tell it to the editors, and you never have to look at it again until you're looking at a cut—if that's the way you want to work as a director.

I think it's really going to speed up the process and make for tighter teams. It's going to shift the emphasis out of the dailies paradigm and to the monitor paradigm, which frankly I think is a good thing.

CDC: You know where everybody is: They'll all be clustered around the monitor....

Cameron: Well, yes, I know. Video village is the heart of the filmmaking community these days anyway, so that shouldn't change too much. But I think there's no reason why eventually the first AC—the focus puller—can't be at the monitors as well. Because you're right there, you see the focus. [Digital 24P HD is] one of the few new tools in terms of smarter-focusing systems; there's no reason for everybody to be at the camera. And half the time the camera's running around on a Steadicam, or up on a crane, or 20 feet in the air anyway, so it's going to actually reinforce and make better a process that's already been there for the last decade or so, which is that people tend to cluster at the monitor in order to control the image.

CDC: Do you want to talk about the underwater test that you're about to do?

Cameron: We did some underwater testing back in September. We went to Truk Lagoon, and basically it was my vacation. It was my dive vacation, but we happened to take a 900 camera and a housing, and I spent the whole week shooting. At the end of that time I took about 10 minutes of the best selects, put them together into a reel, which is the equivalent of a 1000-foot film reel—a small reel—and we did the color correction in about an hour. We went to film using the E-Film laser recorder, and the results were perfect.

It was beautifully timed from end to end. If you think about the amount of time that you'd spend timing a 1000-foot reel of motion-picture film, going through the photochemical timing process, you might spend several days—or as long as a week—going back and forth with multiple prints, subjecting your negative to this potential wear and tear, potential for damage. And at the end of that week you've made some compromises, but basically it looks pretty good, and you've been through hell. With HD we did it in 1 hour.

In theory one could time an entire film in a day or so—finished, timed, done, walk away. That, to me, is huge. Because I always spend the last four or five or six weeks of postproduction color-correcting the film. And it's a nightmare. It's an absolute nightmare because of the vagaries of photochemistry. You make a one-point change, and it will come back two points different just because the "soup" was too hot or something. I think this is going to make a big difference for cinematographers and for directors.

The actual result of the underwater photography from Truk Lagoon projected in 35 mm was absolutely gorgeous. The one thing I can say definitively: Whatever your choice might be for shooting above water, if you have any intention whatsoever of shooting underwater, never shoot film again.

That I can say definitively, and that's not some fine, esthetic consideration. This stuff was beautiful, the blacks were rich, the color reproduction was excellent, the clarity was absolutely perfect, and we had an hour of run time every time we closed up that housing for the camera. And anybody that shoots underwater knows that the most you normally get is 5, 6 minutes off a 400-foot magazine, and then you have to do a reload. That's revolutionary.

And I think any situation in which you have reload issues, whether it's aerial photography inside a space camera or Wescam, or any kind of special-purpose camera applications—you definitely want to be using this stuff.

CDC: Are you saying that, creatively, this is a release, this is a freedom you're going to be allowed here?

Cameron: We're basically done with the age of the meat grinder, of the reciprocating engine. We've moved on to the jet engine. Cameras—film cameras—are piston engines. We've moved on to turbines. Technology always goes through these kind of quantum changes. It doesn't change one's creative process in it's basic sense; you're still going to be working with actors, you're still going to be telling a story, you're still going to be doing shots, you're still going to be thinking editorially, and all those aspects of filmmaking are exactly the same. But I think that these new cameras are really more of an evolutionary change, not a revolutionary change. You're not going to throw out everything you know; it's just going to make what you know better and easier to do.

I don't want to sound like a commercial for these digital 24P HD cameras, but I'm just very excited about the possibilities here. What I want to do is encourage people to start figuring out—to grab the baton and run with it—start figuring out what you can do with these cameras because I want the tools. I can't build all this gear myself; I need people to be building gyro-stabilized systems and better, lighter, smaller camera heads that can go out on the end of much longer telescoping crane arms—things that were physically impossible to do before because of the mass of these big film cameras.

CDC: See any transitional humor in this anywhere?

Cameron: To be perfectly honest I haven't found the process that amusing yet because like with anything you have birth pains; we're going through a lot of stuff right now to figure out, from an engineering standpoint, getting this 3D system up and running, and so on.

I'm sure there'll be some funny moments, like when some DP comes over and puts a .44 Magnum round through a monitor and tells everybody to back off and let him do what he does. But that'll change. People will either get on board or they won't.

And people will continue to shoot film. Why not? It's an art, and an artist can choose their tools. You can still make a clay pot if you want to, you don't have to have a plastic or Fiberglass or ceramic pot; you can make it out of clay if you want to.

So people will choose to use older tools because that's what they're familiar with, and not want to push the envelope. But I think that most filmmakers, most of the time, will try to push themselves further. It's kind of what we do. So I think they're going to run with this stuff.

CDC: Any advice for me as a director or a DP if I'm on the line about this?

Cameron: If you have an aging actress, don't let her look at the monitor. Use a lot of diffusion, back off a bit. That will be an issue because 24P HD is the equivalent of shooting in 65 mm: You're getting more information there. And I think that what people need to look at are ways to soften the image. And DPs will find that; DPs are good at that.

It's lighting, it's makeup, it's all the things we do. It's not going to be a panacea, it's not a magic wand that makes all that stuff go away. In fact it becomes as important as it's ever been to do it properly. If people are visualizing as "Oh now I can shoot a movie like newsgathering, where I throw the camera on my shoulder and just hose the whole thing down"—yeah, you'll be able to do that, but it's still a movie. You're still going to have to take care of business.

GLOSSARY

A Advanced; also the designation of an old British television system simulcast for many years with the current system.

Active Lines The total number of scanning lines devoted to the vertical blanking interval.

Active Line Time The duration of a scanning line minus that period devoted to the horizontal blanking interval.

Acuity *See* Visual Acuity

Adaptive Changing according to conditions.

ADC Analog-to-Digital Converter

Advanced Encoder A device that changes RGB or CAV into NTSC utilizing some form or forms of prefiltering to reduce or eliminate NTSC artifacts. Some advanced encoders also offer image enhancement, gamma correction, and the like.

Advanced Television (ATV) Television better than what is currently broadcast. It includes HDTV, EDTV, IDTV, and artifact elimination schemes.

Alias Something other than what it appears to be. Stairsteps on what should be a smooth diagonal line are an example of a *spatial alias*. Wagon wheels appearing to move backwards are an example of a *temporal alias*. Aliases are caused by sampling and can be reduced or eliminated by prefiltering, which can appear to be a blurring effect.

Analog (Analogue) A signal that is an analogy of a physical process and is continuously variable, rather than discrete. *See also* Digitization.

Anamorphic Squeeze A change in picture geometry to compress one direction (usually the horizontal) more than the other. Anamorphic squeeze lenses made CinemaScope pos-

Mark Schubin, Tracy Swedlow and Jerry Whitaker contributed to this Glossary.

sible. Occasionally, when widescreen movies are transferred to video, an anamorphic squeeze will be used (usually only in credits) to allow the smaller aspect ratio of television to accommodate the larger movie aspect ratio.

Some ATV proponents have suggested a gentle anamorphic squeeze as a technique to assist in aspect ratio accommodation. JVC introduced a video projector in 1989 having an anamorphic expansion lens to convert squeezed widescreen images from a 4:3 aspect ratio back to 16:9. An anamorphic squeeze can be performed either optically or electronically.

Aperture As applied to ATV, the finite size and shape of the point of the electron beam in a camera or picture tube. Because the beam does not come to an infinitesimal point, it affects the area around it, reducing resolution.

Aperture Correction Signal processing that compensates for a loss of detail caused by the aperture. It is a form of image enhancement that adds artificial sharpness; it has been used for many years.

Artifacts Visible (or audible) consequences of various television processes. Artifacts are usually referred to only when they are considered defects. Artifact elimination is often more apparent than quality increases such as resolution enhancement. *See also* Filter Artifacts.

Aspect Ratio Ratio of the width of an image to its height, sometimes expressed as two numbers separated by a colon (e.g., 4:3) or sometimes as one number, with a colon and the number one implied (e.g., 1.85 is the same as 1.85:1). Aspect ratios of ATV systems are sometimes expressed relative to nine (e.g., 16:9, 15:9, 14:9, and 12:9).

Aspect Ratio Accommodation Using these techniques, something shot in one aspect ratio can be presented in another.

Asymmetrical Digital Subscriber Line (ADSL) A type of DSL that provides T1 rates or higher in the downstream (towards the customer) direction and 64 Kbps or higher in the upstream direction.

Asynchronous Transfer Mode (ATM) A high speed data transmission and switching technique that uses fixed-size cells to transmit voice, data, and video, which greatly increases the capacity of transmission paths, both wired and wireless.

Backbone A "fat pipe" within a network. The term is relative to the size of network it serves.

Backchannel Term commonly used to describe the action of sending data back to a host server over a phone wire or cable pipe.

Bandwidth **1.** (Broadcast) The range of frequencies available for signaling. The difference expressed in cycles per second (Hertz) between the highest and lowest frequency of a band (James Martin, 1971). **2.** (Data wire) Analog telephone lines measure capacity in Hertz (the difference in the highest and lowest frequency in the channel). Digital channels measure capacity in bits per second. A T3 connection is approximately 30 times as fast as a T1 connection, which is 50 times as fast as a 28.8 modem, which is twice as fast as a 14.4 modem.

Broadcast: 6 MHz analog
Cable: 450 MHz analog signal
Cable Modems: up to 27 mps downstream–shared
Internet data: 14.4k–26 MBPS with VDSL
VBI: 4 Mbps
Digital Signal: 19.2 Mbps

Bezel The frame that covers the edge of the picture tube in some TV sets and can therefore hide edge information transmitted in an ATV system (such as ACTV I) not meant for the viewer to see. *See also* Overscanning.

Bit Error Rate Similar to signal-to-noise ratio (SNR) for a digital signal.

Bit Rate The digital equivalent of bandwidth. The rate at which the compressed bit stream is delivered from the channel to the input of the decoder.

Bit Rate Reduction Schemes for compressing high bit-rate signals into channels with much lower bit rates.

Blur A state of reduced resolution. Blur can be a picture defect, as when a photograph is indistinct because it was shot out of focus or the camera was moved during exposure. Blur can also be a picture improvement, as when an unnaturally jagged-edged diagonal line or jerky motion is blurred to smoothness.

Broadband A network capable of delivering high bandwidth. Broadband networks are used by Internet and cable television providers. For cable, they range from 550 MHz to 1GHz. A single regularly broadcast TV channel requires 6MHz, for example. In the Internet domain, bandwidth is measured in bits-per-second (BPS). *See* Digital Subscriber Line (DSL).

Broadcast Television Conventional terrestrial television broadcasting, the most technically constrained delivery mechanism for ATV, faced with federal regulations and such potential problems as multipath distortion and cochannel interference.

Burn An image or pattern appearing so regularly on the screen of a picture tube that it ages the phosphors and remains as a ghost image even when other images are supposed to be shown. On computer terminals, the areas occupied by characters are frequently burned, particularly in the upper left corner. In television transmission centers, color bars are sometimes burned onto monitors. There is some concern that some ATV schemes will burn a widescreen pattern on ordinary TV sets because of increased vertical blanking or will burn a non-widescreen pattern on ATV sets because of reception of non-ATV signals. In production, burn refers to long-term or permanent image retention of camera pickup tubes when subjected to excessive highlights.

C Chrominance. Also Compatible.

Cable Modem A device that permits one-way or two-way high speed data communication over a cable television system for purposes such as Internet access at speeds of around 1.5 MBPS. Download rate is 27 Mbps.

Cable Plant The central equipment and broadcasting headquarters of a cable operator. All initial broadcasts from the content providers are sent to the cable plant, aggregated, re-encoded, and broadcast to its set-top box network.

Cable Television The system network for the distribution of the television signal and now digital data by cable (coaxial, twisted pair, or fiber optic).

Carrier An electromagnetic wave modulated with information to be recorded or transmitted; also a term used for such organizations as telephone companies, who carry signals from place to place. *See also* Modulation.

CATV Community Antenna Television or cable television; a delivery mechanism for ATV not necessarily as technically constrained as broadcast television.

CAV Component Analog Video; refers to the handling of unencoded color signals (separate components), a form of signal recording and distribution not subject to cross-color and cross-luminance artifacts and offering potentially greater chroma resolution. When used to ultimately feed NTSC receivers, CAV signals must pass through an NTSC encoder; but, because that might be the only encoder used in production or postproduction, it can be an advanced encoder.

CD Compact Disc. The audio quality some ATV proponents feel is the minimum performance level acceptable for ATV. Others feel CD quality might be excessive.

Channel Stuffing Technique for adding information to an NTSC channel without increasing its bandwidth or eliminating its receiver compatibility.

Characteristic An aspect or a parameter of a particular television system that is different from another system's, but is not necessarily a defect. Characteristics include aspect ratio, colorimetry, resolution, and sound bandwidth.

Chroma Short for chrominance.

Chroma Crawl An NTSC artifact (also sometimes referred to as moving dots); a crawling of the edges of saturated colors in an NTSC picture. Chroma crawl is a form of cross-luminance, a result of a television set decoding color information as high detail luminance information (dots).

Chroma-Key *See* Electronic Matting.

Chroma Resolution The amount of color detail available in a television system, separate from any brightness detail. In almost all television schemes, chroma resolution is lower than luminance resolution, thus matching visual acuity. Horizontal chroma resolution is only about 12 percent of luminance resolution in NTSC; in advanced schemes it is usually 50 percent. *See also* Resolution.

Chrominance Color information signal or signals.

CinemaScope The first modern widescreen movie format, achieving a 2.35:1 aspect ratio through the use of a 2:1 anamorphic squeeze.

Cochannel Interference Interference caused by two or more television broadcast stations utilizing the same transmission channel in different cities. It is a form of interference that affects only broadcast television.

Codec Coder/decoder.

Coder Short for encoder.

COFDM (Coded Othogonal Frequency Division Mulitplexing) A modulation scheme for digital television broadcasting that competes with 8VSB.

Colorimetry Characteristics of color reproduction, including the range of colors that a television system can reproduce.

Color Subcarrier A subchannel created by the second NTSC to carry color information within the original NTSC's black-and-white signal with minimal interference to black-and-white sets. The color subcarrier, an interlaced carrier at 3.579545 MHz, is quadrature modulated with a wideband (I) and a narrowband (Q) signal.

Comb Filter A filter designed to deal with periodic signals (signals that are repetitive) such as scanning lines or frames. Comb filters are so named because a chart of their effect on a signal resembles the teeth of a comb. They are most often used to separate luminance and chrominance signals in NTSC. Depending on their implementation, comb filters can change moving dots to hanging dots and can reduce vertical and diagonal resolution. Some comb filters tend to be incompatible with CCF.

Common Sharing characteristics with another television system. The more that systems share in common, the easier it is to mass manufacture equipment for them.

Component A television system in which chrominance and luminance are distributed separately.

Composite A television system in which chrominance and luminance are combined into a single signal, as they are in NTSC; any single signal comprised of several components.

Compression The reduction in the number of bits used to represent an item of data.

Conditional Access Technology Technology embedded on the set-top box and satellite receiver that enables the cable or satellite broadcaster to filter out content the subscriber has not paid for or provide them with movies or special programs they have purchased on a pay-per-use system.

Constant Luminance Principle A rule of composite color television that states that any change in color not accompanied by a change in brightness should not have any effect on the brightness of the image displayed on a picture tube. The constant luminance principle is generally violated by existing NTSC encoders and decoders. *See also* Gamma.

Crawl An appearance of motion in an image where there should be none. *See also* Chroma Crawl and Line Crawl.

Cross-Color An NTSC artifact occurring when fine-detail luminance is displayed by a TV set as chrominance, resulting in a colorful moiré pattern where there should be none.

Cross-Luminance An NTSC artifact resulting from the interpretation of chrominance as fine-detail luminance in a TV set. The most objectionable form of cross-luminance is chroma crawl or moving dots. Other forms are hanging dots and visible subcarrier.

CRT Cathode Ray Tube or picture tube (CRTs are also used as camera tubes). There are three forms of display CRTs in color television: tricolor (a color picture tube), monochrome (black and white), and single color (red, green, or blue, used in projection television systems).

D Definition.

DAC Digital-to-Analog Converter.

Decoders Devices that change NTSC signals into component signals; sometimes devices that change digital signals to analog (*see* DAC); sometimes devices that descramble a signal. All color TV sets must include an NTSC decoder. Because sets are so inexpensive, such decoders are often quite rudimentary.

Definition Primarily resolution, although some use the term to mean quality.

Delivery Getting television signals to a viewer. Delivery might be physical (e.g., cassette or disc) or electronic (e.g., broadcast, CATV, DBS, optical fiber).

Detail Resolution. Finer detail means higher resolution.

Diagonal Resolution Amount of detail that can be perceived in a diagonal direction. Although diagonal resolution is a consequence of horizontal and vertical resolution, it is not automatically equivalent to them. In fact, ordinary television systems usually provide about 40 percent more diagonal resolution than horizontal or vertical. *See also* Resolution.

Digital Expressable numerically. *See also* Digitization.

Digital Subscriber Line (DSL) Modem telecommunications technology that enables broadband, digital data to be transmitted over ordinary telephone line. DSL comes in many flavors, known collectively as xDSL. *See* Asymmetric DSL, High Bit Rate DSL, SymmetricDSL, Very High Bit Rate DSL.

Digital Video Recorder (DVR) A high-capacity hard drive that is embedded in a set-top box, which records video programming from a television set. These DVRs are operated by personal video recording software, which enables the viewer to pause, fast forward, and manage all sorts of other functions and special applications. TiVo, ReplayTV, and UltimateTV are commercial examples of a DVR.

Digital Video Server A robust, dedicated computer at a central location that receives command requests from the television viewer through a video-on-demand application. Once it receives this request, it then instantly broadcasts specific digital video streams to that viewer. nCUBE, SeaChange, and Concurrent are examples of companies that provide this kind of equipment and software services.

Digitization The process of changing an electronic signal that is an analogy (analog) of a physical process such as vision or hearing into a discrete numerical form. Digitization is subdivided into the processes of *sampling* the analog signal at a moment in time, *quantizing* the sample (assigning it a numerical level), and *coding* the number in binary form. The advantage of digitization includes improved transmission; the disadvantage includes a higher bit rate than the analog bandwidth. Bit rate reduction schemes work to reduce that disadvantage.

Direct Broadcast Satellite (DBS) **1.** A distribution scheme involving transmission of signals directly from satellites to homes. It does not carry the burden of terrestrial broadcasting's restricted bandwidth and regulations and so is thought by many to be an ideal mechanism for the introduction of high base bandwidth ATV. DBS is the most effective delivery mechanism for reaching most rural areas; it is relatively poor in urban areas and in mountainous terrain, particularly in the north. Depending on the frequency band used, it can be affected by factors such as rain. **2.** Satellites powerful enough (approximately 120 watts on the Ku-band) to transmit a signal directly to a medium or small receiving dish (antenna) at 18 inches to 3 feet in diameter. DBS does not require reception and distribution by an intermediate broadcasting facility and transmits directly to the end user.

Direct-To-Home (DTH) Term used to describe satellite broadcasting to the home via 18-inch dishes. *See also* Direct Broadcast Satellite.

Direct-View A CRT watched directly, as opposed to one projecting its image on a screen.

Display The ultimate image presented to a viewer; the process of presenting that image; the device that presents that image.

Distribution The process of getting a television signal from point to point; also the process of getting a television signal to the viewer from the point at which it was last processed.

Dot Crawl *See* Chroma Crawl.

Dot Pitch The distance between phosphor dots in a tricolor direct-view CRT. It can be the ultimate determinant of resolution.

Downlink The action of transmitting analog or digital signal to a satellite dish receiver on earth via a transponder on a satellite.

Downstream To send information from the network to the user.

Dynamic Resolution The amount of spatial resolution available in moving pictures. In most television schemes, dynamic resolution is considerably less than static resolution. *See also* Motion Surprise, Spatial Resolution, and Temporal Resolution.

E Enhanced or Extended.

EBR *See* Electron Beam Recording.

Edge A boundary in an image. The apparent sharpness of edges can be increased without increasing resolution. *See also* Sharpness.

Electron Beam Recording A technique for converting television images to film using direct stimulation of film emulsion by a very fine long focal-length electronic beam.

Electronic Cinematography Photographing motion pictures with television equipment. Electronic cinematography is often used as a term indicating that the ultimate product will be seen on a motion picture screen, rather than a television screen.

Electronic Matting The process of electronically creating a composite image by replacing portions of one image with another. One common, if rudimentary form of this process is *chromakeying*, where a particular color in the foreground scene (usually blue) is replaced by the background scene.

Electronic matting is commonly used to create composite images where actors appear to be in places other than where they are being shot. It generally requires more chroma resolution than vision does, causing contribution schemes to be different from distribution schemes.

Electronic Programming Guide (EPG) A application that allows the viewer to interactively select their television programming. The development of applications to enhance the EPG, such as dynamic video selection, recording options, and more, is a high growth area. Currently, the EPG allows the viewer to also access summaries of shows, the ability to set recording times, show program length and names of crew members, as well as the ability to select content via categories. More advanced EPG (also called Interactive Programming Guides [IPGs]) applications enable the viewer to select shows to record over several weeks every time a show or a selected movie star appears on the schedule. Ultimately, EPGs will enable the TV set to learn the viewing habits of its user and suggest viewing schedules.

Emphasis A boost in signal level that varies with frequency, usually used to improve SNR in FM transmission and recording systems (wherein noise increases with frequency) by applying a pre-emphasis before transmission and a complementary de-emphasis at the receiver.

Encoders Devices that change component signals to composite signals; sometimes devices that change analog signals to digital (*see* ADC); sometimes scramblers.

All NTSC cameras include an encoder. Because many of these encoders omit many of the advanced techniques that can improve NTSC. CAV facilities can use a single, advanced encoder prior to creating a final NTSC signal.

Encryption Scrambling; considered important for signal security and pay-TV.

Enhanced Television (ETV) A type of interactive television technology favored by network broadcasters. This technology allows content producers to send HTML data and graphical "enhancements" through a small part of the regular (US) NTSC analog broadcast signal called the Vertical Blanking Interval (see below). These enhancements appear as overlays on the video and allow viewers to click on them if they are watching TV via special set-top box/software services like WebTV, Wink, WorldGate, and more to come. One major problem associated with enhanced TV today is that producers must pay close attention to timing information. When the digital signal is more widely available, content producers won't have to worry about timing informaton.

European Standards Currently, 625 scanning lines and 50 fields per second as opposed to NTSC's 525/59.94. This led to the opposition to the 1125/60 HDTV proposal when it was presented to the CCIR as a proposed worldwide standard. European producers are developing their own ATV systems.

Even Number The number of scanning lines per frame possible in a progressively scanned television system. An interlaced scan system must use an odd number of lines so that sequential fields will be displaced by one scanning line.

With a "kick" signal, however, even an interlaced system can use an even number of scanning lines.

Extensible Markup Language (XML) A language that acts as a "metalanguage;" XML allows programmers to use it to create their own markup languages for specific uses. XML is written in Standard Generalized Markup Language (SMGL).

Eye Tracking The pattern in which eyes follow a person or object across a television screen. *See also* Dynamic Resolution.

Fiber Optics *See* Optical Fiber.

Field One complete vertical scan of an image. In a progressive scanning system, all of the scanning lines comprising a frame also comprise a field. In an interlaced scanning system, all of the odd-numbered or all of the even-numbered scanning lines comprise a field, and two sequential fields comprise a frame.

Field Alias An alias caused by interlaced scanning. *See also* Interlace Artifacts.

Field Rate Number of fields per second.

Filter A device that allows certain parts of signals to pass through and stops others, in the same way that a coffee filter allows water and certain parts of the coffee to pass through while stopping the grounds. Filters are always used in decoders and usually in encoders.

Filter Artifacts Artifact introduced by filters. The most common visual artifacts introduced by filters are reduced resolution and ringing.

Flash Downloading The ability to automatically send software upgrades to a set-top box network.

Flicker A rapid visible change in brightness, not part of the original scene. *See also* Flicker Frequency, Judder, Large-Area Flicker, and Twitter.

Flicker Frequency The minimum rate of change of brightness at which flicker is no longer visible. The flicker frequency increases with brightness and with the amount of the visual field being stimulated. In a recent study, a still image flashed on and off for equal amounts of time was found to have a flicker frequency of 60 flashes per second at a brightness of 40 foot lamberts (fL) and 70 at 500. Television sets generally range around 100 fL in peak brightness (although some new ones claim over 700). SMPTE recommends 16 fL for movie theater screens (although this is measured without film, which reduces the actual scene brightness by at least 50 percent). One reason for interlaced scanning is to increase television's flashing pictures to the flicker frequency without increasing bandwidth.

FM Frequency Modulation; the form of modulation used for television sound transmission in most of the world, for satellite video transmission, and for videotape recording. Noise in an FM system is greater at higher frequencies, thus usually demanding emphasis.

Footprint A term used to define a logistical area in a region covered by a cable or satellite operator, although not necessarily served directly by them. This term is also used to define the amount of space a particular piece of software or hardware takes up inside a set-top box.

Frame Rate The number of frames per second.

Frequency Repetition rate. In electronics, almost invariably the number of times a signal changes from positive to negative (or vice versa) per second. Only very simple signals (sine waves) have a single constant frequency; the concept of instantaneous frequency therefore applies to any transition, the frequency said to be the frequency that a sine wave making the same transition would have. Images have spatial frequencies, the number of transitions from dark to light (or vice versa) across an image, or per degree of visual field.

Frequency Interleaving The process by which color brightness signals are combined in NTSC.

Frequency Multiplex *See* Multiplex.

Gamma A measure of how much a graph of display brightness versus scene brightness varies from a straight line. Picture tubes have gammas of more than 1; cameras, therefore, use gammas of less than 1 so that the overall brightness-in/brightness-out characteristic will be linear (a straight line). Unfortunately, current CRT gammas do not match those in the NTSC standard, creating nonlinear gammas. Furthermore, because camera gamma is introduced prior to encoding (which lowers chroma resolution), the constant-luminance principle is violated.

Gamma Correction Compensation for gamma problems.

GCR Ghost Cancelling Reference signal. *See* Multipath Distortion.

Geometry The shape of objects in a picture, as opposed to the picture itself (aspect ratio). With good geometry, a picture of a square is square. With poor geometry, a square might be rectangular, trapezoidal, pillow-shaped, or otherwise distorted. Some ATV schemes propose minor adjustments in geometry for aspect ratio accommodation.

Ghosts *See* Multipath Distortion.

GHz Gigahertz; billions of cycles per second.

Group Delay The delay of certain frequencies relative to others; a common characteristic of filtering. Because color is carried as a high frequency subcarrier in NTSC, group delay can cause color to become shifted relative to brightness, creating a ghostlike effect.

H High; also Horizontal.

Hanging Dots A form of cross-luminance created by simple comb filters. It appears as a row of dots hanging below the edge of a highly saturated color. *See also* Cross-Luminance.

HBI *See* Horizontal Blanking Interval.

HD High Definition. A frequently used abbreviation for HDEP and sometimes HDTV. The term "high definition," applied to television, is almost as old as television itself. In its earliest stage, NTSC was considered high definition (previous television systems offered from 20 to 405 scanning lines per frame).

HDEP High Definition Electronic Production. Ronald Reagan was the first U.S. president to be recorded using HDEP, at the 1988 NAB Convention.

Headend The electronic control center of a cable television system, generally located at the antenna site of CATV system. The headend takes incoming signals and amplifies, converts, processes, and combines them into a common coaxial or optical cable for transmission to cable subscribers.

High Bit Rate Digital Subscriber Line (HDSL) A type of DSL that transmits 2 Mbps bidirectional signals over one or two twisted copper pairs. HDSL is used in applications such as corporate internetworking, videoconferencing, and remote data center access.

High Line Rate More than 525 scanning lines per frame (or, in Europe, 625).

Horizontal Blanking Interval That portion of the scanning line not carrying a picture. In NTSC, the HBI carries a synchronizing pulse and a color reference signal. Some scrambling and other systems add sound and/or data signals to the HBI.

Horizontal Resolution Detail across the screen, usually specified as the maximum number of alternating white and black vertical lines (lines of resolution) that can be individually perceived across the width of a picture, divided by the aspect ratio. This number is usually expressed as TV lines per picture height. The reason for dividing by the aspect ratio and expressing the result per picture height is to be able to easily compare horizontal and vertical resolution.

Horizontal chroma resolution is measured between complementary colors (rather than black and white) but can vary in some systems (such as NTSC), depending on the colors chosen. Horizontal resolution in luminance and/or chrominance can vary in some systems between stationary (static resolution) and moving (dynamic resolution) pictures. It is usually directly related to bandwidth.

Host Any computer on a network that offers services or connectivity to other computers on the network. A host has an IP address associated with it.

Hypertext Markup Language (HTML) A collection of tags typically used in the development of Web pages.

Hypertext Transfer Protoco (HTTPl) A set of instructions for communication between a server and a World Wide Web client.

HVS Human Visual System (eyes and brain).

HVT Horizontal, Vertical, and Temporal, the three axes of the spatiotemporal spectrum.

Hybrid Fiber–Coaxial (HFC) A local cable TV or telephone distribution network. An HFC consists of fiber optic trunks ending at neighborhood nodes, with coaxial cable feeders and drop lines downstream of the nodes.

Hypertext A term invented by Ted Nelson in 1965, hypertext was at first a concept and then a real hyperlinking technology enabling users of computers and the Internet to link from one page to another by simply clicking from word to word. Hypertext technologies evolved over many years and through the inventions of many people.

Hypertext Markup Language (HTML) A collection of tags typically used in the development of Web pages.

Hypertext Transfer Protocol (HTTP) A set of instructions for communication between a server and a World Wide Web client.

I Improved or Increased; also the in-phase component of the NTSC color subcarrier, authorized to have more than twice as much horizontal resolution as the Q, or quadrature, component.

IF Intermediate Frequency; the first stage in converting a broadcast television signal into baseband video and audio.

Image Enhancement Technique for increasing apparent sharpness without increasing actual resolution. This usually takes the form of increasing the brightness change at edges. Because image enhancement has advanced continuously for nearly 50 years, ordinary NTSC pictures sometimes look better than the NTSC pictures derived from an HDEP source, particularly when these derived pictures are designed to be augmented by other signals in an ATV receiver. It is very difficult to enhance pictures for NTSC receivers and then unenhance them for receivers with augmentation.

Impulsive Noise High level, short duration unwanted signals that tend to cause a sparkling effect in the picture and/or a percussive effect in the sound. Impulsive noise is often caused by motorized appliances and tools.

Interface A set of textual or graphical symbols that allow a computer user to communicate to underlying software. Computer interfaces work in many ways. Some are text-based and communicate only in letters, numbers, and other keyboard symbols. Others are graphical and require the use of a mouse. Still others are touchscreen.

Interference Defect of signal reproduction caused by a combination of two or more signals that must be separated, whether all are desired or not.

Interlace Technique for increasing picture repetition rate without increasing base bandwidth by dividing a frame into sequential fields. When first introduced, it also had the characteristic of making the scanning structure much less visible. NTSC uses 2:1 interlace (two fields per frame).

Interlace Artifacts Picture defects caused by interlace. These include twitter, line crawl, loss of resolution, and motion artifacts. In addition to causing artifacts, interlaced scanning reduces the self-sharpening effect of visible scanning lines and makes vertical image enhancement more difficult to perform.

Interlace Coefficient A number describing the loss of vertical resolution caused by interlace, in addition to any other loss. It is sometimes confused with the Kell factor.

Interlaced Carrier A television subcarrier at a frequency that is an odd multiple of one half the line rate (for example, the NTSC color subcarrier), and is therefore not likely to be objectionably visible under normal viewing conditions.

Interlaced Scanning Television scanning employing interlace.

Interline Flicker *See* Twitter.

Internet Protocol (IP) A protocol telling the network how packets are addressed and routed.

Internet Service Provider (ISP) Telecommunications companies that sell Internet access. Users either dial-up to an ISP server or have a broadband connection such as DSL. Once connected, they can branch out onto the Web.

ISDN Integrated Services Digital Network, a telecommunications concept and series of standards whereby one, universal, digital transmission network can carry computer data, voice, still, and moving pictures (at any quality level), and high fidelity sound. ISDN is in the process of being slowly implemented by telephone companies worldwide.

Jitter A rapid small shift in image position characteristic of film projection. Projection jitter can reduce the apparent resolution of film.

JND Just Noticeable Difference; a measure of the minimum perceptible change in quality. A one JND change is accurately detected 75 percent of the time; a three JND change is accurately detected 99 percent of the time. There is a large number of JNDs of difference between NTSC as it is now received in U.S. homes and HDEP.

Judder Jerkiness of motion associated with presentation rates below the fusion frequency.

Kell Factor A number describing the loss of vertical resolution from that expected for the number of active scanning lines; named for Ray Kell, a researcher at RCA Laboratories. Many researchers have come up with different Kell factors for progressively scanned television systems. These differences are based on such factors as aperture shape, image content, and measurement technique. A generally accepted figure for the Kell factor is around 0.68, which, multiplied by the 484 active NTSC scanning lines, yields a vertical resolution of 330 lines, matched by NTSC's 330 lines of horizontal resolution per picture height (See Square Pixels). It is important to note that most studies of the Kell factor measure resolution reduction in a progressive scanning system. Interlaced scanning systems suffer from both a Kell factor and an interlace coefficient.

Kick Signal A signal that allows an interlaced television system to use an even number of scanning lines.

kHz Kilohertz, thousands of cycles per second.

Large Area Flicker Flicker of the overall image or large parts of it. *See also* Flicker Frequency and Twitter.

Laser Beam Recording A technique for recording video on film.

Lechner Distance Named for Bernard Lechner, researcher at RCA Laboratories; the Lechner distance is nine feet, the typical distance that Americans sit from television sets, regardless of their screen size. The Jackson distance, 3 meters, named for Richard Jackson, a researcher at Philips in Britain, is similar. There is reason to believe that the Lechner and Jackson distances are why HDTV research was undertaken sooner in Japan (where viewing distances are shorter) than elsewhere. *See also* Viewing Distance.

Letterbox Term generally used for the form of aspect ratio accommodation involving increasing vertical blanking. It is called letterbox because the shape is that of a wall-mounted home mail box.

Line Crawl Tendency of the eyes to follow the sequentially flashing scanning lines of interlaced scanning up or down the screen in the same way that the eyes follow the sequentially flashing light bulbs on a movie theater marquee. Line crawl tends to reduce vertical resolution.

Line-Doubling Any number of schemes to convert interlaced scanning to progressive scanning at the display, the simplest of which simply doubles each scanning line. More elaborate schemes use line interpolation and motion compensation or median filtering.

Line Interpolation An advanced mechanism, used in some line-doublers, which calculates the value of scanning lines to be inserted between existing ones.

Line Kicker *See* Kick Signal.

Line Pairing A reduction in vertical resolution caused when a display (or camera) does not correctly space fields; this results in an overlap of odd and even numbered scanning lines.

Line Pairs A measure of resolution often used in film and print media. In television, lines are used instead, creating confusion when comparing film and video.

Line Rate The rate at which scanning lines appear per second (the number of scanning lines per frame times the frame rate); sometimes used (nonquantitatively) as an indication of the number of scanning lines per frame (e.g., a high line-rate camera).

Lines Scanning lines or lines of resolution. The latter are hypothetical lines alternating between white and black (or, in the case of chroma resolution, between complementary colors). The combined maximum number of black and white lines that might be perceived in a particular direction is the number of lines of resolution. Vertical resolution is measured with horizontal lines; horizontal resolution is measured with vertical lines; diagonal resolution is measured with diagonal lines (no current television system or proposal favors one diagonal direction over the other, so the direction of the diagonal lines doesn't really matter). *See also* PPH.

Line Structure Visibility The ability to see scanning lines. Seeing them makes it harder to see the image (like looking out a window through Venetian blinds or not being able to see the forest for the trees). Some ATV schemes propose blurring the boundary between scanning lines for this reason.

Luminance The brightness detail signal. The human visual system is more sensitive to luminance detail than to chrominance detail.

Matched Resolution A term sometimes used to describe matching the resolution of a television system to the picture size and viewing distance (visual acuity); more often a term used to describe matching of horizontal and vertical (and sometimes diagonal) resolutions. There is some evidence that the lowest resolution in a system (e.g., vertical resolution) can restrict the perception of higher resolutions in other directions. *See also* Square Pixels.

Mb or Megabit 10^6 bits of information (usually used to express a data transfer rate; as in, 1 megabit/second = 1Mbps).

Mbps Megabits Per Second; a data transmission rate in millions of binary digits per second.

Median Filter An averaging technique used by PCEC in its IDTV line interpolation scheme to take an average of lines in the current and previous fields to optimize resolution and avoid motion artifacts without using motion compensation.

MHz Megahertz; millions of cycles per second. A normal U.S. television transmission channel is 6 MHz. The base bandwidth of the video signal in that channel is 4.2 MHz. SMPTE 240M calls for 30 MHz each for red, green, and blue channels.

Modulation The process (or result) of changing information (audio, video, data, etc.) into information-carrying signals suitable for transmission and/or recording. In NTSC-M television transmission, video is modulated onto a picture carrier using AM-VSB, and audio is modulated onto a sound carrier using FM.

Moiré A wavy pattern, usually caused by interference. When that interference is cross-color, the pattern is colored, even if the picture is not.

Monochrome Literally *single color*, usually used to indicate black and white. There have been monochrome high line-rate cameras and displays for many years. The EIA has standardized rates of up to 1,225 scanning lines per frame. NHK developed a monochrome HDTV system with 2,125 scanning lines per frame. Even higher numbers of scanning lines are used in conjunction with the lower frame rates of cathode ray tube scanners used in printing and in film. These extremely high rates are possible because monochrome picture tubes have no triads.

Motion Adaptive An ATV scheme that senses motion and changes the way it functions to avoid or reduce motion artifacts.

Motion Artifacts Picture defects that appear only when there is motion in the scene. Interlaced scanning has motion artifacts in both the vertical and horizontal directions. There is a halving of vertical resolution at certain rates of vertical motion (when the detail in one field appears in the position of the next field 1/60th of a second later), and horizontally moving vertical edges become segmented (reduced in resolution) by the sequential fields. This is most apparent when a frame of a motion sequence is frozen and the two fields flash different information.

Motion Compensation Correction for motion to avoid or reduce motion artifacts. *See also* Motion Adaptive.

Motion Pictures Expert Group (MPEG) A proposed International Standards Organization (ISO) standard for digital video and audio compression for moving images. Responsible for creating standards 1, 2, and 4.

Motion Resolution *See* Dynamic Resolution and Temporal Resolution.

Motion Surprise A major shift in the quality of a television picture in the presence of motion that is so jarring to the viewer that the system might actually appear better if it had continuously lower quality, rather than jumping from a high quality static image to a lower quality dynamic one.

Moving Dots *See* Chroma Crawl.

MPEG-1 1/4 broadcast quality, which translates to 352 × 240 pixels. Typically compressed at 1.5 Mbs.

MPEG-2 Similar to MPEG-1, but includes extensions to cover a wider range of applications. MPEG-2 translates to 704 × 480 pixels at 30 frames per second in North America and 704 × 576 fps at 25 fps in Europe. Typically compressed at higher than 5 Mbs. The primary application targeted during the MPEG-2 definition process was the all-digital transmission of broadcast TV quality video.

Multipath Distortion A form of interference caused by signal reflections. Signals that are reflected take a longer path to reach the receiver than those that are not. The receiver synchronizes to the strongest signal, with the weaker signals travelling via different paths and causing ghostly images superimposed on the main image.

Multiple System Operator (MSO) Cable operators who own a number of different networks and services.

Multiplex To combine multiple signals, usually in such a way that they can be separated. There are three major multiplexing techniques. *Frequency division multiplex* (FDM) assigns each signal a different frequency. This is how radio and television stations in the same metropolitan area can all transmit through the same air space and be individually tuned in. *Time division multiplex* (TDM) assigns different signals different time slots. Different programs can be broadcast over the same channel using this technique.

Narrowband Relatively restricted in bandwidth.

NTSC National Television System Committee; two standardization groups, the first of which established the 525 scanning line per frame/30 frame per second standard and the second of which established the color television system currently used in the United States. Also the common name of the NTSC-established color system.

NTSC is used throughout North America and Central America, except for the French islands of St. Pierre & Miquelon. It is also used in much of the Caribbean and in parts of South America, Asia, and the Pacific. It is also broadcast at U.S. military installations throughout the world and at some oil facilities in the Middle East.

Barbados was the only country in the world to transmit NTSC color on a non–525-line system; they have since switched to 525 lines. Brazil remains the only 525-line country to transmit color TV that is not NTSC; their system is called PAL-M. M is the CCIR designation for 525 line/30 frame television.

NTSC Standard Documentation of the characteristics of NTSC. NTSC is defined primarily in FCC Part 73 technical specifications. Many of its characteristics are defined in EIA-170A. NTSC is also defined by the CCIR.

NTSC is a *living standard*. As problems with it are discovered, they are corrected. For example, a former EIA standard, RS-170, omitted any phase relationship between luminance and chrominance timing, resulting in blanking problems. EIA- 170A defines that relationship (called SC/H for subcarrier to horizontal phase relationship).

NVOD or Near Video on Demand The service of providing a movie to subscribers on multiple channels and staggering its start time (for example every fifteen minutes). Subscribers can then tune in to the next available showing.

Nyquist Nyquist Filter, Nyquist Limit, Nyquist Rule, and Harry Nyquist, for whom they are named. The *Nyquist rule* states that in order to be able to reconstruct a sampled signal without aliases, the sampling must occur at a rate of more than twice the highest desired frequency. The Nyquist rule is usually observed in digital systems. For example, CDs have a sampling frequency of 44.1 kHz to allow signals up to 20 kHz to be recorded. However, it is frequently violated in the vertical and temporal sampling of television, resulting in aliases. (*See also* Alias.)

The *Nyquist limit* is the highest frequency that can be captured without alias for any given sampling rate, essentially half the sampling rate. The *Nyquist filter* is commonly used in the IF stage of a television receiver to separate the desired television channel from potential interference.

Odd Number The number of scanning lines per frame necessary in an interlaced scanning system. One line is split between fields to ensure proper spacing between scanning lines from different fields. A progressively scanned system may use an even number of scanning lines. With the addition of a kick signal, an interlaced system can have an even number of scanning lines.

On-Demand The ability to request video, audio, or information to be sent to the screen immediately by clicking an icon or text on the screen referring to that choice.

Optical Fiber A glass strand designed to carry light in a fashion similar to the manner in which wires carry electrical signals. Because light is electromagnetic radiation of tremendously high frequency, optical fibers can carry much more information than can wires, although multiple paths through the fiber place an upper limit on transmission over long distances because a characteristic called *pulse dispersion*. CATV and telephone companies propose connecting optical fibers directly to homes.

Orientation A direction of presentation affecting resolution requirements. Horizontal lines become vertical lines when their orientation is rotated by 90 degrees; a pattern of dots appearing to be in horizontal and vertical rows may not appear to be diagonally aligned when its orientation is rotated 45 degrees because of certain characteristics of the human visual system.

Orthagonal Sampling Picture sampling arranged in horizontal rows and vertical columns, as opposed to, say, *quincunx sampling*.

Overscanning Displaying less than the complete area of an image to a viewer (i.e., scanning beyond the visible area). All

TV sets are overscanned at least slightly, so that viewers do not see blanking (the size of which can vary). The restricted areas of the SMPTE Safe Action Area and the SMPTE Safe Title Area take this overscanning into account. Newer TV sets have less overscanning than their predecessors.

Packet A header followed by a number of contiguous bytes from an elementary data stream; a layer in the ATSC DTV system coding syntax.

Packet Data Contiguous bytes of data from an elementary data stream present in the packet.

Pan and Scan A process for aspect ratio accommodation wherein an area with the display's aspect ratio is moved across the source's area to follow action across the screen. Pan and scan may be implemented either by introducing panning that was not in the original scene or by repositioning the area between frames, thus introducing edits that were not in the original scene.

Panels The picture area left over when a narrow aspect ratio is superimposed on a wider one. Also sometimes called *side panels* or *widescreen panels*.

Pay-Per-Use A fee for every service, product, and download; often on a tiered basis.

Pay TV Television programming that requires payment up front, usually on a monthly basis as a subscription fee. Cable and satellite operators bundle content programming under packaged names like "Gold," "Silver," and so on.

Pels Picture elements; the smallest dots that can be transmitted for any given bandwidth and scanning system.

Periodic Noise An interfering signal of a particular frequency, which causes a stationary or moving pattern to appear on a screen.

Personal Computer (PC) The device that enables anyone to compute, word process, or perform more complicated functions.

Personal Video Recording (PVR) Software and data services combinations that allow the viewer to interactively select programming choices from an electronic programming guide to either watch or record on their digital video recorder. Data services are provided on a daily basis from the PVR provider. Companies leading this technology include ReplayTV, TiVo, and others like WebTV and DirecTV, which integrate hard drives in their boxes and digital receivers.

Phase Alternation by Line (PAL) A common composite color transmission system (like NTSC) used in many countries of the world. There are 625 scanning lines per frame and 25 frames per second. Brazil is the only country in the world transmitting PAL on a 525/30 system.

Phosphor Material that glows when struck by an electron beam in a CRT; the source of the light emitted by TV screens. Phosphor composition determines the maximum range of colors that can be displayed.

Phosphor Aging Reduction in the amount of light emitted by phosphors caused by deterioration with use.

Picture Carrier The transmitted signal onto which video information is modulated in television broadcasting.

Picture Heights A measurement of viewer distance from a screen in terms of the height of a picture. For example, a viewer sitting 5 feet from a 1-foot high picture would be said to be viewing from a distance of 5 picture heights.

Picture-In-Picture The ability to view a television broadcast in a small window on top of another broadcast or within a larger interactive interface.

Pixel Count A disputed measure of the spatial resolution of a television system in terms of perceivable pixels.

Pixels Picture elements. *See* Pels.

Portal *See* Walled Garden.

PPH Per Picture Height; the standard method of specifying horizontal resolution so that it can be compared directly to vertical resolution. It is obtained by dividing the total horizontal resolution by the aspect ratio, thus obtaining a figure in TV lines per picture height; the number of lines that can be perceived in an area of width equal to the height of the picture.

PPW Per Picture Width; a method of specifying horizontal resolution sometimes used when a large number is desired. *See also* PPH.

Precombing A form of prefiltering used to prevent interference when chrominance and luminance are combined in an encoder. Some forms of precombing can restrict the performance of a receiver equipped with some forms of advanced decoders. Precombing can be considered cleaning a hole and the subcarrier that will be inserted into it.

Progressive Scanning A television scanning system in which each scanning line follows its predecessor in a progressive fashion, rather than skipping intermediate lines to be filled in by the next field. In a progressive scanning system, the field rate and the frame rate are identical. Progressive scanning is also called *sequential scanning*.

Protocol The "language" spoken between computers to help them exchange information. More technically, it's a formal description of message formats and rules that two computers must follow to communicate.

Quantizer A processing step that intentionally reduces the precision of DCT coefficients.

Random Access The process of beginning to read and decode the coded bit stream at an arbitrary point.

Random Noise Also called *thermal noise*, a transmission or recording impairment that manifests itself as snow in a picture and hiss in sound. A number of techniques have been developed to reduce random noise in a picture through signal averaging.

Raster A set of scanning lines; also the type of image sampling using scanning lines (as in raster scanning).

Recommendation 601 A CCIR recommendation (standard) for digital component video, equally applicable to 525 and 625 scanning lines, also called 4:2:2. Digital component video is about as close in quality as current 525 scanning line equipment can come to ATV.

Resolution Perceivable detail. *See also* Chroma Resolution, Diagonal Resolution, Dynamic Resolution, Horizontal Resolution, Spatial Resolution, Static Resolution, and Temporal Resolution.

RF Radio Frequency, a term used for modulated signals, as opposed to baseband. *See also* IF.

RGB Red, Green, and Blue; the most basic color components.

Ringing A common filter artifact, manifesting itself in television pictures as ghostlike images of sharp edges.

Robust A transmission or recording scheme that can tolerate significant impairments without catastrophic failure (severe degradation).

RS-170A Now called EIA170A; the EIA standard for NTSC television production equipment.

Sampling The process of dealing with something continuous in discrete sections. Sampling is probably best known as the first step in the process of *digitization*, wherein an analog (continuous) signal is divided into discrete moments in time. Yet, even analog television signals have already been sampled twice: once temporaly (time being sampled in discrete frames) and once vertically (the vertical direction being divided into discrete scanning lines). If these initial sampling processes are not appropriately filtered (and they rarely are in television), they can lead to aliases. *See also* Alias, Digitization, and Nyquist.

Sampling Rate The rate at which samples are taken of a signal in the process of digitizing it. Places an upper limit on horizontal resolution.

Saturated Color A color as far from white, black, or grey as it can be (e.g., vermillion rather than pink).

Scanning Circuitry Camera or display subsystems designed for moving an electron beam around to form a raster.

Scanning Lines Horizontal or near-horizontal lines sampling a television image in the vertical direction. In tube-type cameras and displays equipped with CRTs, the scanning lines are caused by electron beam traces.

Scanning Structure A term sometimes used to describe a number of scanning lines per frame, interlace ratio, and frame rate; also sometimes used to describe what appears when scanning lines are visible.

Scrambling Usually used as a synonym for *encryption*; controlled disordering of a signal to prevent unauthorized reception; sometimes used to describe controlled disorganization of a signal to improve its robustness. The latter form is more often called *shuffling*.

Sequential Scanning *Progressive scanning*, so named because scanning lines are transmitted in numerical sequence, rather than in odd or even numbered fields, as in *interlaced scanning*.

Set-Top Box (STB) An electronic device that sits on top of a TV set and allows it to connect to the Internet, game systems, or cable systems.

Sharpness Apparent image resolution. High sharpness may be the result of high resolution, or it might be an optical illusion caused by image enhancement or by visible edges in a display, such as the vertical stripes of an aperture grille CRT (e.g., Trinitron). Visible scanning lines can actually increase perceived sharpness.

Shoot and Protect A concept of aspect ratio accommodation central to the selection of the 16:9 aspect ratio for the SMPTE HDEP standard. In a shoot and protect system, in production the action is confined to certain bounds (the shoot range) but a larger area (the protect range) is kept free of microphone booms, lights, and other distracting elements. Shoot and protect has been used for years in film, where the shoot aspect ratio is the 1.85:1 used in U.S. theatrical projection, and the protect aspect ratio is the 1.33:1 used in NTSC.

The 16:9 aspect ratio was selected mathematically as that requiring the least area to protect both 1.33:1 television and 2.35:1 widescreen film. In such a system, both the shoot and the protect aspect ratios are 16:9. A rectangle of shoot width and protect height would be 1.33:1 (12:9); a rectangle of shoot height and protect width would be 2.35:1 (about 21:9). The concept of 3-perf film conflicts strongly with 1.85:1 shoot and protect.

Sideband A signal that is a consequence of some forms of modulation. When modulation forms two sidebands, one can sometimes be filtered out to increase efficiency without sacrificing information.

Signal-to-Noise Ratio *See* SNR.

Simulcast Simultaneous broadcast. Prior to the advent of multichannel television sound broadcasting, the only way to transmit a stereo television show to homes was by simultaneous broadcasting on TV and radio stations.

SMPTE 240M The first SMPTE standard rejected by the American National Standards Institute, on the basis of an appeal by ABC. *See also* Society of Motion Picture and Television Engineers.

SNR Signal-to-Noise Ratio; it may not be possible to directly compare SNRs for ATV and for NTSC because the eye's sensitivity to noise varies with the detail of the noise.

Society of Motion Picture and Television Engineers (SMPTE) A television standards organization. The Society has current standards for more than 10 different videotape recording formats, with more pending.

Spatial Relating to the area of an image.

Spatial Resolution Usually referred to as resolution; the linearly measurable detail in an image, in the vertical, horizontal, or diagonal directions.

Spatiotemporal Filtering Filtering in both space and time.

Spatiotemporal Spectrum A three-dimensional representation of the energy distribution of a television signal. The three dimensions are horizontal, vertical, and time.

Spectrum Usually, a range of frequencies. The electromagnetic spectrum includes power frequencies, radio, TV, infrared, light, ultraviolet, X-rays, and so on.

Spectrum Allocation Designation of certain bandwidths at certain frequencies for certain purposes. For example, channel 2 has been allocated 6 MHz of bandwidth from 54 MHz to 60 MHz for television broadcasting.

Square Pixels Pixels generated in a television system having the same horizontal and vertical resolution. There is some evidence that a large mismatch between horizontal and vertical resolution prevents the higher resolution from being fully perceived by the human visual system. NTSC was created with square pixels with a resolution of approximately 330 × 330 lines. Square pixels also ease digital image manipulations, such as rotation.

Squeeze A change in aspect ratio. Anamorphic lenses sometimes squeeze a widescreen scene by a factor of two horizontally so it will fit on a 1.33:1 aspect ratio frame. In projection, another anamorphic lens "expands" the squeeze (squeezes vertically) to restore the original aspect ratio. When a widescreen film is presented on television without being expanded, it is said to be squeezed. An unexpanded film print is said to be a *squeeze print* (the opposite is *flat*).

Static Resolution Detail in a stationary image. Any amount of bandwidth is sufficient for the transmission of HDTV images with high static resolution—even a telephone line; the smaller the bandwidth, the longer it takes to transmit all of the resolution.

Standard A set of rules or characteristics defining a particular television system. Some standards (such as those contained in FCC rules and regulations) are mandatory. Most (including those of the EIA, IEEE, and SMPTE) are voluntary. The establishment of a standard often freezes development at a certain level but allows users and manufacturers to deal with a much larger array of products than might be available without a standard.

Standards Converter A device for converting signals from one standard to another. Converting between different color schemes with the same scanning structure is called *transcoding*. Converting between different scanning structures requires line and field interpolation, which usually introduces artifacts. Standards conversion between 525 scanning line and 625 scanning line signals is performed regularly. Conversion from HDEP to either NTSC or a receiver-compatible ATV system will require standards conversion.

Subcarrier An auxiliary information carrier added to the main baseband signal prior to modulation. The most common example in television is the NTSC color subcarrier.

Subchannel A transmission path within the main transmission path. Subcarriers are examples of subchannels, but there are others. Quadrature modulation of the picture carrier provides a subchannel; so does blanking stuffing.

Subsampling Sampling within samples. For example, dividing an NTSC pixel into three or four subpixels is an example of subsampling.

S-VHS Super VHS; a consumer videotape format offering horizontal resolution somewhat greater than that offered by NTSC broadcasting and allowing the component recording and playback without cross-luminance or cross-color artifacts through a four-pin S-Video connection.

S-Video The separate luminance and chrominance (Y/C) connection found on all S-VHS and ED-Beta VCRs, many 8 mm video products, and many television sets and cameras, allow recording and playback of signals without cross-color or cross-luminance. The "S" stands for "separate," not "super."

Symmetrical Digital Subscriber Line (SDSL) A type of DSL that uses only one of the two cable pairs for transmission. SDSL allows residential or small office users to share the same telephone for data transmission and voice or fax telephony.

Sync Buzz A noise containing harmonics of 59.94 Hz, heard on television set speakers under certain signal and transmission conditions. One such condition is the transmission of electronically generated characters of high level and resolution greater than can be carried in NTSC. The ringing resulting when those signals hit an NTSC filter causes the television carrier to momentarily disappear. Because the characters are within a television field, the rate of appearance and disappearance is a multiple of the field rate, 59.94 Hz.

System Gamma The overall light-in/light-out characteristic of a television system, from camera through receiver. In an ideal system, the gamma should be 1. In practice, it is about 1.4.

T Tele- (as in television) or time.

T-Axis Time axis of the spatiotemporal spectrum.

Tcommerce A word based on "e-commerce," this term describes interactive commerce on television. The word "shopping" also suffices.

Teletext A series of static pages on the television that offer all sorts of text news and community services. Invented by BBC engineers in the 1970s, teletext remains popular today across all UK television networks. Text pages are offered in carousel to keep things fresh. A viewer can request different pages using the number pad on their remote control. Only recently has this technology been upgraded to digital.

Temporal Relating to time.

Temporal Alias An alias caused by violation of the Nyquist limit on sampling in time with frames. The most common temporal alias is in images of spinning wheels that appear to be turning backwards.

Temporal Resolution The finest moments of time that can be perceived in a particular system. It is not the same as *dynamic resolution*, which is spatial resolution when an image is changing. As an example, suppose a spoked wheel is turning. If the spokes are a blur when the wheel is not turning, the system has poor static resolution; if they are clearly visible, it has good static resolution (for the spokes). If they're a blur when the wheel is turning, the system has poor dynamic resolution and poor temporal resolution. If they're clearly visible when the wheel is turning, the system has good dynamic resolution. If, though clearly visible, they appear to be stationary, or turning in the wrong direction, or turning at the wrong speed, or flashing rapidly in different positions so that it is impossible to tell which way or at what speed they're turning (a temporal blur), the system has poor temporal resolution. A great deal of evidence indicates that the human visual system cannot simultaneously perceive high spatial resolution and high temporal resolution.

Terminal A device that allows user to send commands to a computer located somewhere else.

Terrestrial Broadcasting analog or digital signal via a large antenna that stands on the ground.

Three-Dimensional Spectrum *See* Spatiotemporal Spectrum.

Time Multiplex *See* Multiplex.

Transmission Aperture A number used to compare amounts of light passed through an optical system, such as a camera lens. Transmission aperture takes into consideration both the *f*-stop (geometric aperture) and the amount of light absorbed or reflected in the optical system.

Transmission Standard A standard used for transmitting signals to the home, not necessarily for producing them. The scanning structure of NTSC is identical for both production and transmission, but this need not be the case in DTV.

Transponder Located on a space satellite, the transponder receives programming uplinked by the content provider and then downlinks it to the broadcaster. Signal space on a transponder is very expensive.

Triad Three-colored phosphor dots on the faceplate of a tricolor CRT. Some tricolor CRTs use vertical stripes of different color phosphors or vertically oriented oblong dots. These dots or stripes are the ultimate determinants of maximum horizontal resolution. When the dots are round, they are also the maximum determinants of vertical resolution. The finer the dot pitch, the higher the resolution, because it is not possible to reduce the size of a black-and-white pixel below the size of one triad. Triad spacing also cannot be optimized for all numbers of scanning lines. Thus, a tube optimized for 1,125 scanning lines will not yield optimum performance with a signal of 1,050 scanning lines, or vice versa.

Neither black-and-white CRTs nor the three single-color CRTs used in most projection TV sets suffer from these limitations because their faceplates are uniformly covered with a layer of phosphor, and resolution is ultimately determined by the size of the electron beam and the projection optics. Picture tubes with striped apertures can deal effectively with multiple scanning rates, but still restrict horizontal resolution to the width of three stripes.

Triggers "Transport A" and "Transport B" A command from the host server that notifies the viewer's set-top box that interactive content is available at this point. A trigger shows up seamlessly as an icon or some kind of clickable text within the television broadcast program. Once clicked by using the remote control, the trigger disappears and more content or a new interface appears on the TV screen. Transport A triggers travel over the VBI first and then send the following content through the Internet (mostly in use now). Transport B triggers and content travel over the digital broadcast signal.

TV Lines Measure of resolution. A TV line is either black or white, so two TV lines (a black and a white) form one cycle of spatial resolution. TV lines are often confused with scanning lines. For vertical resolution, scanning lines multiplied by the Kell factor (and, when appropriate, by the interlace coefficient) yield TV lines.

TV Tuner Card The TV tuner card enables a PC to receive television signals, which are then converted to digital format for viewing on screen.

Twinkle A sparkling effect that can be caused by subsampling, because the finest detail is transmitted at a rate below the flicker frequency (and sometimes even below the fusion frequency).

Twitter A flickering of fine horizontal edges caused by interlaced scanning. A fine line appearing in only one field is presented below the flicker frequency; therefore, it flickers. Twitter is eliminated in line-doubling schemes that change from interlaced to progressive scanning, as most of the IDTV schemes do. Interestingly, twitter was much less of a problem in the early days of NTSC than it is today, because cameras and displays didn't have sufficient detail to confine an edge to one scanning line.

Two-Way Used to describe data as it travels both from the broadcaster's headend (cable) or central office (telecommunications) to the viewers location and then back to the headend of central office. Two-way always connotes the presence of an interactive infrastructure.

UHF Ultra-High Frequency; the band from 300 MHz to 3 GHz. In television, UHF refers to a subset of that band, the range from 470 MHz to 890 MHz, once allocated to TV channels 14 through 83. Demands of other transmission services (such as police radios) have eaten into both the lower and the upper reaches of the UHF TV allocations.

Uniform Resource Locator (URL) The address of a document or other resource available on the Internet by clicking a link. A URL has three components, the protocol (http:), server domain name (intel.com), and the file location on their server.

Uplink To transmit analog or digital signal to a satellite so that it can then be transmitted back down to earth.

Upstream Information that travels from the user to the Internet or to a central office or headend.

Vertical Alias An alias caused by unfiltered sampling in the vertical direction by scanning lines. Vertical aliasing is frequently noticed when reading vertical resolution on a resolution chart. The wedgelike lines get finer and finer until they reach the limit of the vertical resolution of the system, but then they may appear to widen or to change position. This is caused by lines on the chart sometimes falling between scanning lines and sometimes on them. In a properly filtered television system, detail finer than the vertical resolution of the system is a smooth blur.

Vertical Blanking Interval (VBI) A period during which the electron beam in a display is blanked out while it travels from the bottom of the screen to the top. It is the black bar that becomes visible when the vertical hold on a television set isn't correctly adjusted. The VBI is usually measured in scanning lines. When the VBI is subtracted from the total number of scanning lines, the result is the number of active scanning lines.

In NTSC, the VBI has a duration of 20.5 or 21 lines (depending on the field), of which nine lines are devoted to the vertical synchronizing signal that lets television sets know when a field has been completed. The remaining lines have long been used to carry auxiliary information, such as test and reference signals, time code, and encoded text, such as captions for the hearing impaired.

Vertical Resolution The amount of detail that can be perceived in the vertical direction; the maximum number of alternating white and black horizontal lines that can be counted from the top of the picture to the bottom. It is not the same as the number of scanning lines. It is the number of scanning lines *minus* the VBI *multiplied by* the Kell factor (and, where appropriate, the interlace coefficient).

Vertical-Temporal Pre-Filtering Filtering at the camera or transmission end to eliminate vertical and temporal aliases. When a high line rate, progressively scanned camera is prefiltered to NTSC rates, the resulting image is not only alias-free but can also be used by an advanced receiver to provide vertical and temporal resolution beyond that normally found in NTSC. The Kell factor of such a system can be close to one.

Vertical-Tempered Sampling Sampling that occurs in every television signal because of individual frames (which sample in time) and individual scanning lines (which sample in the vertical direction). This sampling can cause aliases unless properly prefiltered.

Vestigial Sideband The vestige of a sideband left after filtering.

Very High Bit Rate Digital Subscriber Line (VDSL) A type of DSL that is primarily intended to be used as the last transmission system section in a network. VDSL can serve as the primary transmission element for video-on-demand (VOD) and Asynchronous Transfer Mode (ATM) applications over the existing infrastructure of twisted copper pairs in the local plant. It is used to carry multiple television channels, HDTV, and ATM to the home for interactive services (home banking, shopping, remote medical care).

VHF Very High Frequency; the range from 30 MHz to 300 MHz, within which are found U.S. television channels 2 through 13.

Videophone This future device promises to incorporate real-time video transmissions with telephony. Although this technology was promised and available many years ago, it has never been fully realized. Today, Internet Telephony or Cable Telephony may offer this service.

Video-On-Demand (VOD) The ability to interactively select videos whenever one wants to watch them. Videos are streamed in MPEG format.

Viewing Distance Distance between an image and a viewer's eyes. In television, the distance is usually measured in picture heights. In film it is sometimes measured in picture widths. As a viewer gets closer to a television set from a distance, the amount of detail perceptible on the screen continually increases until, at a certain point, it falls off rapidly.

At that point, scanning line or triad visibility is interfering with the viewer's ability to see all of the detail in the picture (sort of not being able to see the forest for the trees). The finer the triad or scanning structure, the closer to the screen this point can be (in picture heights).

Therefore, high-definition screens allow either closer viewing for the same size screen or larger screens for the same physical viewing distance (not in picture heights). Recent research indicates that subjective quality is not strictly a function of viewing angle. Viewers appreciate large screens even if they are so far from them that they cannot perceive all of their resolution.

When the effects of scanning lines and triads are reduced, other artifacts (such as a temporal alias of panning called *strobing*) may become more obvious. From far enough away, it is impossible to tell high-definition resolution from NTSC resolution.

Visible Scanning Lines Normally considered a defect that affects perception of fine vertical detail. Scanning line visibility can also have an apparent sharpness increasing effect, however. *See also* Sharpness and Viewing Distance.

Visible Subcarrier The most basic form of cross-luminance.

Visual Acuity The amount of detail perceptible by the human visual system. It depends on many factors, including brightness, color, orientation, and contrast. Optimum viewing distance depends on visual acuity.

Walled Garden A term that appeared in the mid- to late-90s to define interactive content offerings contained or walled-off from direct access to Internet users. Walled garden users may link to the Internet from walled gardens, but not vice versa. America Online is an example of a very successful walled garden. Certain interactive TV middleware software solutions enable cable and satellite providers to create their own walled gardens or "portals." Inside an ITV walled garden an interface allows the viewer to have access to news, sports information, email on TV, and other applications.

Wideband Relatively wide in bandwidth.

Widescreen An image with an aspect ratio greater than 1.33:1.

X-Axis The horizontal axis of a graph. When a television signal is examined in one dimension, the x-axis is usually time. When it is examined in two or three dimensions, the x-axis is usually horizontal resolution.

Y Luminance; so named because it is the y-axis of the chart of the spectral sensitivity of the human visual system.

Y-Axis The vertical axis of a graph. When a television signal is examined in one dimension, the y-axis is usually signal strength. When it is examined in two or three dimensions, the y-axis is usually vertical resolution.

Y/C Connections Connections between videotape recorders and between videotape recorders and cameras, monitors, and other devices that keep luminance and chrominance separate and thus avoid cross-color and cross-luminance. *See also* S-Video.

Z-Axis An axis of a three-dimensional graph, which, when printed on a flat piece of paper, is supposed to be perpendicular to the plane of the paper. When a television signal is examined in three dimensions, the z-axis is usually time.

INDEX

Note: Boldface numbers indicate illustrations.

ABOUT THE AUTHORS

JOHN RICE has over 25 years experience in the video industry as a producer, writer, consultant, and journalist and is the former publisher/editorial director of *Videography* and *Corporate Television* magazines.

BRIAN MCKERNAN is the editor of *Digital Cinema* magazine and the former Editorial Director of *Videography*, *Television Broadcast*, and *Government Video*.